Milestones in Drug Therapy
MDT

Combination Therapy of AIDS

Edited by E.D.A. De Clercq and A.-M.I. Vandamme

Springer Basel AG

Editors
Erik D.A. De Clercq / Anne-Mieke I. Vandamme
Katholieke Universiteit Leuven, Rega Institute
Minderbroedersstraat 10
B-3000 Leuven, Belgium

Library of Congress Cataloging-in-Publication Data
Combination therapy of AIDS / edited by E.D.A. De Clercq and A.-M.I. Vandamme.
p. ; cm. -- (Milestones in drug therapy)
Includes bibliographical references and index.
ISBN 978-3-0348-9604-7 ISBN 978-3-0348-7869-2 (eBook)
DOI 10.1007/978-3-0348-7869-2
1. AIDS (Disease)--Chemotherapy. 2. AIDS (Disease)--Treatment. 3. Chemotherapy, Combination. 4. Antiviral agents. I. De Clercq, Erik. II. Vandamme, Anne-Mieke, 1960- III. Series.
[DNLM: 1. HIV Infections--drug therapy. 2. Anti-HIV Agents--therapeutic use. 3. Antiretroviral Therapy, Highly-Active. 4. Treatment Outcome. WC 503.2 C7314 2003]
RC606.6.C64 2003
616.97'92061--dc22

2003052372

Bibliographic information published by Die Deutsche Bibliothek
Die Deutsche Bibliothek lists this publication in the Deutsche Nationalbibliografie; detailed bibliographic data is available in the internet at http://dnb.ddb.de

ISBN 978-3-0348-9604-7

Originally published by Birkhäuser Verlag in 2004
Softcover reprint of the hardcover 1st edition 2004

Printed on acid-free paper produced from chlorine-free pulp. TFC ∞
Cover illustration: provided by Prof. De Clercq.

NRTIs: nucleoside reverse transcriptase inhibitors (zidovudine, didanosine, zalcitabine, stavudine, lamivudine, abacavir, emtricitabine)
NtRTIs: nucleotide reverse transcriptase inhibitors (tenofovir, disoproxil fumarate)
NNRTIs: non-nucleoside reverse transcriptase inhibitors (nevirapine, delaviridine, efavirenz)
FIs: fusion inhibitors (enfuvirtide)
PIs: protease inhibitors (saquinavir, ritonavir, indinavir, nelfinavir, amprenavir, lopinavir, atazanavir)

ISBN 978-3-0348-9604-7

9 8 7 6 5 4 3 2 1 www.birkhauser-science.com

Contents

List of contributors

Ume L. Abbas, University of Pittsburgh, Division of Infectious Diseases, Falk Medical Building, Suite 3-A, 3601 Fifth Ave, Pittsburgh, PA 15213, USA; e-mail: abbasu@msx.dept-med.pitt.edu

Brigitte Autran, Laboratoire d'Immunologie Cellulaire et Tissulaire, CNRS UMR 7627, Hôpital Pitié-Salpétrière, 47-83, bld de l'hôpital, Bâtiment CERVI, 75013 Paris, France; e-mail: brigitte.autran@psl.ap-hop-paris.fr

Marie-Charlotte Bernard, Division of Infectious Diseases and Laboratory of Virology, Geneva University Hospital, 1211 Geneva 14, Switzerland

Charles C.J. Carpenter, Department of Medicine, The Miriam Hospital, 164 Summit Avenue, Providence, RI 02906, USA; e-mail: ccjc@lifespan.org

Erik De Clercq, Rega Institute for Medical Research, Katholieke Universiteit Leuven, B-3000 Leuven, Belgium; e-mail: Erik.DeClercq@rega.kuleuven.ac.be

Bonaventura Clotet, HIV Unit, Hospital Universitari Germans Trias I Pujol, 08916 Barcelona, Spain

Nathan Clumeck, Saint-Pierre University Hospital, Department of Infectious Diseases, 322 rue Haute, 1000 Brussels, Belgium, e-mail: maladiesinfectieuses@stpierre-bru.be

Lisa M. Frenkel, University of Washington, Department of Pediatrics and Laboratory of Medicine, 4800 Sand Point Way, NE, Seattle, WA 98105, USA; e-mail: lfrenkel@u.washington.edu

Felipe García, Clinic Institute of Infectious Diseases and Immunology, IDIBAPS, Hospital Clinic, Faculty of Medicine, University of Barcelona, Villarroel, 170, 08036 Barcelona, Spain; e-mail: fgarcia@medicina.ub.es

José M. Gatell, Clinic Institute of Infectious Diseases and Immunology, IDIBAPS, Hospital Clinic, Faculty of Medicine, University of Barcelona, Villarroel, 170, 08036 Barcelona, Spain; e-mail: gatell@medicine.ub.es

Roy M. Gulick, Weill Medical College of Cornell University, Cornell Clinical Trials Unit, 525 East 68th Street, New York, NY 10021, USA; e-mail: rgulick@med.cornell.edu

Marianne Harris, British Columbia Centre for Excellence in HIV/AIDS, 667-1081 Burrard Street, Vancouver, BC V6Z 1Y6, Canada; e-mail: mharris@hivnet.ubc.ca

Martin S. Hirsch, Massachusetts General Hospital, Harvard Medical School, 65 Landsdowne St, Room 419, Cambridge, MA 02139, USA; e-mail: mshirsch@partners.org

Christopher Holtzer, University of California, San Francisco, Gilead Sciences Inc., Foster City, CA, USA

Joan Joseph, Clinic Institute of Infectious Diseases and Immunology, IDIBAPS, Hospital Clinic, Faculty of Medicine, University of Barcelona, Villarroel, 170, 08036 Barcelona, Spain; e-mail: jjoseph@clinic.ub.es

Deborah Konopnicki, Saint-Pierre University Hospital, Department of Infectious Diseases, 322 rue Haute, 1000 Brussels, Belgium; e-mail: maladiesinfectieuses@stpierre-bru.be

Michael Kurowski, Therapia GmbH, Rubensstrasse 125, 12157 Berlin, Germany: e-mail: labor-kurowski@t-online.de

John W. Mellors, University of Pittsburgh, Division of Infectious Diseases, School of Medicine, Scaife Hall, Suite 818, 3550 Terrace Street, Pittsburgh, PA 15261, USA; e-mail: mellors@msx.dept-med.pitt.edu

Julio S.G. Montaner, University of British Columbia/St. Paul's Hospital, British Columbia Centre for Excellence in HIV/AIDS, 667-1081 Burrard Street, Vancouver, BC V6Z 1Y6, Canada; e-mail: jmontaner@hivnet.ubc.ca

Roger Paredes, HIV Unit, Hospital Universitari Germans Trias I Pujol, 08916 Barcelona, Spain; e-mail: 34120rpd@comb.es

Luc Perrin, Division of Infectious Diseases and Laboratory of Virology, Geneva University Hospital, 1211 Geneva 14, Switzerland; e-mail: luc.perrin@hcuge.ch

Lidia Ruiz, HIV Unit, Hospital Universitari Germans Trias I Pujol, 08916 Barcelona, Spain; e-mail: lruiz@ns.hugtip.scs.es

Nicole H. Tobin, University of Washington, Department of Pediatrics and Laboratory of Medicine, 4800 Sand Point Way, NE, Seattle, WA 98105, USA; e-mail: nicole.tobin@seattlechildrens.org

Cécile L. Tremblay, C.H.U.M., Hôpital Hôtel-Dieu de Montréal, Pavillon Jeanne-Mance, Bureau 7-355, 3840, rue Saint-Urbain, Montréal Qc, H2W 1 T8, Canada, e-mail: c.tremblay@umontreal.ca

Mike Youle, HIV Clinical Research, Royal Free Centre for HIV Medicine, Royal Free Hospital, Pond Street, London NW3 2QG, UK; e-mail: myoule@rfh1.demon.co.uk

Preface

HIV infection has been a greater challenge to current medicine than any other viral disease of modern times. HIV leads to a persistent infection and the virus has an immense genetic flexibility under selective pressure. During its replicative cycle in patients, HIV accumulates mutations at such a high rate that the selective pressure inflicted on the immune system, or generated by antiviral drugs rapidly triggers the appearance of escape mutants. Currently available drugs, when used singly, are not capable of suppressing virus replication in patients to such a level that the generation of mutations, from which a variant resistant to immune attack or antiviral drugs can be selected, is prevented. This is the main reason why combination therapy, usually of three drugs, has become the standard procedure for the treatment of AIDS.

It is obvious that virus eradication will not readily be achievable, so that drugs have to be taken for a prolonged time or even lifelong so as to keep the viral load as low as possible. Whether the currently used drug combinations will be able to control virus replication in a particular patient for such a prolonged period of time depends on many factors, most of which are addressed in the different chapters of this book.

The aim of antiviral drug combination therapy for AIDS is ultimately to restore full function of the immune system. The expected immunological benefit is one of the major determinants driving the decision when to start therapy. In principle, treatment should be started sufficiently early, so as not to compromise too much immune competence. However, in the past the traditional drug regimens have been compounded by a number of side-effects which often caused patients to not adhere to drug intake. It could be argued therefore that therapy should not be started too early, so as not to weaken motivation for lifelong compliance in the face of more harm from side-effects of the drugs than of the disease itself. Shifting treatment guidelines reflects this difficult balance. In addition, the potency of the individual drugs also contributes to the choice of the drugs to be incorporated in the combination regimens.

Current clinical tools to evaluate the success of antiretroviral combination therapies are still based on viral load, reflecting virus replication, and CD4 count, reflecting the status of the immune system. A new tool has recently emerged: drug resistance testing. It is logical that knowledge of drug resistance will help in therapy decisions; however, scientific evidence was only recently firmly established, so that guidelines now include resistance testing. Other parameters such as drug level monitoring and drug adherence are considered very valuable, but have not been evaluated sufficiently thoroughly in prospective studies to warrant their use in routine clinical practice.

In the introduction chapter of the book, Erik De Clercq provides an overview of the four different classes of compounds that are now available for the treatment of AIDS, the so-called NRTIs (nucleoside reverse transcriptase inhibitors), NtRTIs (nucleotide reverse transcriptase inhibitors), NNRTIs (non-nucleoside reverse transcriptase inhibitors) and PIs (protease inhibitors). A fifth class, that of the virus-cell fusion inhibitors, is represented by a single compound (enfuvirtide) that has recently been approved and that, together with the NRTIs, NtRTIs, NNRTIs and PIs, could be included, when available and desirable, into combination drug regimens.

How antiviral drug therapy of AIDS evolved from a single drug (zidovudine) therapy in 1987 to what today is commonly referred to as highly active antiretroviral therapy (HAART) is chronicled by Deborah Konopnicki and Nathan Clumeck.

Cécile L. Tremblay and Martin S. Hirsch then describe the basic principles for combining different drugs in the treatment of HIV infection. Simply stated, triple drug combinations do better than dual drug combinations, which, in turn, score better than single drug monotherapy. Roger Paredes, Bonaventura Clotet and Lidia Ruiz have assessed different forms of multiple-drug regimens from a comparative viewpoint.

The drug levels achieved at the site(s) of virus replication are of paramount importance to predict the efficacy of any drug regimen, and this should become clear from the chapter of Michael Kurowski on the pharmacokinetics and pharmacodynamics of the therapeutic regimens used.

Luc Perrin and Marie-Charlotte Bernard discuss HAART treatment for primary HIV infection, an indication which today is recommended only within the scope of formal clinical trials. This chapter is followed by the chapter of Marianne Harris and Julio S.G. Montaner on salvage therapy for which, as for first-line therapies, sustained immunological and clinical benefit has been demonstrated only when complete suppression of plasma viral load is achieved.

Felipe García, Joan Joseph and José M. Gatell tell us what to expect from structural therapy interruptions (STIs) as a strategy to re-activate or reconstitute immunity, a provocative approach towards the therapeutic management of HIV infections. The kinetics of immune reconstitution and the current limitations thereof are further analyzed by Brigitte Autran.

Specific aspects have to be considered when dealing with HAART in children infected with HIV-1, and suggestions on how to treat such pediatric HIV-1 infections are proposed by Nicole H. Tobin and Lisa M. Frenkel.

Treatment failure continues to occur commonly because of a multitude of reasons and, when assessing remedies to prevent or overcome treatment failures, discontinuation of drug therapy for toxicity, non-adherence or other reasons must be distinguished from virologic, immunologic and clinical treatment failures, as pointed out by Roy M. Gulick.

One of the major reasons for treatment failure is the development of HIV resistance towards the antiretroviral medications. The sensitivity of HIV-1 to antiretrovirals can be monitored by either genotyping or phenotyping, or both. The economic impact, i.e., costs of medical care, for HIV-1 drug resistance testing is analyzed by Christopher Holtzer and Michael Youle.

Taken all the foregoing considerations together, a number of issues could be raised, as to when antiretroviral therapy should be indicated, which drug regimens are likely to be the most efficacious, how antiretroviral therapy, once it has been initiated, should be monitored, and when drug regimens need to be changed in individual patients. Charles C.J. Carpenter has formulated the guidelines for optimal antiretroviral therapy, ultimately aimed at providing the most effective and durable virologic, immunologic and clinical response.

In the final chapter, Ume L. Abbas and John W. Mellors envision the future of antiretroviral therapy. They describe a number of new promising antiretrovirals that are in development, addressing, in particular, new NRTIs (among which emtricitabine has in the mean time been approved for clinical use (Emtriva™)), NNRTIs, PIs (such as atazanavir, now also approved for clinical use (Reyataz™)), viral entry and integrase inhibitors, as well as some important procedures, such as therapeutic drug monitoring (TDM) and strategies, such as immune-based therapies (including therapeutic vaccines), cytokines (i.e. IL-2) and intermittent therapy with once-daily dosing regimens, that should keep pace with globally growing AIDS pandemic.

Combination therapy of AIDS is a complex matter. Many things can go wrong, and clinicians need sufficient training and expertise to be able to make the wisest decision in any particular setting. Continued education is a must, and comprehensive coverage is needed to bring this complex matter to the clinical and medical practice in a sufficiently elaborated way. Also researchers, whether basic or clinical, should familiarize themselves with all aspects of treating HIV patients, in order to grasp where the needs are for continued progress and extended knowledge. Drug designers should be aware of the complexities associated with the ultimate use of their potential new drugs. This is why this book will fill the needs of clinicians, healthcare workers, researchers and drug designers. It aims at delineating the complexity of combination therapy for AIDS, thereby providing guidance on the optimal use of the anti-HIV drugs.

Erik De Clercq
Anne-Mieke Vandamme July 2003

Combination Therapy of AIDS
Edited by E.D.A. De Clercq and A.-M.I. Vandamme

Anti-HIV agents to be used in drug combination regimens

Erik De Clercq

Rega Institute for Medical Research, Katholieke Universiteit Leuven, B-3000 Leuven, Belgium

Introduction

Antiviral therapy and, particularly, anti-HIV therapy has definitely come of age [1]. There are now more than 30 compounds that have been formally approved for the therapy of virus infections. The nucleoside reverse transcriptase inhibitors (NRTIs) zidovudine, didanosine, zalcitabine, stavudine, lamivudine and abacavir are used in the treatment of human immunodeficiency virus (HIV) infections and lamivudine is also licensed for the treatment of hepatitis B virus (HBV) infections. The nucleotide reverse transcriptase inhibitor (NtRTI) tenofovir has been approved for the treatment of HIV infections and is also effective against HBV infections. The non-nucleoside reverse transcriptase inhibitors (NNRTIs) nevirapine, delavirdine and efavirenz are exclusively used in the treatment of HIV infections, and so are the HIV protease inhibitors (PIs) saquinavir, ritonavir, indinavir, nelfinavir, amprenavir and lopinavir.

The drugs currently used in the treatment of HIV infections (AIDS), with their generic and brand names, are listed in Table 1. They can be catalogued according to their molecular target of action: reverse transcriptase (substrate binding site (NRTIs and NtRTIs), non-substrate binding site (NNRTIs) or protease (PIs)).

Virus-drug resistance

Mutations, engendering resistance (or reduced susceptibility) to the different NRTIs, NtRTIs, NNRTIs and PIs, have been described at, respectively, the reverse transcriptase (RT) and protease level. These mutations follow characteristic patterns, which differ for the different classes of HIV inhibitors (NRTIs, NtRTIs, NNRTIs and PIs) but show remarkable similarities within each class of inhibitors (Tab. 2) [2]. For tenofovir, only one mutation (K65R) has been associated with resistance, which makes it clearly different from the NRTIs, which, in general, give rise to multiple mutations in the RT.

Table 1. Drugs currently used to treat HIV infections (AIDS)

Drug	Brand name
Nucleoside reverse transcriptase inhibitors (NRTIs)	
Zidovudine (AZT)	Retrovir
Didanosine (ddI)	Videx
Zalcitabine (ddC)	Hivid
Stavudine (d4T)	Zerit
Lamivudine (3TC)	Epivir
Abacavir (ABC)	Ziagen
Emtricitabine [(–)FTC]	Emtriva
Nucleotide reverse transcriptase inhibitors (NtRTIs)	
Tenofovir disoproxil fumarate	Viread
Non-nucleoside reverse transcriptase inhibitors (NNRTIs)	
Nevirapine	Viramune
Delavirdine	Rescriptor
Efavirenz	Sustiva, Stocrin
Protease inhibitors (PIs)	
Saquinavir	Fortovase, Invirase
Ritonavir	Norvir
Indinavir	Crixivan
Nelfinavir	Viracept
Amprenavir	Agenerase
Lopinavir (with ritonavir)	Kaletra
Atazanavir	Reyataz
Viral fusion inhibitors	
Enfuvirtide	Fuzeon

As the emergence of resistance towards current anti-HIV drugs is an important determinant in eventual therapy failure, attempts have been focussed on the development of new anti-HIV compounds that either interact with targets other than the HIV reverse transcriptase or protease, or are capable of suppressing HIV strains that are resistant to the current HIV RT and protease inhibitors.

Virus adsorption inhibitors

Numerous polyanionic compounds, i.e., polysulfates (i.e., dextran sulfate, dextrin sulfate, heparin, heparan sulfate, polyvinylalcohol sulfate (PVAS), ...), polysulfonates (i.e., suramin, poly(4-styrene)sulfonate, polyvinylsulfonate (PVS), ...), polynucleotides (such as zintevir (a 17-mer capable of forming a double guanine quartet)), polyoxometalates, negatively charged albumins, and polycarboxylates (i.e., aurintricarboxylic acid (ATA)) have been reported to block HIV replication through interference with the interaction of the viral

Table 2.

Mutations in the reverse transcriptase gene associated with reduced susceptibility to reverse transcriptase inhibitors

Nucleoside reverse transcriptase inhibitors

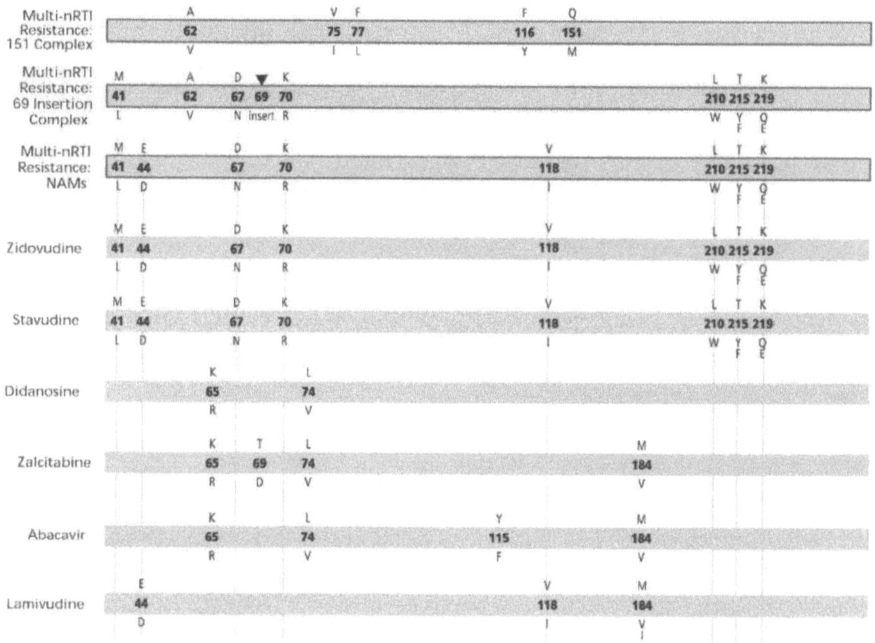

Nucleotide reverse transcriptase inhibitor

Nonnucleoside reverse transcriptase inhibitors

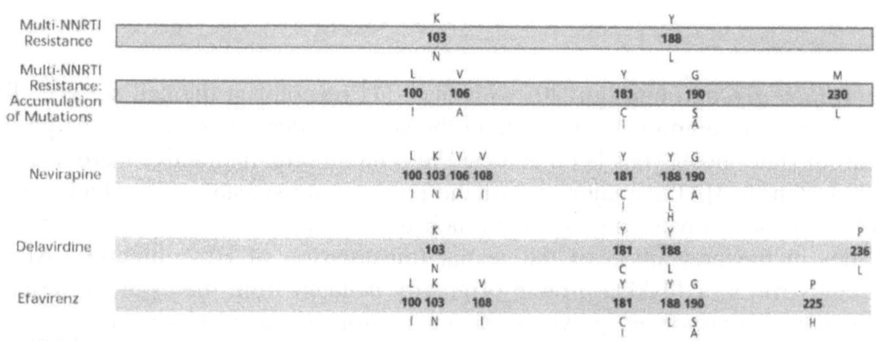

(Continued on next page)

Table 2. (Continued)

Mutations in the protease gene associated with reduced susceptibility to protease inhibitors

Protease inhibitors

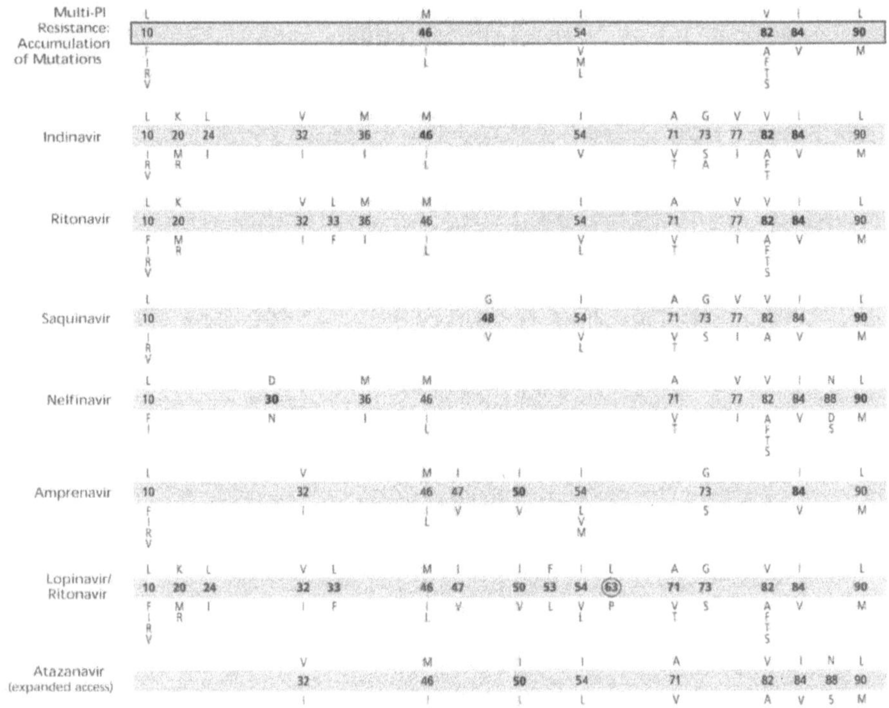

Mutations in the gp41 envelope gene associated with resistance to entry inhibitors

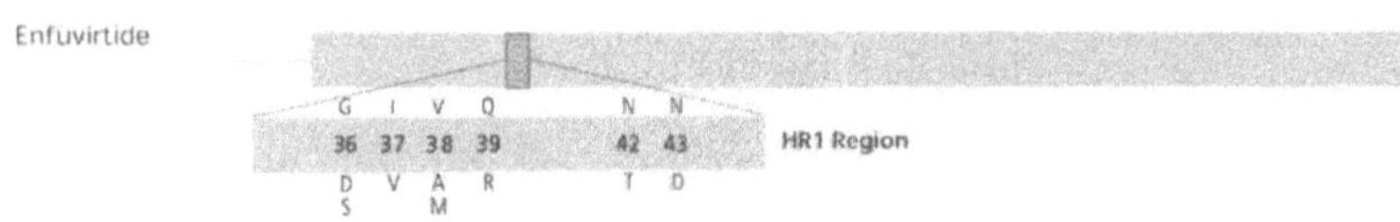

envelope glycoprotein (gp120) with the CD4 receptor at the cell surface. To this class of compounds also belong the cosalane derivatives where the polycarboxylate moiety has been attached to a membrane-interactive steroid (i.e., cholestane) [3]. The major role of the polyanionic substances, including the cosalane derivatives, may reside in their use as topical (i.e., vaginal) microbicides in the prevention of the sexual transmission of HIV infection. Also, cyanovirin-N, a 11-kDa protein originally isolated from the cyanobacterium *Nostoc ellipsosporum*, prevents the interaction of the viral envelope gp120 with its target cell receptors and thus qualifies as a potential microbicide to prevent the transmission of HIV [4, 5].

The attachment of the HIV surface glycoprotein gp120 to $CD4^+$ cells, the first step in virus entry, can also be blocked by CD4-immunoglobulin G2 (CD4-IgG2, PRO 542), a recombinant antibody-like fusion protein wherein the heavy- and light-chain variable domains of human IgG2 have been replaced with the D1D2 domains of human CD4 [6]. Preliminary evidence of antiviral activity, as monitored by a reduction in plasma viral load, was noted after (single) intravenous infusion of PRO 542 in HIV-infected individuals [7].

Viral co-receptor antagonists

Following their binding to the CD4 receptor, the HIV particles must interact, again through their envelope glycoprotein gp120, with either CXCR4 (the receptor for the CXC chemokine SDF-1 (stromal-cell derivatived factor)) or CCR5 (the receptor for the CC chemokine RANTES (regulated upon activation, normal T-cell expressed and secreted)) before they can enter the cells: CXCR4 acts as the co-receptor for HIV-1 strain that infect T-cells (T-tropic or X4 strains) and CCR5 functions as the co-receptor for HIV-1 strains that infect macrophages (M-tropic or R5 strains). Viral entry into the cells can be blocked at the CCR5 level, i.e., by TAK-779, a quaternary ammonium derivative, and at the CXCR4 level, i.e., by the bicyclam AMD3100. TAK-779 and AMD3100 have been shown to inhibit, in the nanomolar concentration range, the replication of R5 HIV-1 strains [8] and X4 HIV-1 strains [9], respectively. So as to simultaneously block X4 and R5 HIV strains, future developmental plans should focus on the combination of CXCR4 and CCR5 antagonists, or, if feasible, on the use of compounds that antagonize both receptors.

Following TAK-779, a number of piperidinylpiperidine derivatives, i.e., SCH 351125 [10], have been described as potent and selective CCR5 antagonists. They block the binding of RANTES to CCR5, as well as the replication of R5 HIV-1 strains, within the nanomolar concentration range. SCH 351125 (SCH-C) also strongly inhibits the replication of R5 HIV-1 in SCID-hu Thy/Liv mice, shows a favorable pharmacokinetic profile (i.e., good oral bioavailability in rats, dogs and monkeys), and has been selected as the drug candidate for further human clinical trials [11]. Preliminary clinical data, based on the oral administration of 25 mg SCH-C twice daily for 10 days to 12 adults infected with HIV-1 and currently on no antiretroviral agents, indicated that SCH-C is able to achieve a 0.5–1.0 $\log_{10}$ reduction in viral load [12].

The bicyclams had been known as potent and selective HIV inhibitors for a number of years before their target of action was identified as the CXCR4 coreceptor [9]. The bicyclam AMD3100 inhibits the replication of X4 HIV-1 strains within the nanomolar concentration range [13, 14]. The inhibitory effects of AMD3100 on the T-tropic and dual (X4/R5)-tropic HIV-1 strains have been demonstrated in a wide variety of cells expressing CXCR4, including peripheral blood mononuclear cells (PBMCs). AMD3100 has proved efficacious, alone and in combination with other anti-HIV drugs, in achieving a marked reduction in viral load in the SCID-hu Thy/Liv mouse model [15]. Following a phase I clinical trial for safety in normal healthy volunteers [16], AMD3100 entered phase I/II clinical trials in HIV-infected individuals, where it was found to effect a dose-dependent reduction of the T-tropic X4 HIV plasma viral load [17]. Notably, in eight of 19 patients with dual (X4/R5) or mixed (X4 + R5) virus at baseline, a complete loss of X4 virus was observed by day 11 of treatment with an AMD-3100 dose as low as 5 μg/kg/h.

Viral fusion inhibitors

The interaction of the X4 or R5 HIV envelope glycoprotein gp120 with the coreceptors CXCR4 and CCR5, respectively, is followed by a spring-loaded action of the viral glycoprotein gp41 that then anchors through its amino terminus into the target cell membrane and initiates the fusion of the two lipid bilayers (viral envelope and cellular membrane). This fusion process can be blocked by enfuvirtide (T-20, also called pentafuside and DP-178), a synthetic 36-amino acid peptide corresponding to residues 127–162 of the ectodomain of gp41 (or residues 643–678 in the gp120 precursor) (Fig. 1). Proof-of-concept that HIV fusion inhibitors are able to reduce virus replication *in vivo* has been provided by an initial clinical trial, where T-20 was found to achieve a 1.5- to 2.0-fold reduction in plasma viral load at the highest dose used (100 mg, twice daily) [18].

In a phase III, open-label, study, the so-called T-20 *versus* Optimized Regimen Only Study 1 (TORO 1) [19], patients from 48 sites in North and South America, with at least 5000 copies of HIV-1 RNA per ml of plasma were randomly assigned in a 2:1 ratio to receive enfuvirtide plus an optimized background regimen of three to five antiretroviral drugs or such a regimen alone (control group). Enfuvirtide (90 mg) was administered twice daily by subcuta-

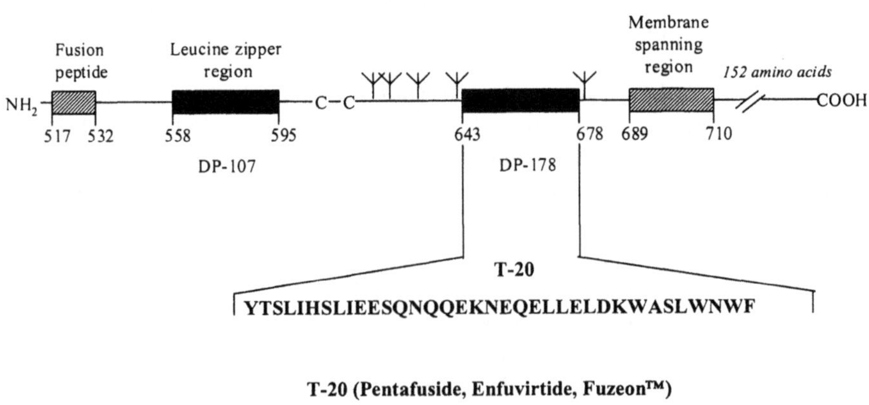

Figure 1. Enfuvirtide.

neous injection. At 24 weeks, the mean change from base line in the viral load (intention-to-treat) was a decrease of 1.696 $\log_{10}$ copies per ml in the enfuvirtide group, and a decrease of 0.764 $\log_{10}$ copies per ml in the control group ($p < 0.001$). The mean increases in $CD4^+$ cell count were 76 cells per µl and 32 cells per µl, respectively ($p < 0.001$). It was concluded that addition of enfuvirtide to an optimized antiretroviral drug regimen provided significant virologic and immunologic benefit through 24 weeks in patients who had previously received multiple antiretroviral drugs and had developed multi-drug HIV-1 resistance [19]. Enfuvirtide has been recently approved by the US Food and Drug Administration (FDA) for clinical use in the treatment of HIV infections.

Sensitivity to T-20 is strongly influenced by coreceptor specificity: the EC_{50} for R5 HIV-1 isolates is, at an average, 0.8 $\log_{10}$ higher than the EC_{50} for X4 HIV-1 isolates, this coreceptor specificity being defined by the gp120 V3 loop [20]. Baseline sensitivity to T-20 is determined by the N-terminal heptad repeat or first helical region (HR1) of gp41, where mutations G36D and V38A confer reduced sensitivity [20, 21]. These and other mutations seem to emerge promptly in patients upon monotherapy with T-20 [21]. Given the variability in T-20 sensitivity between different virus strains depending on mutations in the HR1 domain of gp41, as well as modifications in the gp120 V3 loop that determine coreceptor specificity [22], clinical use of fusion inhibitors such as T-20 may be made more efficient if combined with coreceptor antagonists.

Viral uncoating inhibitors

The ultimate capsid protein to be dissociated (“uncoated”) from the HIV RNA genome, before the latter can be submitted to the “reverse” transcription to proviral DNA, is the nucleocapsid protein (NCp7). NCp7 is firmly attached to the HIV genome, and, as it contains two zinc fingers motifs, targeting these zinc figures with, for example, zinc-ejecting compounds, such as azodicarbonamide (ADA) [23], has been considered as an attractive approach towards anti-HIV therapy. In principle, NCp7-targeted compounds should be able to interfere with both early (i.e., uncoating) and late stages (i.e., assembly) of the retroviral replicative cycle. Of the NCp7-targeted compounds, ADA has been the first to proceed to clinical trials: some preliminary evidence of efficacy was witnessed with add-on ADA in advanced AIDS patients failing current antiretroviral therapy [24]. These studies should be further extended before drawing any conclusions on the potential of ADA, or any other putative NCp7 zinc-finger inhibitors, in the treatment (or prevention) of HIV infections.

Nucleoside reverse transcriptase inhibitors (NRTIs)

The substrate (dNTP) binding site of the HIV reverse transcriptase (RT) is the target for a large variety of NRTI analogues, which have since many years [25] been recognized as efficacious drugs, and which are currently used in the treatment of HIV infections: i.e., zidovudine (AZT) (Fig. 2), didanosine (ddI) (Fig. 3), zalcitabine (ddC) (Fig. 4), stavudine (d4T) (Fig. 5), lamivudine (3TC) (Fig. 6) and abacavir (ABC) (Fig. 7). In addition to these formally approved drugs, emtricitabine [(–)2'-deoxy-3'-thia-5-fluorocytidine or (–)FTC] has been recently submitted for FDA approval and several other NRTIs such as (±)2'-

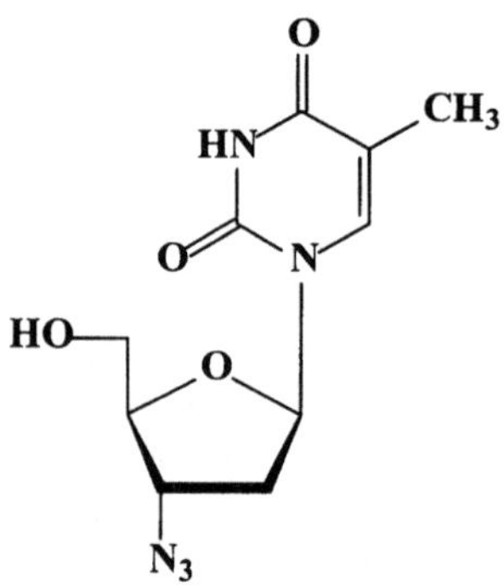

Zidovudine
3'-Azido-2',3'-dideoxythymidine
AZT
Retrovir®

Figure 2. Zidovudine.

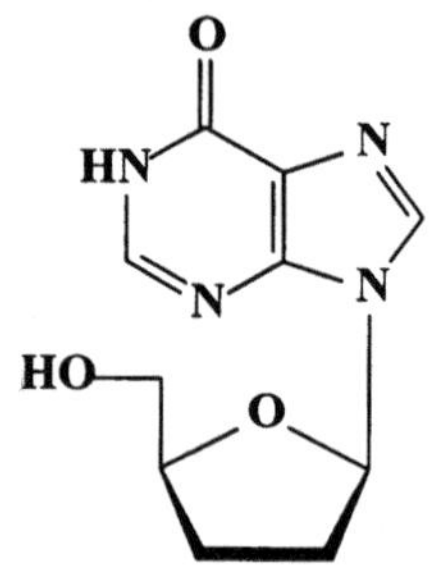

Didanosine
2',3'-Dideoxyinosine
DDI
Videx®

Figure 3. Didanosine.

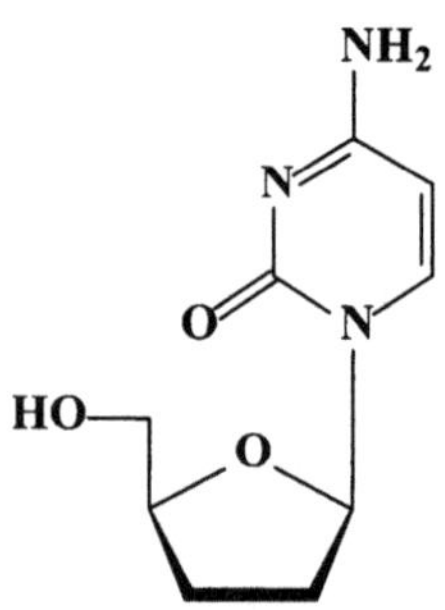

Zalcitabine
2',3'-Dideoxycytidine
DDC
Hivid®

Figure 4. Zalcitabine.

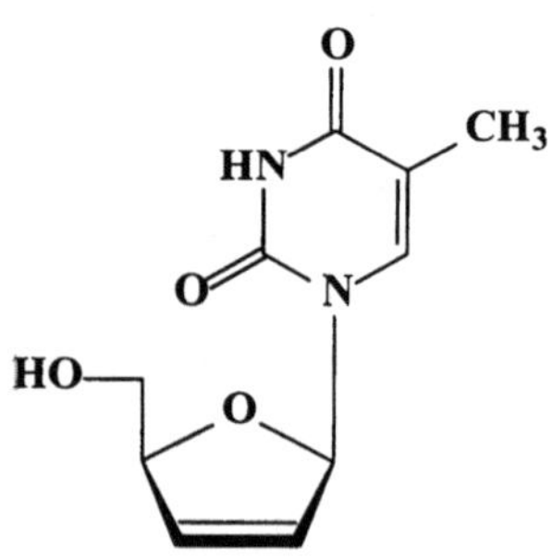

Stavudine
2',3'-Didehydro-2',3'-dideoxythymidine
D4T
Zerit®

Figure 5. Stavudine.

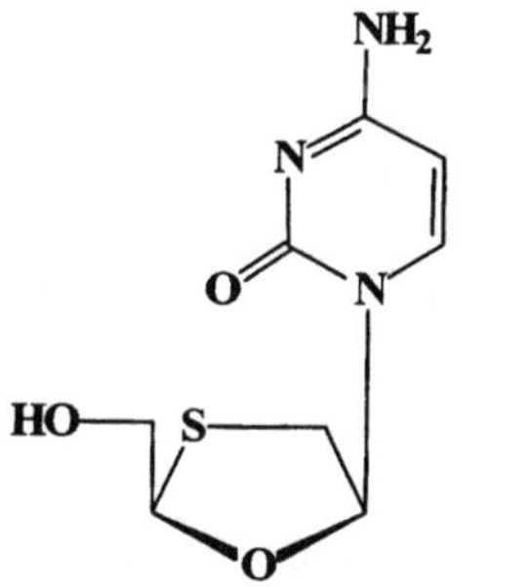

Lamivudine
2',3'-Dideoxy-3'-thiacytidine
3TC
Epivir®

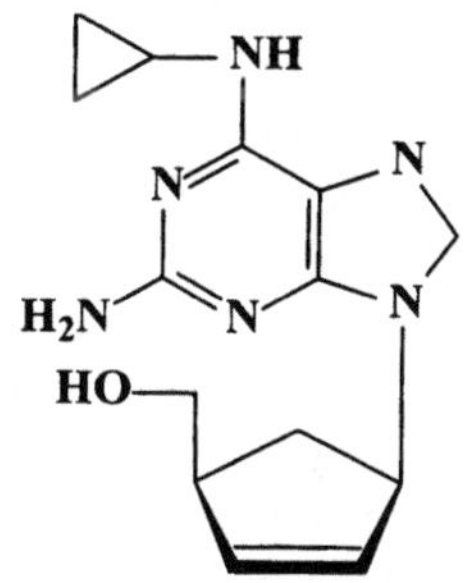

Abacavir
1592U89
Ziagen®

Figure 6. Lamivudine.

Figure 7. Abacavir.

deoxy-3'-oxa-4'-thiocytidine (dOTC), amdoxovir [(–)-β-D-2,6-diaminopurine dioxolane or DAPD], racivir [(±)2'-deoxy-3'-oxa-4'-thiocytidine (FdOTC)], reverset [RVT, DPC 817, β-D-2',3'-didehydro-2',3'-dideoxy-5-fluorocytidine (β-D-d4FC)], elvucitabine [ACH-126443, β-L-2',3'-didehydro-2',3'-dideoxy-5-fluorocytidine (β-L-d4FC)], and alovudine [MIV-310, 3'-fluoro-2',3'-dideoxythymidine (FddThd)] are in different stages of clinical development.

As a rule, these newer NRTIs retain activity against AZT- and/or 3TC-resistant HIV-1 strains, as has been specifically demonstrated for dOTC [26] and DAPD [27]. All NRTIs (whether old or new) have to be converted intracellularly, through three consecutive phosphorylations, to their 5'-triphosphate form before the latter can act as competitive inhibitors/substrate analogues/chain terminators at the reverse transcriptase level. The first phosphorylation step is the rate-limiting step in the intracellular metabolism of the NRTIs, and several nucleotide prodrugs have been designed that directly deliver the 5'-monophosphate form inside the cells: i.e., d4T aryloxyphosphoramidate [28] and *cyclo*saligenyl d4TMP [29]. Both these constructs release the d4T 5'-monophosphate within the cells and thus circumvent or by-pass the thymidine kinase needed for the first phosphorylation step.

Nucleotide reverse transcriptase inhibitors (NtRTIs)

The acyclic nucleoside phosphonates 9-(2-phosphonylmethoxyethyl)adenine (PMEA) and (*R*)-9-(2-phosphonylmethoxypropyl)adenine (PMPA) can be considered as nucleotide analogues in which the phosphate group has been built in as an enzymatically stable phosphonate moiety. These compounds thus bypass the first phosphorylation step, and require only two phosphorylations to be converted to their active metabolites, the diphosphate derivatives PMEApp and PMPApp. In this form they act as competitive inhibitors/substrates, with respect to dATP, in the HIV and HBV reverse transcriptase reaction, and when incorporated into the viral DNA chain, they obligatorily terminate further chain elongation. As neither PMEA (adefovir) nor PMPA (tenofovir) are readily bioavailable by the oral route, they have been converted to their oral prodrugs bis(pivaloyloxymethyl)-PMEA (adefovir dipivoxil) and bis(isopropyl-oxycarbonyloxymethyl)-PMPA (tenofovir disoproxil). The former has been licensed (Hepsera®) for the treatment of chronic HBV infections, whereas the latter, tenofovir disoproxil fumarate (Viread®) (Fig. 8), has been approved for the treatment of HIV infections.

Clinical trials have demonstrated that tenofovir (disoproxil fumarate) is well tolerated, reduces the viral load at non-toxic doses, and induces little, if any, virus-drug resistance even after prolonged (>1 year) treatment [30, 31]. In rhesus macaques infected with the highly pathogenic chimeric virus SHIV, tenofovir treatment initiated 1 week post infection, at a time when disseminated infection and extensive viral replication had already been established and $CD4^+$ T-cell loss had begun, led to prompt, virtually complete suppression of viral replication and long-term stabilization of $CD4^+$ T-cell levels, which were sustained, even after withdrawal of tenofovir (after 12 weeks of treatment) [32]. In a monotherapy trial with tenofovir disoproxil fumarate in 10 chronically HIV-1-infected antiretroviral-naïve individuals, the drug achieved a 1.5 $\log_{10}$ reduction in HIV-1 RNA levels during the 3 weeks of therapy, a response that was considered as robust as that observed with ritonavir monotherapy [33]. Apart from its efficacy, safety and tolerability, tenofovir disoproxil

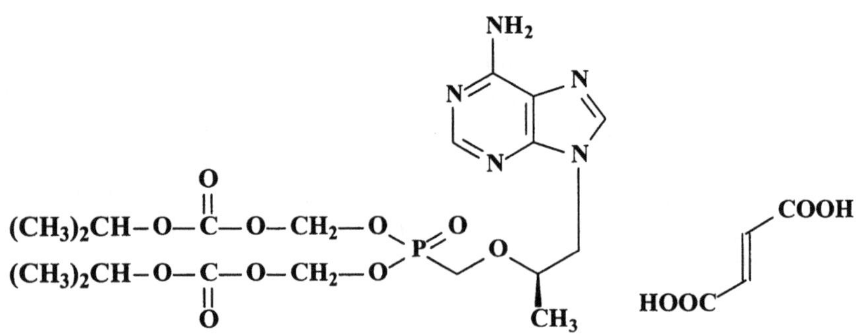

bis(POC)-PMPA
Tenofovir disoproxil fumarate
Viread®

Figure 8. Tenofovir disoproxil fumarate (TDF).

fumarate has a number of unique characteristics. It demonstrates activity against viruses resistant to the "classical" nucleoside analogues (i.e., zidovudine, didanosine and zalcitabine), including multidrug-resistant viruses carrying the Q151M mutation in their reverse transcriptase; and, furthermore, the long intracellular half-life of the active metabolite of tenofovir, its diphosphate PMPApp, which ranges from 12 to 15 h in activated lymphocytes and from 33 to 50 h in resting lymhocytes [34] allows infrequent, i.e., once-daily, dosing.

Non-nucleoside reverse transcriptase inhibitors (NNRTIs)

NNRTIs are defined by the fact that they interact with an allosteric, non-substrate binding site of the HIV-1 reverse transcriptase [35], and, consequently, they are only (or primarily) active against HIV-1, and not any of the other retroviruses. The officially approved NNRTIs nevirapine (Fig. 9), delavirdine (Fig. 10) and efavirenz (Fig. 11) have acquired a definitive "niche" in drug combination regimens for the treatment of HIV infections. As the older NNRTIs are notorious for rapidly eliciting virus-drug resistance (resulting from mutations at amino acid residues surrounding the NNRTI-binding site of HIV-1 RT), attempts have been made to design newer NNRTIs with higher potency and/or resilience to drug resistance mutations. This search has yielded a number of compounds which are indeed active against HIV strains that are resistant to the older NNRTIs. These "second generation" NNRTIs are now in development as potential drugs for the treatment of HIV-1 infections: they include capravirine (S-1153, AG1549) [36], SJ-3366 [37], DPC 083 [38] and dapivirine (TMC 125, R165335) [39]. As a rule, these newer NNRTIs are indeed active against those mutant virus strains that have developed resistance (primarily due to the K103N and Y181C mutations in the HIV-1 reverse transcriptase) to the "first generation" NNRTIs.

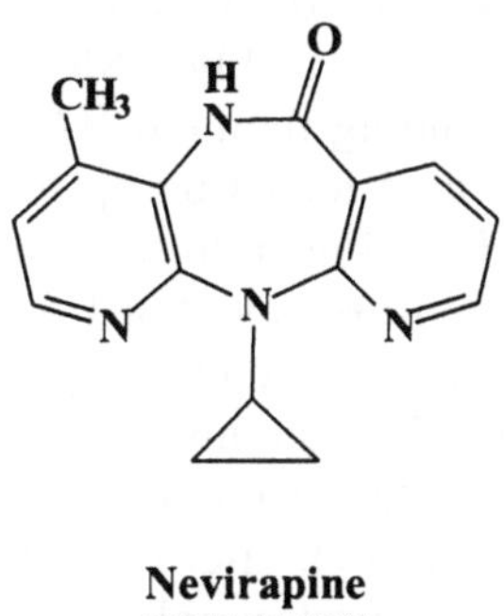

Nevirapine
BI-RG-587
Viramune®

Figure 9. Nevirapine.

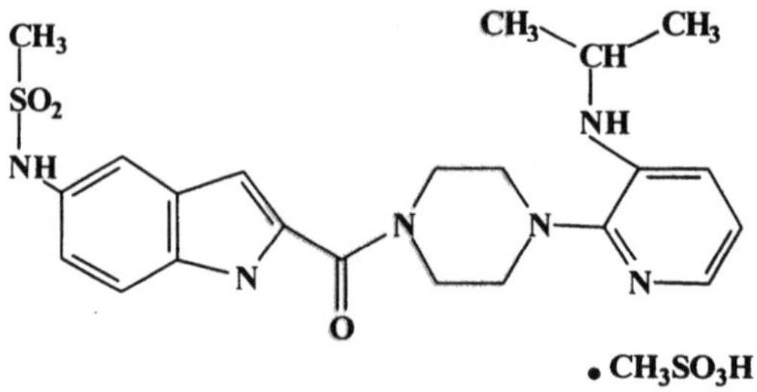

U-90152S
Delavirdine
Rescriptor®

Figure 10. Delavirdine.

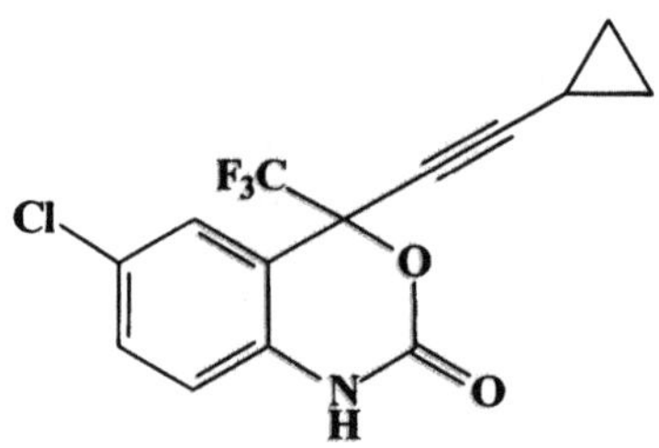

Benzoxazinone DMP 266
Efavirenz
Sustiva®
Stocrin®

Figure 11. Efavirenz.

When administered orally twice daily at a dose of 900 mg for 7 days in treatment-experienced patients with highly NNRTI-resistant virus and currently failing on an NNRTI-containing regimen, TMC125 demonstrated significant antiviral potency (viral load reduction up to 0.9 $\log_{10}$) [40]. In another study, monotherapy with TMC125 in antiretroviral-naïve HIV-1-infected individuals effected, after 1 week of treatment, a 1.5–2.0 $\log_{10}$ reduction of viral load, with a similar initial rate of decline of plasma HIV-1 as seen with a five-drug regimen [41]. DPC 083, administered orally once daily at a dose of 100 mg, in patients who had failed on the current NNRTIs and harbored NNRTI-resistant mutations, effected, after 8 weeks of treatment, a viral load reduction of 1.28 $\log_{10}$ [42]. In antiretroviral-naïve HIV-1-infected individuals, DPC 083 at an oral once-daily drug regime of 50, 100 or 200 mg provided through plasma drug levels that exceeded the IC_{90} for NNRTI-resistant HIV-1 mutants [43].

HIV integrase inhibitors

Gene expression (i.e., transcription to viral RNA) of the (pro)viral HIV DNA is only possible after the latter has been integrated into the host chromosome, and thus integrase has been considered an attractive target for chemotherapeutic intervention with the HIV replicative cycle. In the past, numerous integrase inhibitors have been described. None, however, showed sufficient specificity to be further pursued as an integrase-targeted drug. In some cases, as with L-chicoric acid [44], selective anti-HIV activity was noted in cell culture, but, then, the antiviral activity could be attributed primarily to inhibition of virus adsorption rather than proviral DNA integration. Of the compounds reported thus far, very few qualify as genuine inhibitors of HIV integrase, the first example being the 4-aryl-2,4-dioxobutanoic acid derivatives, or the so-called diketo acids (L-731,988 and L-708,906). These compounds were found to inhibit HIV-1 replication in cell culture, on the one hand, and to inhibit the strand transfer function of the integrase, on the other hand; and these two events could be causally linked, as mutations in the HIV-1 integrase conferred resistance to the inhibitory effects of the compounds on both strand transfer and HIV-1 infectivity [45]. Antiviral resistance to the diketo acid L-708,906 is associated with the mutations T66I, L74M and S230R in the HIV-1 integrase gene [46]: after 30 passages of HIV-1 in the presence of L-708,906, the T66I mutant was detected. On further passaging (up to 70 passages), the virus retained the T66I mutation but, in addition, acquired the L74M and S230R mutations. This multiple-mutant virus proved highly resistant to the diketo acid L-708,906 [46].

Reportedly in clinical development is the diketo derivative S-1360: this compound was recently announced as an HIV-1 integrase inhibitor for oral use [47]. S-1360 would inhibit the HIV-1 integrase at an IC_{50} of 20 nM, and HIV-1 replication at an EC_{50} of 140 nM, while its CC_{50} would be 110 μM, thus achieving a therapeutic index of almost one thousand [47]. Another HIV-1 integrase inhibitor that reportedly entered clinical development is the 1,6-naphthyridine-7-carboxamide L-870810 [48]. The mechanism of action of the diketo acids (i.e., L-731,988), as well as S-1360 and L-870810 may be based on an interaction between the carboxylate of the diketo acid or the isosteric heterocycle in the other compounds and metal ion(s) in the active site of the integrase, resulting in a functional sequestration of these critical metal cofactors [49].

Recently, an entirely new class of HIV integrase inhibitors was identified, namely that of the 5*H*-pyrano[2,3-di:-6,5-d']dipyrimidines (PDPs); the most potent congener of this series, V-165 inhibited the replication of HIV-1 at an EC_{50} of 8.9 μM, which is 14-fold below the cytotoxic threshold [50]. A close correlation was found between the anti-HIV activity observed in cell culture and the inhibitory activity in the integrase strand transfer assays. Time-of-addition experiments confirmed that V-165 interfered with the viral replication cycle at a time point coinciding with integration, a conclusion corroborated by *Alu*-PCR [50]. Whether V-165 or any of the aforementioned HIV integrase inhibitors may have clinical utility as such, or in combination with other anti-HIV agents, remains to be demonstrated.

HIV transcription inhibitors

At the transcription level, HIV gene expression may be inhibited by compounds that interact with cellular factors that bind to the long terminal repeat (LTR) promoter and that are needed for basal level transcription, such as NF-κB inhibitors [51]. Greater specificity, however, can be expected from those compounds that specifically inhibit the transactivation of the HIV LTR promoter by the viral trans-acting transactivator (Tat) protein. Several compounds have been described as inhibitors of the transcription process. For example, the 9-mer peptoid CGP64222, which is structurally reminiscent of the amino acid 48–56 sequence RKKRRQRRR of Tat, was designed to act as a Tat antagonist, and, although this 9-mer peptoid is able to interact with the Tat-driven transcription process, its anti-HIV activity in cell culture could be primarily attributed to an interaction with CXCR4, the co-receptor for X4 HIV strains [52].

Capping and methylation of HIV pre-mRNAs are coupled to the elongation by polymerase II. Binding of the capping enzyme and cap methyltransferase to polymerase II depends on phosphorylation of its carboxyl-terminal domain, and capping and methylation reactions start as soon as the nascent pre-mRNA has attained a chain length of 19–22 nucleotides. It has been recently demonstrated that the cotranscriptional capping of HIV mRNA is stimulated by Tat, consequently to its binding to the capping enzyme [53]. These findings implicate capping as an elongation checkpoint critical to HIV gene expression, and thus corroborate earlier observations that S-adenosylmethionine-dependent methylations play an important role in the Tat-dependent transactivation of transcription from LTR [54]. They also offer an explanation for the inhibitory effects of S-adenosylhomocysteine hydrolase inhibitors, such as neplanocin A and 3-deazaneplanocin A on both Tat-dependent transactivation and HIV replication. The (potential) role that may be reserved for S-adenosylhomocysteine hydrolase inhibitors *in vivo*, in the treatment of HIV infections, remains to be established.

HIV protease inhibitors

Viral proteases play a critical role in the life cycle of many different viruses, i.e., besides HIV, herpes-, picorna- and flaviviruses. As to its role in HIV replication, HIV protease secures the cleavage of the gag and gag-pol precursor proteins to the mature structural proteins (p17, p24, p9, p7) and functional proteins (protease, reverse transcriptase/RNAse H, and integrase), thus arresting maturation and thereby blocking infectivity of the nascent virions [55]. HIV protease inhibitors have been tailored after the target peptidic linkages in the gag and gag-pol precursor proteins that have to be cleaved by the protease. All six protease inhibitors that are currently approved and used for the treatment of HIV infections, namely saquinavir (Fig. 12), ritonavir (Fig. 13), indinavir (Fig. 14), nelfinavir (Fig. 15), amprenavir (Fig. 16) and lopinavir (Fig. 17), share the same structural determinant, i.e., an hydroxyethylene core group (instead of the normal peptidic linkage) that makes them non-scissile, "peptidomimetic", substrate analogues of the HIV protease [56]. The HIV protease inhibitors have proven to be valuable therapeutics in drug combination schedules with NRTIs and NNRTIs, in the treatment of HIV infections. Yet, they are met by a number of compounding factors, including overlapping resistance patterns (for example, mutations L10F/I/R/V, K20M/R, L24I, M46I/L, F53L, I54L/T/V, L63P, A71I/L/T/V, V82A/F/T, I84V and L90M are associated with

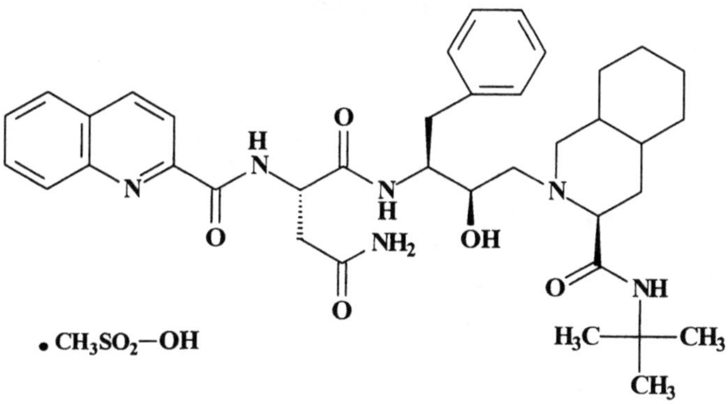

Saquinavir
Invirase®
Fortovase®

Figure 12. Saquinavir.

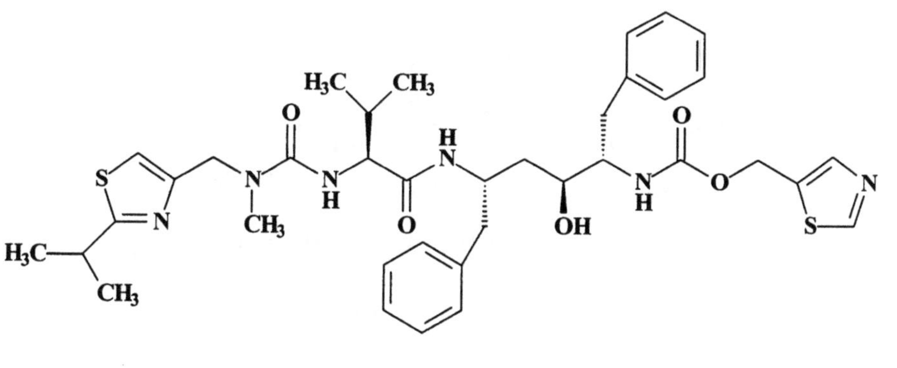

Ritonavir
Norvir®

Figure 13. Ritonavir.

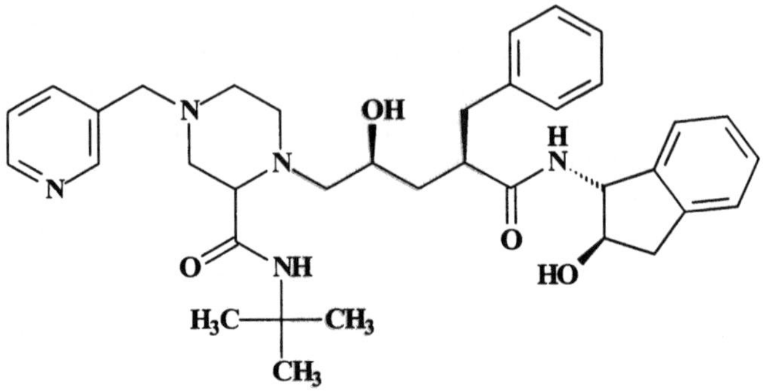

Indinavir
Crixivan®

Figure 14. Indinavir.

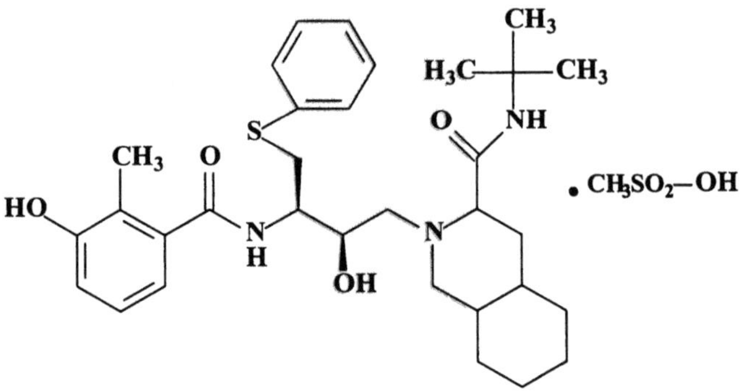

Nelfinavir
Viracept®

Figure 15. Nelfinavir.

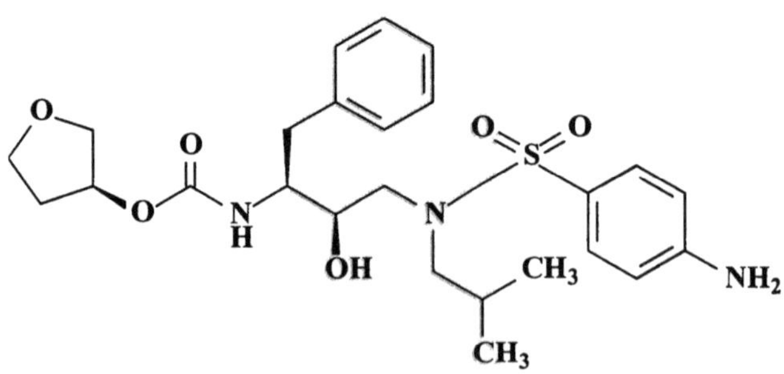

Amprenavir
Agenerase®

Figure 16. Amprenavir.

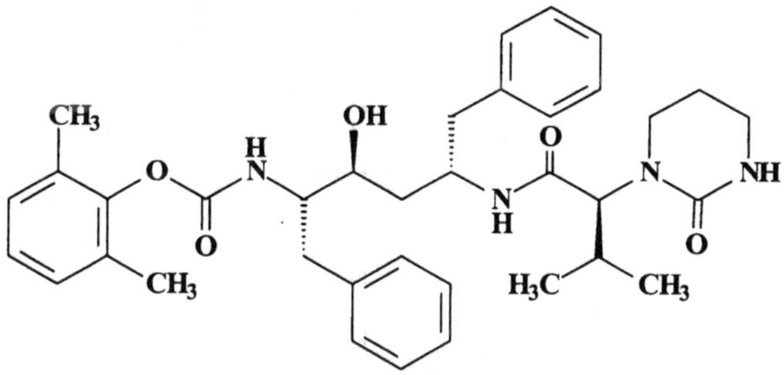

Lopinavir
Kaletra®
Kaletra = Lopinavir/Ritonavir at 4/1 ratio

Figure 17. Lopinavir.

reduced susceptibility to lopinavir [57]) and long-term side-effects (e.g., metabolic disturbances such as lipodystrophy and insulin-refractory diabetes, and cardiovascular complications). This has prompted the search for new, non-peptidic inhibitors of HIV protease, that would, hopefully, combine excellent anti-HIV potency with little or no cross-resistance with the "older" drugs and better tolerability on both short and long term. To these potential new candidate drugs that may fulfill at least part of the expectations belong atazanavir (BMS-232632) [58] and tipranavir (PNU-140690) [59].

Atazanavir is now in phase III clinical trials; it has been accredited with a favorable resistance profile that does not parallel any of the other protease inhibitors currently in clinical use, as well as a favorable pharmacokinetic profile that would allow once-daily dosing. Atazanavir, in combination with either stavudine, didanosine, lamivudine, zidovudine, nelfinavir, indinavir, ritonavir, saquinavir or amprenavir, yielded additive to moderately synergistic antiviral effects [58]. Nelfinavir-, saquinavir-, and amprenavir-resistant HIV-1 strains remained sensitive to BMS-232632, while indinavir- and ritonavir-resistant viruses showed six- to nine-fold changes in BMS-232632 sensitivity. On the other hand, BMS-232632-resistant (N88S, I84V) virus, selected upon virus passage in the presence of the compound, remained sensitive to saquinavir, but showed various levels (0.1- to 71-fold decrease in sensitivity) of cross-resistance to nelfinavir, indinavir, ritonavir and amprenavir [60]. Also tipranavir showed low cross-resistance to HIV strains that were resistant to the established (peptidomimetic) inhibitors of HIV protease [61]. Of 105 clinical HIV-1 samples with more than tenfold resistance to three or four of the peptidomimetic protease inhibitors and an average of six mutations per sample, 90% were susceptible to tipranavir, 8% had four- to 10-fold resistance, and only 2% had more than 10-fold resistance, to tipranavir [62]. Typically HIV-1 isolates with the protease inhibitor resistance mutations L10I, K20M, M36I, L63P, A71V, V82T and L90M retained susceptibility to tipranavir [63]. In addition, tipranavir retains marked activity against HIV-1 isolates derived from patients with multidrug resistance to other protease inhibitors [59], and might therefore be useful in combination regimens with other antiretroviral agents for patients who already failed on other protease inhibitor-containing drug regimens.

Recently, a novel HIV-1 protease inhibitor [UIC-94003 (TMC-126)], containing a *bis*-tetrahydrofuranyl urethane and 4-methoxybenzenesulfonamide, and thus structurally related to amprenavir, was reported to be extremely potent against a wide spectrum of HIV-1 strains (EC_{50}: 0.3–0.5 nM), whether resistant to other PIs or not [64]. Upon selection of HIV-1 in the presence of UIC-94003, mutants carrying a novel active-site mutation, A28S, in the presence of L10F, M46I, I50V, A71V and N88D, appeared. Modeling analysis revealed that close contact of UIC-94003 with the main chain of the protease active-site amino acid residues D29 and D30 differed from that of other PIs and may be important for its potency, particularly against drug-resistant HIV-1 variants [64].

Drug combinations

Given the number of anti-HIV agents that have now been licensed for clinical use and that essentially belong to five different categories (NRTIs, NtRTIs, NNRTIs, PIs and Fusion inhibitors), the number of possible drug combinations (Fig. 18) has become so large (and may be expected to expand in the future)

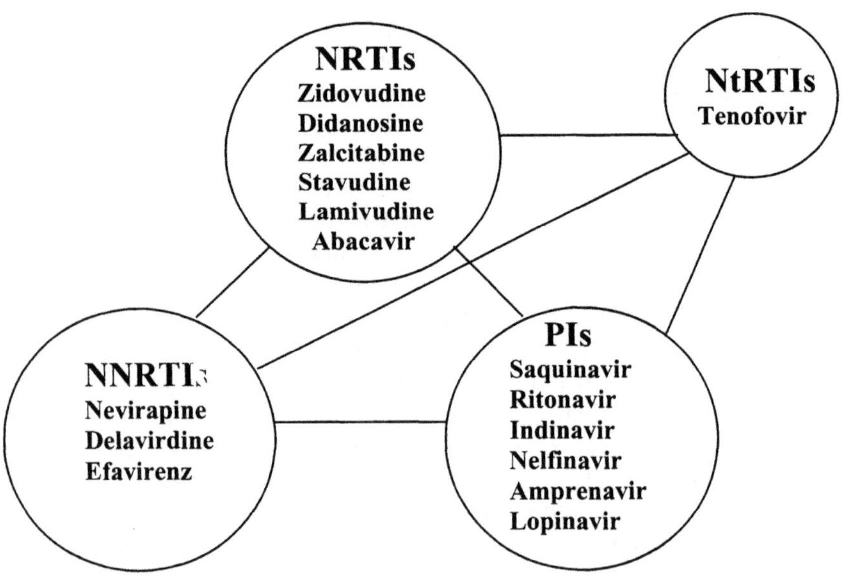

Figure 18. Combinations of NRTIs, NtRTIs, NNRTIs and PIs. These combinations could be extended with viral fusion inhibitors (i.e., enfuvirtide), when available.

that a rational choice of the optimal combination will be hard, if not impossible. It also follows that the optimal drug combination could not be generalized but should be individualized.

From an overview of the anti-HIV drug combinations that have been the subject of several clinical studies (Fig. 19), it appears that the highest success rates, in terms of achieving undetectable HIV loads in the highest percentage of patients, have been achieved with the combination of stavudine (d4T) + lamivudine (3TC) + efavirenz (EFV) and the combination of tenofovir disoproxil fumarate (TDF) + lamivudine (3TC) + efavirenz (EFV) [65, 66]. The latter combination appeared to be the more promising because of the lack of concomitant biochemical side-effects (triglyceride and cholesterol increases) [67].

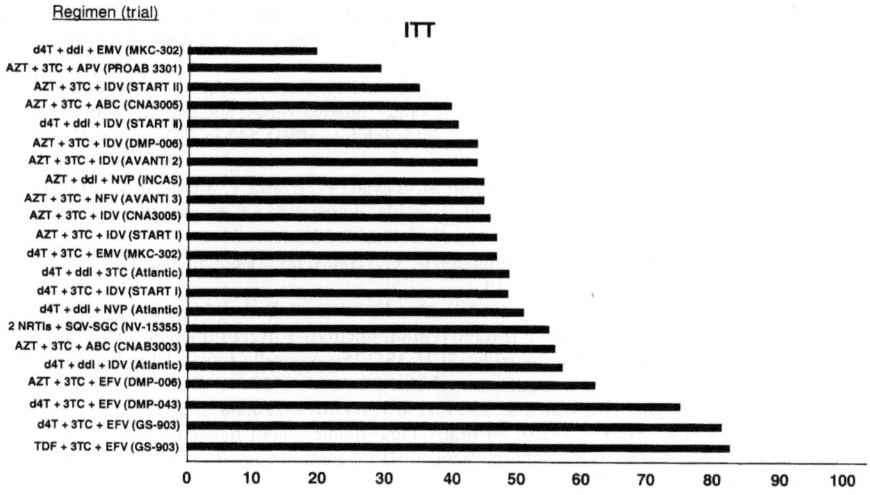

Figure 19. Results from clinical trials of various anti-HIV drug combinations: percent of patients with HIV load RNA ≤ 50 copies/ml at 48 weeks (ITT: intention to treat). Figure adapted from references [65, 66].

Conclusion

Various anti-HIV agents, targeted at a multitude of different steps within the HIV replicative cycle are now available, or under development, as therapeutic agents for the treatment of HIV infection (AIDS). So as to minimize the risks for the emergence of virus-drug resistance, it is obvious that these different compounds will have to be combined in the appropriate drug combination schedules which will have to be adjusted regularly, pending the clinical response to the older drugs, and the availability of newer compounds.

Acknowledgments
I thank Christiane Callebaut for her proficient editorial assistance.

References

1 De Clercq E (2001) Antiviral drugs: current state of the art. *J Clin Virol* 22: 73–89
2 *http://www.iasusa.org/resistance_mutations/index.html*
3 Santhosh KC, Paul GC, De Clercq E, Pannecouque C, Witvrouw M, Loftus TL, Turpin JA, Buckheit RW Jr, Cushman M (2001) Correlation of anti-HIV activity with anion spacing in a series of cosalane analogues with extended polycarboxylate pharmacophores. *J Med Chem* 44: 703–714
4 Boyd MR, Gustafson KR, McMahon JB, Shoemaker RH, O'Keefe BR, Mori T, Gulakowski RJ, Wu L, Rivera MI, Laurencot CM et al. (1997) Discovery of cyanovirin-N, a novel human immunodeficiency virus-inactivating protein that binds viral surface envelope glycoprotein gp120: potential applications to microbicide development. *Antimicrob Agents Chemother* 41: 1521–1530
5 Mori T, Boyd MR (2001) Cyanovirin-N, a potent human immunodeficiency virus-inactivating protein, blocks both CD4-dependent and CD4-independent binding of soluble gp120 (sgp120) to target cells, inhibits sCD4-induced binding of sgp120 to cell-associated CXCR4, and dissociates bound sgp120 from target cells. *Antimicrob Agents Chemother* 45: 664–672
6 Allaway GP, Davis-Bruno KL, Beaudry GA, Garcia EB, Wong EL, Ryder AM, Hasel KW, Gauduin MC, Koup RA, McDougal JS (1995) Expression and characterization of CD4-IgG2, a novel heterotetramer that neutralizes primary HIV type 1 isolates. *AIDS Res Hum Retroviruses* 11: 533–539
7 Jacobson JM, Lowy I, Fletcher CV, O'Neill TJ, Tran DN, Ketas TJ, Trkola A, Klotman ME, Maddon PJ, Olson WC et al. (2000) Single-dose safety, pharmacology, and antiviral activity of the human immunodeficiency virus (HIV) type 1 entry inhibitor PRO 542 in HIV-infected adults. *J Infect Dis* 182: 326–329
8 Baba M, Nishimura O, Kanzaki N, Okamoto M, Sawada H, Iizawa Y, Shiraishi M, Aramaki Y, Okonogi K, Ogawa Y et al. (1999) A small-molecule, nonpeptide CCR5 antagonist with highly potent and selective anti-HIV-1 activity. *Proc Natl Acad Sci USA* 96: 5698–5703
9 De Clercq E (2000) Inhibition of HIV infection by bicyclams, highly potent and specific CXCR4 antagonists. *Mol Pharmacol* 57: 833–839
10 Palani A, Shapiro S, Clader JW, Greenlee WJ, Cox K, Strizki J, Endres M, Baroudy BM (2001) Discovery of 4-[(Z)-(4-bromophenyl)-(ethoxyimino)methyl]-1'-[(2,4-dimethyl-3-pyridinyl)carbonyl]-4'-methyl-1,4'-bipiperidine N-oxide (SCH 351125): an orally bioavailable human CCR5 antagonist for the treatment of HIV infection. *J Med Chem* 44: 3339–3342
11 Strizki JM, Xu S, Wagner NE, Wojcik L, Liu J, Hou Y, Endres M, Palani A, Shapiro S, Clader JW et al. (2001) SCH-C (SCH 351125), an orally bioavailable, small molecule antagonist of the chemokine receptor CCR5, is a potent inhibitor of HIV-1 infection *in vitro* and *in vivo*. *Proc Natl Acad Sci USA* 98: 12718–12723
12 Reynes J, Rouzier R, Kanouni T, Baillat V, Baroudy B, Keung A, Hogan C, Markowitz M, Laughlin M (2002) SCH C: safety and antiviral effects of a CCR5 receptor antagonist in HIV-1 infected subjects. 9th Conference on Retroviruses and Opportunistic Infections, Seattle, Washington, USA, 24–28 February 2002. Abstracts, p 53, no 1
13 De Clercq E, Yamamoto N, Pauwels R, Balzarini J, Witvrouw M, De Vreese K, Debyser Z, Rosenwirth B, Peichl P, Datema R et al. (1994) Highly potent and selective inhibition of human immunodeficiency virus by the bicyclam derivative JM3100. *Antimicrob Agents Chemother* 38: 668–674
14 Schols D, Esté JA, Henson G, De Clercq E (1997) Bicyclams, a class of potent anti-HIV agents, are targeted at the HIV coreceptor fusin/CXCR-4. *Antiviral Res* 35: 147–156
15 Datema R, Rabin L, Hincenbergs M, Moreno MB, Warren S, Linquist V, Rosenwirth B, Seifert J, McCune JM (1996) Antiviral efficacy *in vivo* of the anti-human immunodeficiency virus bicyclam SDZ SID 791 (JM 3100), an inhibitor of infectious cell entry. *Antimicrob Agents Chemother* 40: 750–754
16 Hendrix CW, Flexner C, MacFarland RT, Giandomenico C, Fuchs EJ, Redpath E, Bridger G, Henson GW. (2000) Pharmacokinetics and safety of AMD3100, a novel antagonist of the CXCR-4 chemokine receptor, in human volunteers. *Antimicrob Agents Chemother* 44: 1667–1673
17 Schols D, Claes S, De Clercq E, Hendrix C, Bridger G, Calandra G, Henson GW, Fransen S, Huang W, Whitcomb JM et al. (2002) AMD-3100, a CXCR4 antagonist, reduced HIV viral load and X4 virus levels in humans. 9th Conference on Retroviruses and Opportunistic Infections, Seattle, Washington, USA, 24–28 February 2002. Abstracts, p 53, no 2
18 Kilby JM, Hopkins S, Venetta TM, Dimassimo B, Cloud GA, Lee JY, Alldredge L, Hunter E,

Lambert D, Bolognesi D et al. (1998) Potent suppression of HIV-1 replication in humans by T-20, a peptide inhibitor of gp41-mediated virus entry. *Nature Medicine* 4: 1302–1307

19 Lalezari JP, Henry K, O'Hearn M, Montaner JSG, Piliero PJ, Trottier B, Walmsley S, Cohen C, Kuritzkes DR, Eron JJ Jr, et al. (2003) Enfuvirtide, an HIV-1 fusion inhibitor, for drug-resistant HIV infection in North and South America. *N Engl J Med* 348: May 29

20 Derdeyn CA, Decker JM, Sfakianos JN, Wu X, O'Brien WA, Ratner L, Kappes JC, Shaw GM, Hunter E (2000) Sensitivity of human immunodeficiency virus type 1 to the fusion inhibitor T-20 is modulated by coreceptor specificity defined by the V3 loop of gp120. *J Virol* 74: 8358–8367

21 Wei X, Decker JM, Liu H, Zhang Z, Arani RB, Kilby JM, Saag MS, Wu X, Shaw GM, Kappes JC (2002) Emergence of resistant human immunodeficiency virus type 1 in patients receiving fusion inhibitor (T-20) monotherapy. *Antimicrob Agents Chemother* 46: 1896–1905

22 Reeves JD, Gallo SA, Ahmad N, Miamidian JL, Harvey PE, Sharron M, Pohlmann S, Sfakianos JN, Derdeyn CA, Blumenthal R et al. (2002) Sensitivity of HIV-1 to entry inhibitors correlates with envelope/coreceptor affinity, receptor density, and fusion kinetics. *Proc Natl Acad Sci USA* 99: 16249–16254

23 Rice WG, Turpin JA, Huang M, Clanton D, Buckheit RW Jr, Covell DG, Wallqvist A, McDonnell NB, DeGuzman RN, Summers MF et al. (1997) Azodicarbonamide inhibits HIV-1 replication by targeting the nucleocapsid. *Nature Medicine* 3: 341–345

24 Goebel F-D, Hemmer R, Schmit J-C, Bogner JR, De Clercq E, Witvrouw M, Pannecouque C, Valeyev R, Vandevelde M, Margery H et al. (2001) Phase I/II dose escalation and randomized withdrawal study with add-on azodicarbo-namide in patients failing on current antiretroviral therapy. *AIDS* 15: 33–45

25 De Clercq E (1995) Toward improved anti-HIV chemotherapy: therapeutic strategies for intervention with HIV infections. *J Med Chem* 38: 2491–2517

26 Stoddart CA, Moreno ME, Linquist-Stepps VD, Bare C, Bogan MR, Gobbi A, Buckheit RW Jr, Bedard J, Rando RF, McCune JM (2000) Antiviral activity of 2'-deoxy-3'-oxa-4'-thiocytidine (BCH-10652) against lamivudine-resistant human immunodeficiency virus type 1 in SCID-hu Thy/Liv mice. *Antimicrob Agents Chemother* 44: 783–786

27 Furman PA, Jeffrey J, Kiefer LL, Feng JY, Anderson KS, Borroto-Esoda K, Hill E, Copeland WC, Chu CK, Sommadossi JP et al. (2001) Mechanism of action of 1-β-D-2,6-diaminopurine dioxolane, a prodrug of the human immunodeficiency virus type 1 inhibitor 1-β-D-dioxolane guanosine. *Antimicrob Agents Chemother* 45: 158–165

28 Saboulard D, Naesens L, Cahard D, Salgado A, Pathirana R, Velazquez S, McGuigan C, De Clercq E, Balzarini J (1999) Characterization of the activation pathway of phosphoramidate triester prodrugs of stavudine and zidovudine. *Mol Pharmacol* 56: 693–704

29 Balzarini J, Aquaro S, Knispel T, Rampazzo C, Bianchi V, Perno CF, De Clercq E, Meier C (2000) Cyclosaligenyl-2',3'-didehydro-2',3'-dideoxythymi-dine monophosphate: efficient intracellular delivery of d4TMP. *Mol Pharmacol* 58: 928–935

30 Squires K, Pierone G, Berger D, Steinhart C, Bellos N, Becker SL, Chen SS, Miller MD, Coakley DF, Cheng A (2002) Tenofovir DF: a 48-week final analysis from a phase III randomized, double blind placebo controlled study in antiretroviral experienced patients. 9th Conference on Retroviruses and Opportunistic Infections, Seattle, Washington, USA, 24–28 February 2002. Abstracts, p 209, no 413-W

31 Margot NA, Johnson A, Cheng A, Coakley DF, Miller MD (2002) Final 48-week genotypic and phenotypic analyses of study 907: tenofovir DF (TDF) added to stable background regimens. 9th Conference on Retroviruses and Opportunistic Infections, Seattle, Washington, USA, 24–28 February 2002. Abstracts, p 209, no 414-W

32 Smith MS, Foresman L, Lopez GJ, Tsay J, Wodarz D, Lifson JD, Page A, Wang C, Li Z, Adany I et al. (2000) Lasting effects of transient postinoculation tenofovir [9-R-(2-phosphonomethoxypropyl)adenine] treatment on SHIV(KU2) infection of rhesus macaques. *Virology* 277: 306–315

33 Louie M, Hogan C, Hurley A, Captan B, Flaherty J, Lamy P, Balagtas A, Coakley D, Chung C, Ho DD et al. (2002) Determining the relative efficacy of tenofovir DF using frequent measurements of HIV-1 RNA during a short course of monotherapy in antiretroviral drug naïve individuals. 9th Conference on Retroviruses and Opportunistic Infections, Seattle, Washington, USA, 24–28 February 2002. Abstracts, p 54, no 3

34 Deeks SG, Barditch-Crovo P, Lietman PS, Hwang F, Cundy KC, Rooney, JF, Hellmann NS, Safrin S, Kahn JO (1998) Safety, pharmacokinetics, and antiretroviral activity of intravenous 9-[2-(R)-

(phosphonomethoxy)propyl]adenine, a novel anti-human immunodeficiency virus (HIV) therapy, in HIV-infected adults. *Antimicrob Agents Chemother* 42: 2380–2384
35 De Clercq E (1998) The role of non-nucleoside reverse transcriptase inhibitors (NNRTIs) in the therapy of HIV-1 infection. *Antiviral Res* 38: 153–179
36 Ren J, Nichols C, Bird LE, Fujiwara T, Sugimoto H, Stuart DI, Stammers DK (2000) Binding of the second generation non-nucleoside inhibitor S-1153 to HIV-1 reverse transcriptase involves extensive main chain hydrogen bonding. *J Biol Chem* 275: 14316–14320
37 Buckheit RW Jr, Watson K, Fliakas-Boltz V, Russell J, Loftus TL, Osterling MC, Turpin JA, Pallansch LA, White EL, Lee JW et al. (2001) SJ-3366, a unique and highly potent nonnucleoside reverse transcriptase inhibitor of human immunodeficiency virus type 1 (HIV-1) that also inhibits HIV-2. *Antimicrob Agents Chemother* 45: 393–400
38 Corbett JW, Ko SS, Rodgers JD, Jeffrey S, Bacheler LT, Klabe RM, Diamond S, Lai CM, Rabel SR, Saye JA et al. (1999) Expanded-spectrum nonnucleoside reverse transcriptase inhibitors inhibit clinically relevant mutant variants of human immunodeficiency virus type 1. *Antimicrob Agents Chemother* 43: 2893–2897
39 Andries K, de Béthune M-P; Kukla MJ, Azijn H, Lewi PJ, Janssen PAJ, Pauwels R (2000) R165335-TMC125, a novel non nucleoside reverse transcriptase inhibitor (NNRTI) with nanomolar activity against NNRTI resistant HIV strains. Abstracts of the 40th Interscience Conference on Antimicrobial Agents and Chemotherapy, Toronto, Canada, 17–20 September 2000, no. 1840
40 Gazzard B, Pozniak A, Arasteh K, Staszewski S, Rozenbaum W, Yeni P, van 't Klooster G, De Dier K, Peeters M, de Béthune MP et al. (2002) TMC125, a next-generation NNRTI, demonstrates high potency after 7 days therapy in treatment-experienced HIV-1-infected individuals with phenotypic NNRTI resistance. 9th Conference on Retroviruses and Opportunistic Infections, Seattle, Washington, USA, 24–28 February 2002. Abstracts, p 54, no 4
41 Sankatsing S, Weverling G, van 't Klooste G, Prins J, Lange J (2002) TMC125 monotherapy for 1 week results in a similar initial rate of decline of HIV-1 RNA as therapy with a 5-drug regimen. 9th Conference on Retroviruses and Opportunistic Infections, Seattle, Washington, USA, 24–28 February 2002. Abstracts, p 54, no 5
42 Ruiz N, Nusrat R, Lauenroth-Mai E, Berger D, Walworth C, Bacheler LT, Ploughman L, Tsang P, Labriola D, Echols R et al. (2002) Levy R, and the DPC 083-203 Study Team. Study DPC 083-203, a phase II comparison of 100 and 200 mg once-daily DPC 083 and 2 NRTIs in patients failing a NNRTI-containing regimen. 9th Conference on Retroviruses and Opportunistic Infections, Seattle, Washington, USA, 24–28 February 2002. Abstracts, p 55, no 6
43 Ruiz N, Nusrat R, Lazzarin A, Arasteh K, Goebel F-D, Audagnotto S, Rachlis A, Arribas J, Ploughman L, Fiske W et al. (2002) Labriola D, Levy R, Echols R, and the DPC 083-201 Study Team. Study DPC 083-201: a phase II double-blind (DB) comparison of 3 once-daily doses of the NNRTI DPC 083 vs 600 mg efavirenz (EFV) in combination with 2 NRTIs in HIV antiretroviral (ARV) treatment naïve-patients. 9th Conference on Retroviruses and Opportunistic Infections, Seattle, Washington, USA, 24–28 February 2002. Abstracts, p 55, no 7
44 Pluymers W, Neamati N, Pannecouque C, Fikkert V, Marchand C, Burke TR Jr, Pommier Y, Schols D, De Clercq E, Debyser Z et al. (2000) Viral entry as the primary target for the anti-HIV activity of chicoric acid and its tetra-acetyl esters. *Mol Pharmacol* 58: 641–648
45 Hazuda DJ, Felock P, Witmer M, Wolfe A, Stillmock K, Grobler JA, Espeseth A, Gabryelski L, Schleif W, Blau C et al. (2000) Inhibitors of strand transfer that prevent integration and inhibit HIV-1 replication in cells. *Science* 287: 646–650
46 Witvrouw M, Fikkert V, Van Maele B, Pannecouque C, Neamati N, Burke TR Jr, Pais G, De Clercq E, Debyser Z (2002) Antiviral resistance to diketo acids is associated with the mutations T66I, L74M and S230R in the HIV-1 integrase gene. Fifteenth International Conference on Antiviral Research, Prague, Czech Republic, 17–21 March 2002. *Antiviral Res* 53, A48, no 45
47 Yoshinaga T, Sato A, Fujishita T, Fujiwara T (2002) S-1360: *in vitro* activity of a new HIV-1 integrase inhibitor in clinical development. 9th Conference on Retroviruses and Opportunistic Infections, Seattle, Washington, USA, 24–28 February 2002. Abstracts, p 55, no 8
48 Miller MD (2002) HIV integrase inhibitors: *in vitro* consequences of resistance. XIVth World Congress of Pharmacology, San Francisco, CA, USA, 7–12 July 2002. Abstracts, no 118.2
49 Grobler JA, Stillmock K, Hu B, Witmer M, Felock P, Espeseth AS, Wolfe A, Egbertson M, Bourgeois M, Melamed J et al. (2002) Diketo acid inhibitor mechanism and HIV-1 integrase: implications for metal binding in the active site of phosphotransferase enzymes. *Proc Natl Acad Sci USA* 99: 6661–6666
50 Pannecouque C, Pluymers W, Van Maele B, Tetz V, Cherepanov P, De Clercq E, Witvrouw M,

Debyzer Z. (2002) New class of HIV integrase inhibitors that block viral replication in cell culture. *Curr Biol* 12: 1169–1177

51 Daelemans D, Vandamme A-M, De Clercq E (1999) Human immunodeficiency virus gene regulation as a target for antiviral chemotherapy. *Antiviral Chem Chemother* 10: 1–14

52 Daelemans D, Schols D, Witvrouw M, Pannecouque C, Hatse S, Van Dooren S, Hamy F, Klimkait T, De Clercq E, Vandamme A-M (2000) A second target for the peptoid Tat/transactivation response element inhibitor CGP64222: inhibition of human immunodeficiency virus replication by blocking CXC-chemokine receptor 4-mediated virus entry. *Mol Pharmacol* 57: 116–124

53 Chiu YL, Ho CK, Saha N, Schwer B, Shuman S, Rana TM (2002) Tat stimulates cotranscriptional capping of HIV mRNA. *Mol Cell* 10: 585–597

54 Daelemans D, Esté JA, Witvrouw M, Pannecouque C, Jonckheere H, Aquaro S, Perno CF, De Clercq E, Vandamme A-M (1997) S-Adenosylhomocysteine hydrolase inhibitors interfere with the replication of human immunodeficiency virus type 1 through inhibition of the LTR transactivation. *Mol Pharmacol* 52: 1157–1163

55 Flexner C (1998) HIV-protease inhibitors. *N Eng J Med* 338: 1281–1292

56 Molla A, Granneman GR, Sun E, Kempf DJ (1998) Recent developments in HIV protease inhibitor therapy. *Antiviral Res* 39: 1–23

57 Kempf DJ, Isaacson JD, King MS, Brun SC, Xu Y, Real K, Bernstein BM, Japour AJ, Sun E, Rode RA (2001) Identification of genotypic changes in human immunodeficiency virus protease that correlate with reduced susceptibility to the protease inhibitor lopinavir. *J Virol* 75: 7462–7469

58 Robinson BS, Riccardi KA, Gong YF, Guo Q, Stock DA, Blair WS, Terry BJ, Deminie CA, Djang F, Colonno RJ et al. (2000) BMS-232632, a highly potent human immunodeficiency virus protease inhibitor that can be used in combination with other available antiretroviral agents. *Antimicrob Agents Chemother* 44: 2093–2099

59 Rusconi S, La Seta Catamancio S, Citterio P, Kurtagic S, Violin M, Balotta C, Moroni M, Galli M, d'Arminio-Monforte A (2000) Susceptibility to PNU-140690 (tipranavir) of human immunodeficiency virus type 1 isolates derived from patients with multidrug resistance to other protease inhibitors. *Antimicrob Agents Chemother* 44: 1328–1332

60 Gong YF, Robinson BS, Rose RE, Deminie C, Spicer TP, Stock D, Colonno RJ, Lin PF (2000) *In vitro* resistance profile of the human immunodeficiency virus type 1 protease inhibitor BMS-232632. *Antimicrob Agents Chemother* 44: 2319–2326

61 Poppe SM, Slade DE, Chong KT, Hinshaw RR, Pagano PJ, Markowitz M, Ho DD, Mo H, Gorman RR 3rd, Dueweke TJ et al. (1997) Antiviral activity of the dihydropyrone PNU-140690, a new nonpeptidic human immunodeficiency virus protease inhibitor. *Antimicrob Agents Chemother* 41: 1058–1063

62 Larder BA, Hertogs K, Bloor S, van den Eynde CH, DeCian W, Wang Y, Freimuth WW, Tarpley G (2000) Tipranavir inhibits broadly protease inhibitor-resistant HIV-1 clinical samples. *AIDS* 14: 1943–1948

63 Back NK, van Wijk A, Remmerswaal D, van Monfort M, Nijhuis M, Schuurman R, Boucher CA (2000) *In vitro* tipranavir susceptibility of HIV-1 isolates with reduced susceptibility to other protease inhibitors. *AIDS* 14: 101–102

64 Yoshimura K, Kato R, Kavlick MF, Nguyen A, Maroun V, Maeda K, Hussain KA, Ghosh AK, Gulnik SV, Erickson JW et al. (2002) A potent human immunodeficiency virus type 1 protease inhibitor, UIC-94003 (TMC-126), and selection of a novel (A28S) mutation in the protease active site. *J Virol* 76: 1349–1358

65 Bartlett J, Demasi R, Quinn J, Moxham C, Rousseau F (2000) Meta-analysis of efficacy of triple combination therapy in antiretroviral-naive HIV-1-infected adults. 7th Conference on Retroviruses and Opportunistic Infections, San Francisco, CA, USA, 30 January– February 2000. Abstracts, no 519

66 Staszewski S, Gallant J, Pozniak A, Suleiman JMAH, Dejesus E, Koenig E, Coleman S, Lu B, Cheng AK, Coakley DF (2002) Efficacy and safety of tenofovir disoproxil fumarate (TDF) *versus* stavudine (d4T) when used in combination with lamivudine (3TC) and efavirenz (EFV) in HIV-1 infected patients naïve to antiretroviral therapy (ART): 48-week interim results. XIV International AIDS Conference, Barcelona, Spain, 7–12 July 2002. Abstracts, no Or17

67 Staszewski S, Gallant JE, Pozniak AL, Suleiman JMAH, DeJesus E, Lu B, Sayre J, Cheng A (2003) Efficacy and safety of tenofovir DF (TDF) *versus* stavudine (d4T) when used in combination with lamivudine and efavirenz in antiretroviral naïve patients: 86-week preliminary interim reslts. 10th Conference on Retroviruses and Opportunistic Infections, Boston, MA, USA, 10–14 February 2003. Abstracts, no 564b

Combination Therapy of AIDS
Edited by E.D.A. De Clercq and A.-M.I. Vandamme

A perspective of the history of HAART

Deborah Konopnicki and Nathan Clumeck

Saint-Pierre University Hospital, Department of Infectious Diseases, 322 rue Haute, 1000 Brussels, Belgium

Introduction

The history of HIV treatment is characterised by a rapid development. In 1987, 6 years after the discovery of AIDS, the US Food and Drug administration approved the first drug against HIV. Since then three different classes of antiretroviral drugs including 16 drugs are currently available, others have reached phase III studies and immunotherapy and vaccines are in development.

This history is also characterised by new approaches to treatment research and drug development. Patient associations in USA and Europe have for the first time in medical history contributed to accelerating and changing the process of access to new drugs using pressure on politicians, researchers, clinicians and industry. The regulatory authorities have had to adapt their administrative requirements in order to accelerate the availability of new compounds. The pharmaceutical industry has invested in treatment research but the development of new drugs has mostly been directed towards the developed world. Finally clinicians have learned to design clinical trials for new drugs or strategies and to switch from supportive care to an early highly active treatment which is aimed to control virological and/or immunological parameters. This chapter will focus on the history of therapy in chronically HIV-infected adults.

Nucleoside reverse transcriptase inhibitors

The first class of antiretroviral drugs to be approved in the United States and in Europe were the nucleoside reverse transcriptase inhibitors or NRTIs. These non-competitive inhibitors of the reverse transcriptase need to be activated by host cell phosphorylation into a triphosphate form to compete with normal nucleoside substrates for incorporation into the viral genome, resulting in chain termination and interruption of viral replication.

NRTI monotherapy

Zidovudine (3'-azido-3'-deoxythymidine, AZT) a structural analogue of thymidine was the first agent approved for the treatment of HIV infection after an initial phase II double-blind placebo-controlled trial showed in 1986 a significant survival advantage in patients with advanced HIV disease receiving zidovudine [1]. The study compared oral zidovudine 250 mg q4h to placebo in 282 patients with recent *Pneumocystis carinii* pneumonia or advanced AIDS-related complex. The trial was terminated early when an intermediate analysis after 127 days showed a significant decrease in mortality in the zidovudine arm (1 *versus* 19 deaths in the placebo arm), a decrease in the frequency and severity of opportunistic infections and an increase in the Karnofski performance score. However the dosage of 1500 mg daily was associated with frequent and severe hematotoxicity: 24% of patients had haemoglobin levels less than 7.5 mg/dl and 21% of patients required red cells transfusion [2]. Further studies with patients at the same advanced stage of disease (ACTG 002) or mildly symptomatic (ACTG 016) demonstrated similar short term efficacy of zidovudine at lower doses with better tolerance [3–4].

ACTG 019 a large placebo-controlled trial compared zidovudine 500 mg qd or 1500 mg qd to placebo in the earlier stages of HIV infection: patients with less than 500 CD4 cells/μL after a median follow up of 51 to 61 weeks (w) had clinical and laboratory benefits when receiving zidovudine with a two- to three-fold decrease in the progression rate over placebo. Toxicity was reduced in the 500-mg arm [5].

However, the benefit of zidovudine monotherapy was not confirmed in a longer study which included 1749 asymptomatic patients followed for a median of 3.3 years. Indeed, in the Concorde trial comparison between immediate *versus* deferred therapy with zidovudine 250 mg q6h showed no difference in both arms in the 3-year estimated survival rate (92–94%) and probability of progression to AIDS defining illness (29–32%) [6]. In the following years other NRTIs were developed and given as monotherapy: zalcitabine (ddC), didanosine (ddI), lamivudine (3TC) and stavudine (d4T).

Rapidly, after zidovudine use, resistance induced by mutations of the reverse transcriptase were noted [7, 8]. Clinical trials (ACTG 116 and 117) with symptomatic patients or patients with less than 200 CD4 cells/μL compared continuing zidovudine *versus* switching to didanosine either soon or several months to years after beginning zidovudine therapy [9, 10]. Patients with no prior or less than 2 months of zidovudine use had the same benefits in terms of clinical progression or survival with either zidovudine or didanosine. However, prior zidovudine therapy greater than 2 months was associated with a clear benefit of didanosine over zidovudine, implying that resistance to zidovudine monotherapy appeared rapidly and earlier than expected from previous studies.

NRTI monotherapy trials have allowed to distinguish toxicity profiles for each drug and with their long-term use mitochondrial toxicity due to mitochondrial DNA polymerase inhibition by NRTIs was demonstrated. Different side-effects such as myopathy, peripheral neuropathy, lactic acidosis and steatosis could be explained by this common mitochondrial toxicity [11].

Dual NRTI therapy

The next step was to compare NRTIs mono- to dual therapy. Rapidly it appeared that higher and more sustained increase in CD4 levels and decrease in viral load (–1 to –1.5 $\log_{10}$ cp/mL) resulted from dual therapy [12, 13]. In the ACTG 175 trial which included 2467 patients with CD4 cells count between 200 and 500/µL, didanosine monotherapy or a combination of zidovudine plus didanosine or zidovudine plus zalcitabine showed slower clinical progression, CD4 cells count decrease and lower mortality compared to zidovudine monotherapy over 143 weeks [14]. This study demonstrated also that the progression to AIDS or death or loss of >50% of CD4 cells was associated with the baseline plasma viral load (VL) and the decrease of VL under therapy [15, 16]. These results were confirmed by the European Delta study. Therapy with zidovudine and zalcitabine or didanosine had substantial benefits in naïve patients in terms of survival (relative reduction in mortality of 42% in the zidovudine plus didanosine arm and 32% in zidovudine plus zalcitabine arm) compared to zidovudine alone [17]. In zidovudine-experienced patients the addition of didanosine improved survival providing a relative reduction in mortality of 23%. Increase in CD4 cell count was the greatest with zidovudine plus didanosine at least during the first year and was superior in the dual therapy groups throughout the study.

Other comparisons of zidovudine alone with combined zidovudine plus lamivudine have shown both in naïve and experienced patients a more sustained decrease in plasma VL and increase in CD4 cell count under dual NRTI therapy [18–20].

Various double-NRTI therapy regimens have also been compared but during short follow-up periods. In naïve patients didanosine plus stavudine reduced plasma VL and increased CD4 cells count (median: –2.23 $\log_{10}$ copies/mL and +125 cells/µL) more effectively than zidovudine plus lamivudine after 6 months (median: –1.28 $\log_{10}$ copies/mL and +62 cells/µL) [21]. After 1 year stavudine plus lamivudine was comparable to zidovudine plus lamivudine in terms of virologic effect, but lamivudine with didanosine resulted in a less important VL decrease [22].

Protease inhibitors

A major breakthrough occurred in the mid-1990s when protease inhibitors (PIs) were used in a triple therapy combination with NRTIs.

PIs prevent the packaging of mature virions by inhibiting the cleavage of the gag and gag-pol polyprotein precursors. In early dose escalation studies, ritonavir, saquinavir or indinavir monotherapy affected plasma VL and CD4 cells count markedly, although this effect was transient at lower doses and associated with the emergence of drug resistances [23–27].

In a double-blind study with 302 patients published in1996, triple therapy with saquinavir 1800 mg plus zidovudine 600 mg plus zalcitabine 2.25 mg provided significantly greater reduction in plasma VL (–1 $\log_{10}$ copies/mL at week 4) than dual therapy with zidovudine plus zalcitabine or monotherapy [28]. Gulick et al. in a pivotal study published in 1997, compared indinavir 800 mg tid to zidovudine 200 mg tid plus lamivudine 150 mg bid to a combination of the three drugs [29]. After 24 w 90% of patients receiving the triple therapy had undetectable HIV-RNA levels (threshold: 500 copies/mL) *versus* 43% in the indinavir arm and 0 in the dual-NRTI arm. These results were maintained in patients reaching 52 weeks of follow up. Mean increase in CD4 cell counts was within the same range for the combination of three drugs (+86 cells/μL) and indinavir (+100) but inferior in the zidovudine plus lamivudine arm (+46).

At that same period, the concept of triple therapy providing sustained viral suppression was reinforced by the demonstration that measures of VL alone or combined to CD4 cell counts were surrogate markers of HIV-1 progression to AIDS or death and that the level of viral reduction predicts clinical outcome [15, 30].

The association of two antiretroviral classes including three different drugs leading to sustained viral suppression has become today the standard of care for HIV infection under the acronym of "Highly Active Antiretroviral Therapy" or HAART. Soon after the introduction of HAART in the developed world, mortality and morbidity dramatically decreased with fewer hospital admissions, lower rates of opportunistic diseases and deaths [31].

In clinical trials which are usually performed in selected centres with highly motivated patients, HAART including ritonavir, indinavir, nelfinavir or amprenavir showed durable and profound viral suppression (VL at 24–52 weeks <50 copies/mL in 48–69% of patients, depending on the PI [32–35]). However, in day-to-day practice few patients are reaching or maintaining viral suppression, probably because of the initial PI requirement. Numerous daily intakes because of short half-life, fasting or dietary constraints related to variability in oral absorption or heavy pill burden and complicated schedules of administration made initial HAART difficult to support on the mid or longer term [36, 37]. In addition frequent digestive tolerance and toxicity problems such as indinavir-induced nephrolithiasis, PI-induced insulin resistance, hyperlipemia or lipodystrophy jeopardise patient compliance and adherence to the treatment and add physical and psychological burdens to the therapy regimes.

Different strategies have been developed to try to decrease these obstacles linked to the PI use, while maintaining viral suppression. Induction of therapy with PI-based HAART followed by a maintenance by dual-NRTI therapy studies (Trilege, ACTG 343) have been disappointing because viral suppression was lost under the maintenance therapy. On the contrary, initial PI sparing regimens associating NRTIs and NNRTI or switching from a PI regimen to a simpler therapy (3 NRTIs or 2 NRTIs and 1 NNRTI) after reaching viral suppression are good alternatives that will be discussed in the NNRTI section.

PI tolerance profile has also been improved thanks to ritonavir boosting, which allows a decrease of the number of daily pills. In addition new molecules have recently been developed such as atazanavir, a once-daily PI with a promising tolerance and toxicity profile inducing less metabolic disturbances [38].

Dual and boosted protease inhibitors therapy

The inhibition of the cytochrome p450 isoenzyme 3A4 by ritonavir allows lowering of doses and a decrease in dosing frequency of other PIs. This has led to improved adherence and therefore to increased antiretroviral efficacy. The ritonavir association encloses two different concepts. Ritonavir can be used at the 400 mg bid dose acting as the second PI in a dual PI therapy and can at the same time be used as a boost allowing a lowering of the other PI dose and intake frequency. Ritonavir 100 mg bid low dose only plays the role of a booster to simplify and improve other PI pharmacodynamic profiles. These low ritonavir doses are better tolerated in terms of gastrointestinal toxicity. The first PI combination to be clinically evaluated was ritonavir-saquinavir [39]. When saquinavir alone provided a weak viral response (–1 $\log_{10}$ copies/mL at week 4), its association with ritonavir 400 mg bid has been able to provide better saquinavir plasma concentrations with sustained viral response.

The changes in pharmacokinetic profile may also increase the toxicity as is the case with indinavir/ritonavir 800/100 mg bid, which induces a higher rate of nephrolithiasis because of a higher plasma indinavir peak [40]. Combinations of indinavir/ritonavir 600/100 or 400/100 mg bid are now under evaluation.

PIs such as indinavir, amprenavir, and saquinavir or recently lopinavir (a fixed dose of ritonavir 33 mg plus lopinavir 133 mg) are currently given together with ritonavir.

Non-nucleoside reverse transcriptase inhibitors

The non-nucleoside reverse transcriptase inhibitors (NNRTIs) are non-competitive inhibitors of the reverse transcriptase, which becomes unable to function after the NNRTI binds to a unique site in a reversible and non-competitive manner. NNRTIs are only active against HIV-1 (not HIV-2) and they have a low resistance threshold: a single key mutation in the HIV-1 reverse transcriptase codon can induce a high level of phenotypic resistance to the entire class. They were discovered in the late 1980s just before the PIs, but their potential efficacy was underestimated by their initial misuse in monotherapy. Indeed it was only in the mid-1990s that their use in combination therapy was initiated. A good bioavailability and a long half-life allowing a twice or once daily intake are good arguments for their use as part of a triple-drug therapy. In the INCAS trial, a randomised study comparing zidovudine plus didanosine plus nevirapine to two NRTI dual regimens, 50% of patients under triple therapy reached HIV RNA <400 copies/mL after 52 weeks *versus* 22% for zidovudine plus didanosine and 0% for zidovudine plus nevirapine [41].

Another open-label randomised key study compared indinavir to efavirenz in association with zidovudine and lamivudine in 450 naïve patients. Over 48 weeks the efavirenz arm provided a better viral response than the indinavir arm (70% *vs* 48% <400 copies/mL, $p < 0.001$) [42]. This could be explained by fewer adverse events and by an easier treatment provided by the NNRTI-based therapy. These results contributed to the present success of PI-sparing regimens. Indeed such a regimen is efficient in naïve patients even in case of high baseline VL. It allows the preservation of PI for future therapeutical options, while PI-induced toxicity (insulin resistance and hyperlipemia increasing cardiovascular risks on the long term, body changes) can be avoided.

Among the NNRTIs, delavirdine is no longer used because of frequent drug-drug interactions and a high pill burden. There is a debate about the respective efficacy of efavirenz and nevirapine. A recent trial (the 2 NN study) comparing both drugs seems to indicate that they are equivalent in terms of efficacy [43].

Switch therapy and PI-sparing regimens

In contrast to NRTIs or PIs, NNRTIs have not been associated with long-term toxicity and in particular they do not induce lipodystrophy and cause less metabolic disturbances which explains why they have been extensively used as "switch strategy". This strategy consists in changing the class of drugs supposed to be more toxic when viral load has been suppressed for at least 6 months with initial therapy.

Switching from a PI-based regimen to a NNRTI-based regimen successfully maintains undetectable viral loads, increases in CD4 cell counts, while both adherence and quality of life are improved. The impact on established lipodystrophy is so far quite disappointing as switching from a PI to a NNRTI does not seem to improve greatly the body changes [44–47]. From one study to another, there are discrepancies as far as improvement of metabolic changes are concerned. Lack of standardised definition of lipodystrophy or of hyperlipemia evaluation and concomitant use of NRTIs (which are also implicated in lipodystrophy) probably explain these differences.

Switching from a PI-based regimen to a triple NRTIs therapy is another easy and convenient (two pills per day) option that was developed with Trizivir®, a fixed combination of zidovudine, lamivudine and abacavir, a NRTI with higher potency. This approach with a simplified regimen provides sustained virological response when patients do not harbour archived mutations due to prior treatment with NRTI mono- or dual therapy [48, 49]. In a study enrolling patients with previous lipodystrophy, the cholesterol and triglycerids improved 6 months after switching but lipodystrophy worsened [50].

When used as first line therapy, triple therapy with zidovudine plus lamivudine plus abacavir are less efficient than indinavir-based therapy in patients with baseline plasma VL higher than 100,000 copies/mL [51]. For most physicians, the present opinion is that PI-sparing regimens preserve therapeutic options while inducing little or no metabolic disorders, therefore reducing the potential for atherogenic risk associated with PI use. However, the concept of mitochondriopathy and lipodystrophy associated with NRTIs has emerged recently, raising the need for the appraisal of new strategies such as "NRTI-sparing regimens" aiming to avoid these complications.

Lipodistrophy, NRTI-sparing regimens and structured therapy interruption

PI and NRTIs respectively induce through separated and specific pathways lipoaccumulation, hyperlipemia and insulin resistance or mitochondrial toxicity and lipodystrophy [52, 53]. In addition the association of NRTIs and PIs induces more lipodystrophy than PIs alone [54, 55]. The high prevalence of HAART mid or long term induced toxicity associated with our better understanding of HIV physiopathology have led to a change in the recommendations for initiating HIV therapy from "hit hard and early" to "wait and see", depending on the CD4 level and VL thresholds in order to prevent unnecessary patient exposure to potentially toxic drugs [56, 57].

In this setting during the late 1990s, a new strategy called "structured treatment interruption" or STI was developed. In addition to the goal of reducing exposure to drugs, it was postulated that STI might stimulate the patient's specific immunity against HIV and induce specific helper and cytotoxic responses that could control viral replication without antiviral therapy on a long-term basis. So far preliminary encouraging trials on STI in chronically infected patients included only a few patients [58–59]. A larger trial showed that only a minority of patients did not need to restart therapy (18% at week 52 and 11% at week 96 had plasma VL <5,000 copies/mL without therapy) suggesting that the impact on specific immunity against HIV is disappointing and do not favour the auto-vaccination hypothesis [60].

Pharmacological progress, new compounds and once daily therapy

During recent years, most progress in HIV therapy has been related to pharmacological development of more convenient drugs with lower pill burden, once or twice daily schedules and a more favourable tolerance profile. Zidovudine plus lamivudine and zidovudine plus lamivudine plus abacavir combined in a fixed tablet twice daily have been available for several years. More recently didanosine pills have been enteric coated so that the buffer responsible for gastrointestinal intolerance is not required and instead of four chewable tablets, only one pill is needed once daily. Nevirapine studies have shown that it can be taken at 400 mg or two pills once daily. A single tablet of lamivudine 300 mg is now available and can also be taken once daily. Stavudine extended release 100 mg in one pill qd has reached final phases of development. Nelfinavir pills have been covered with a new layer to allow easier swallowing. Fosamprenavir, the amprenavir pro-drug is currently being developed with the same efficacy using only two pills bid instead of six bid with amprenavir. Atazanavir, a two-tablet once-daily PI, inducing very few lipid disturbances, has reached phase IV development. Recently tenofovir (300 mg or one pill qd), a nucleotide analogue, has become available.

The access to eight drugs that can be taken once daily (efavirenz, nevirapine, lamivudine, stavudine, didanosine, tenofovir, atazanavir, saquinavir/ritonavir 1600/100) has opened the clinical field of once-daily regimens. These regimens have similar efficacy compared to bid therapy, while patients' compliance and quality of life can be improved, making chronic therapy easier and more achievable in the longer term [61].

Resistance testing, salvage therapy and new drugs

Since the very early development of HIV therapy, concerns have been raised with regard to viral resistance to treatment. Indeed a significant proportion of patients (between 20 and 40%) are failing their current therapy [36] and the numbers of naïve patients infected with resistant virus are increasing [62]. Failure of therapy may be due to poor patient compliance, pharmacological problems (drug interaction, poor absorption, etc...) or extensive antiretroviral experience. In this setting resistance testing with genotypic assay has constituted great progress in the clinical evaluation of the failing patient. Recently, the Havana study has shown that genotype testing in addition to expert advice is more efficient in predicting which medications are likely to fail than standard of care or genotype testing or expert advice alone [63]. Genotype resistance testing is therefore recommended before changing treatment in chronically infected patients with viral failure [56].

Salvage strategy for failing patients includes "mega-" or "giga-HAART" therapy with more than six drugs including NRTIs, NNRTIs and PIs and combining the most potent drugs such as abacavir, efavirenz or amprenavir. These heavy regimens have sometimes been initiated after drug holidays aiming to release the selection pressure induced by HAART letting the wild-type HIV replace the mutant strains. Favourable effects have been suggested in a 52 weeks trial including patients with very low CD4 cells levels (27/μL) [64].

Unfortunately this was not confirmed in less advanced patients probably because drug holidays induce severe immune loss that are more harmful in less immunosuppressed patients.

Today multiresistant patients can benefit from new drugs in existing or new antiretroviral classes. Among the existing classes new and more potent PIs such as lopinavir are currently available; TMC-114 and tipranavir are still in development. Tenofovir is a nucleotide analogue that has recently become available. TMC-125, a NNRTI active against HIV strains that are highly resistant to other NNRTIs is still under evaluation.

As a new promising class of drugs, the fusion inhibitors inhibit the entry of the virus in any cell harbouring the CD4+ receptor by binding to the viral gp41 thus preventing the virus and cell membranes from fusing. Fuzeon (enfuvirtide, T-20) a twice daily subcutaneously injected fusion inhibitor in association with optimised background therapy based on genotype and phenotype testing has given promising results after 24 weeks in two phase III pivotal studies (Toro I and II) including 995 highly experienced patients (–1.43 to –1.70 $\log_{10}$ copies/mL in T-20 arm *versus* –0.65 to –0.76 in optimised background alone) [65]. Fuzeon is not active against HIV-2 and resistant strains have already been described, but T-1249, another fusion inhibitor, active against HIV 1 and 2, more potent *in vitro* and less prone to induce viral resistance is currently under phase I-II development.

HIV-1 integrase inhibitors prevent the integration of HIV proviral DNA into the host genome and are currently in phase I-II development. Several agents inhibiting virus fusion or entry in the CD4+ cells and interacting at the level of the CD4+ chemokine co-receptors (CCR5 and CXCR4) are also now being studied in phases I and II.

Evolution of understanding HIV physiopathology and treatment strategies

After the revolution induced by the introduction of HAART, estimations of the HIV turnover kinetics led to the consideration that eradication of HIV could be possible leading to recommendation of HAART very early in the infection course. However later studies demonstrated that resting memory CD4 lymphocytes had integrated the virus forming a latent persistent reservoir that would need at least 60 years to eradicate with HAART and that the human body harboured sanctuaries such as the brain where drug penetration could be more problematic [66]. Further immunological works demonstrated that HAART induces sustained immune recovery even in severely immunosuppressed patients and that this recovery is depending on the amplitude and duration of the viral load suppression [67]. This progress in the understanding of the HIV physiopathology and the perspective of HIV infection as a chronic infection that should be controlled lifelong contributed to change from the "hit hard and early" strategy to a more conservative approach of initiation of therapy together with the consideration of alternative and complementary strategies such as immune-based therapy.

Immune-based therapy

Administration of interleukin-2, a pro-inflammatory cytokine that activates T-cells and the latently infected resting lymphocytes is one of the most studied approaches aiming to decrease or ultimately eliminate the reservoir of HIV-infected T-cells [68]. Since 1995, numerous trials have shown in both naïve or antiretroviral experienced patients receiving HAART plus interleukin-2 a substantial qualitative and quantitative increase in CD4 T-cells to levels significantly higher than those achieved with HAART alone [69–73]. In these trials tolerance was acceptable and patients had baseline CD4 cell counts above 200/μL. In the Ilstim study including patients with CD4 cell counts between 25 and 100/μL, the effect of interleukin-2 was also dramatic resulting in 81% of patients achieving more than 200 CD4 cells/μL after 80 weeks of interleukin-2 *versus* 33% with HAART alone ($p < 0.0001$) [74]. In order to evaluate the clinical benefit (progression to AIDS or death) of interleukin-2, two major multinational randomised trials enrolling a very large population of chronically infected patients are ongoing: the ESPRIT trial for patients with more than 300 cells/μL and the SILCAAT study for subjects with 50 to 300 cells/μL. Another strategy which is presently evaluated is to use intermittent interleukin-2 as a boost for immunity, maintaining CD4 levels above a certain threshold and thus allowing the postponement of the indication for continuous HAART.

Other immune-stimulating or -suppressive agents include granulocyte-colony stimulating factor or G-CSF, hydroxyurea, cyclosporine or mycophenolate. Hydroxyurea decreases intracellular deoxynucleoside triphosphates that are required for DNA synthesis, resulting in a synergistic activity with didanosine *in vitro*, but because of its toxic profile (CD4 response blunting, neuropathy, severe pancreatitis and mitochondrial toxicity) it is not used anymore. G-CSF may lead to an increase in CD4 T-cell counts as well as a decrease in viral load when added to HAART [75]. Mycophenolate mofetil selectively inhibits lymphocyte division and proliferation. It reduces the pool of activated CD4 T-lymphocytes that support productive HIV infection but it also inhibits viral replication by depleting dGTP, an important substrate for the reverse transcriptase. A randomised trial (MOCHA) comparing HAART alone or HAART with mycophenolate is underway.

Vaccination

Therapeutic vaccines have been under investigation for several years. They include DNA vaccines (using various vectors of delivery systems such as canarypox virus, ALVAC) which encode various HIV immunogenic peptides such as gp120, whole killed virus preparation (Remune), HIV core antigens (p24VLP), recombinant HIV envelop proteins (VaxSyn), DNA plasmids or anti-TAT vaccination associated with different adjuvants.

Several studies have shown the appearance of specific cytotoxic or antibody responses after using certain types of vaccination, but this has not been so far associated with a clinical benefit in terms of clinical progression, significant viral load control or ability to stop antiretroviral therapy [76]. Moreover, a recent report of a subject super-infected with a second closely related HIV strain despite persistent recognition of multiple CD8 epitopes has blunted the enthusiasm for a potential therapeutic vaccine [77].

Conclusion

Since the arrival of the first efficient therapy in developed countries, major progress has been accomplished in the fields of virology, pharmacology, immunology, physiopathology and resistance understanding leading to the expansion of new therapy strategies, immune-based therapy and vaccination research. The deadly HIV infection has turned into a chronic disease which allows a professional and family life but imposes the burden of a treatment that can be toxic in the long term. Still numerous questions such as drug-drug interactions, toxicity management, long-term adherence, efficacy, and treatment costs need to be answered. But most of all, access to therapy which is currently not available for 90% of the HIV-infected population in the world remains our main challenge today.

References

1 Fischl MA, Richman DD, Grieco MH, Gottlieb MS, Volberding PA, Laskin OL, Leedom JM, Groopman JE, Mildvan D, Schooley RT et al. (1987) The efficacy of azidothymidine (AZT) in the treatment of patients with AIDS and AIDS-related complex. A double-blind, placebo-controlled trial. *N Engl J Med* 317: 185–191

2 Richman DD, Fischl MA, Grieco MH, Gottlieb MS, Volberding PA, Laskin OL, Leedom JM, Groopman JE, Mildvan D, Hirsch MS et al. (1987) The toxicity of azidothymidine (AZT) in the treatment of patients with AIDS and AIDS-related complex. A double-blind, placebo-controlled trial. *N Engl J Med* 317: 192–197

3 Fischl MA, Parker CB, Pettinelli C, Wulfsohn M, Hirsch MS, Collier AC, Antoniskis D, Ho M, Richman DD, Fuchs E et al. (1990) A randomized controlled trial of a reduced daily dose of zidovudine in patients with the acquired immunodeficiency syndrome. The AIDS Clinical Trials Group. *N Engl J Med* 323: 1009–1014

4 Fischl MA, Richman DD, Hansen N, Collier AC, Carey JT, Para MF, Hardy WD, Dolin R, Powderly WG, Allan JD et al. (1990) The safety and efficacy of zidovudine (AZT) in the treatment of subjects with mildly symptomatic human immunodeficiency virus type 1 (HIV) infection. A double-blind, placebo-controlled trial. The AIDS Clinical Trials Group. *Ann Intern Med* 112: 727–737

5 Volberding PA, Lagakos SW, Koch MA, Pettinelli C, Myers MW, Booth DK, Balfour HH Jr, Reichman RC, Bartlett JA, Hirsch MS et al. (1990) Zidovudine in asymptomatic human immunodeficiency virus infection. A controlled trial in persons with fewer than 500 CD4-positive cells per cubic millimeter. The AIDS Clinical Trials Group of the National Institute of Allergy and

Infectious Diseases. *N Engl J Med* 322: 941–949
6 Concorde Coordinating Committee (1994) Concorde: MRC/ANRS randomised double-blind controlled trial of immediate and deferred zidovudine in symptom-free HIV infection. *Lancet* 343: 871–881
7 Larder BA, Kellam P, Kemp SD (1991) Zidovudine resistance predicted by direct detection of mutations in DNA from HIV-infected lymphocytes. *AIDS* 5: 137–144
8 Fitzgibbon JE, Howell RM, Schwartzer TA, Gocke DJ, Dubin DT (1991) *In vivo* prevalence of azidothymidine (AZT) resistance mutations in an AIDS patient before and after AZT therapy. *AIDS Res Hum Retroviruses* 7: 265–269
9 Kahn JO, Lagakos SW, Richman DD, Cross A, Pettinelli C, Liou SH, Brown M, Volberding PA, Crumpacker CS, Beall Get al. (1992) A controlled trial comparing continued zidovudine with didanosine in human immunodeficiency virus infection. The NIAID AIDS Clinical Trials Group. *N Engl J Med* 327: 581–587
10 Shafer RW, Kozal MJ, Winters MA, Iversen AK, Katzenstein DA, Ragni MV, Meyer WA 3rd, Gupta P, Rasheed S, Coombs R et al. (1994) Combination therapy with zidovudine and didanosine selects for drug-resistant human immunodeficiency virus type 1 strains with unique patterns of pol gene mutations. *J Infect Dis* 169: 722–729
11 Moyle G (2000) Toxicity of antiretroviral nucleoside and nucleotide analogues: is mitochondrial toxicity the only mechanism? *Drug Saf* 23: 467–481
12 Collier AC, Coombs RW, Fischl MA, Skolnik PR, Northfelt D, Boutin P, Hooper CJ, Kaplan LD, Volberding PA, Davis LG et al. (1993) Combination therapy with zidovudine and didanosine compared with zidovudine alone in HIV-1 infection. *Ann Intern Med* 119: 786–793
13 Fischl MA, Stanley K, Collier AC, Arduino JM, Stein DS, Feinberg JE, Allan JD, Goldsmith JC, Powderly WG (1995) Combination and monotherapy with zidovudine and zalcitabine in patients with advanced HIV disease. The NIAID AIDS Clinical Trials Group. *Ann Intern Med* 122: 24–32
14 Hammer SM, Katzenstein DA, Hughes MD, Gundacker H, Schooley RT, Haubrich RH, Henry WK, Lederman MM, Phair JP, Niu M et al. (1996) A trial comparing nucleoside monotherapy with combination therapy in HIV-infected adults with CD4 cell counts from 200 to 500 per cubic millimeter. AIDS Clinical Trials Group Study 175 Study Team. *N Engl J Med* 335: 1081–1090
15 Katzenstein DA, Hammer SM, Hughes MD, Gundacker H, Jackson JB, Fiscus S, Rasheed S, Elbeik T, Reichman R, Japour A et al. (1996) The relation of virologic and immunologic markers to clinical outcomes after nucleoside therapy in HIV-infected adults with 200 to 500 CD4 cells per cubic millimeter. AIDS Clinical Trials Group Study 175 Virology Study Team. *N Engl J Med* 335: 1091–1098
16 Lathey JL, Hughes MD, Fiscus SA, Pi T, Jackson JB, Rasheed S, Elbeik T, Reichman R, Japour A, D'Aquila RT et al. (1998) Variability and prognostic values of virologic and CD4 cell measures in human immunodeficiency virus type 1-infected patients with 200–500 CD4 cells/mm^3 (ACTG 175). AIDS Clinical Trials Group Protocol 175 Team. *J Infect Dis* 177: 617–624
17 Delta Coordinating Committee (1996) Delta: a randomised double-blind controlled trial comparing combinations of zidovudine plus didanosine or zalcitabine with zidovudine alone in HIV-infected individuals. *Lancet* 348: 283–291
18 Eron JJ, Benoit SL, Jemsek J, MacArthur RD, Santana J, Quinn JB, Kuritzkes DR, Fallon MA, Rubin M (1995) Treatment with lamivudine, zidovudine, or both in HIV-positive patients with 200 to 500 CD4+ cells per cubic millimeter. North American HIV Working Party. *N Engl J Med* 333: 1662–1669
19 Katlama C, Ingrand D, Loveday C, Clumeck N, Mallolas J, Staszewski S, Johnson M, Hill AM, Pearce G, McDade H (1996) Safety and efficacy of lamivudine-zidovudine combination therapy in antiretroviral-naïve patients. A randomized controlled comparison with zidovudine monotherapy. Lamivudine European HIV Working Group. *JAMA* 276: 118–125
20 Staszewski S, Loveday C, Picazo JJ, Dellarnonica P, Skinhoj P, Johnson MA, Danner SA, Harrigan PR, Hill AM, Verity L et al. (1996) Safety and efficacy of lamivudine-zidovudine combination therapy in zidovudine-experienced patients. A randomized controlled comparison with zidovudine monotherapy. Lamivudine European HIV Working Group. *JAMA* 276: 111–117
21 Molina JM, Chêne G, Ferchal F, Journot V, Pellegrin I, Sombardier MN, Rancinan C, Cotte L, Madelaine I, Debord T et al. (1999) The ALBI trial: a randomized controlled trial comparing stavudine plus didanosine with zidovudine plus lamivudine and a regimen alternating both combinations in previously untreated patients infected with human immunodeficiency virus. *J Infect Dis* 180: 351–358

22 Kuritzkes DR, Marchner I, Johnson VA, Bassett R, Eron JJ, Fischl MA, Murphy RL, Fife K, Maenza J, Rosandich ME et al. (1999) Lamivudine in combination with zidovudine, stavudine, or didanosine in patients with HIV-1 infection. A randomized, double-blind, placebo-controlled trial. *AIDS* 13: 685–694
23 Danner SA, Carr A, Leonard JM, Lehman LM, Gudiol F, Gonzales J, Raventos A, Rubio R, Bouza E, Pintado V et al. (1995) A short-term study of the safety, pharmacokinetics, and efficacy of ritonavir, an inhibitor of HIV-1 protease. European-Australian Collaborative Ritonavir Study Group. *N Engl J Med* 333: 1528–1533
24 Markowitz M, Saag M, Powderly WG, Hurley AM, Hsu A, Valdes JM, Henry D, Sattler F, La Marca A, Leonard JM et al. (1995) A preliminary study of ritonavir, an inhibitor of HIV-1 protease, to treat HIV-1 infection. *N Engl J Med* 333: 1534–1539
25 Kitchen VS, Skinner C, Ariyoshi K, Lane EA, Duncan IB, Burckhardt J, Burger HU, Bragman K, Pinching AJ, Weber JN (1995) Safety and activity of saquinavir in HIV infection. *Lancet* 345: 952–955
26 Steigbigel R, Berry P, Teppler H et al. (1996) Treatment with indinavir, zidovudine, and lamivudine in adults with human immunodeficiency virus infection and prior antiretroviral therapy. *11th International Conference on AIDS,* 7–12 July 1996, Vancouver, BC, Canada, Abstract Mo.B.412
27 Molla A, Korneyeva M, Gao Q, Vasavanonda S, Schipper PJ, Mo HM, Markowitz M, Chernyavskiy T, Niu P, Lyons N et al. (1996) Ordered accumulation of mutations in HIV protease confers resistance to ritonavir. *Nat Med* 2: 760–766
28 Collier AC, Coombs RW, Schoenfeld DA, Bassett RL, Timpone J, Baruch A, Jones M, Facey K, Whitacre C, McAuliffe VJ et al. (1996) Treatment of human immunodeficiency virus infection with saquinavir, zidovudine, and zalcitabine. AIDS Clinical Trials Group. *N Engl J Med*; 334: 1011–1017
29 Gulick RM, Mellors JW, Havlir D, Eron JJ, Gonzalez C, McMahon D, Richman DD, Valentine FT, Jonas L, Meibohm A et al. (1997) Treatment with indinavir, zidovudine, and lamivudine in adults with human immunodeficiency virus infection and prior antiretroviral therapy. *N Engl J Med* 337: 734–739
30 Mellors JW, Muñoz A, Giorgi JV, Margolick JB, Tassoni CJ, Gupta P, Kingsley LA, Todd JA, Saah AJ, Detels R et al. (1997) Plasma viral load and CD4+ lymphocytes as prognostic markers of HIV-1 infection. *Ann Intern Med* 126: 946–954
31 Palella FJ Jr, Delaney KM, Moorman AC, Loveless MO, Fuhrer J, Satten GA, Aschman DJ, Holmberg SD (1998) Declining morbidity and mortality among patients with advanced human immunodeficiency virus infection. HIV Outpatient Study Investigators. *N Engl J Med* 338: 853–860
32 Mathez D, Bagnarelli P, Gorin I, Katlama C, Pialoux G, Saimot G, Tubiana P, De Truchis P, Chauvin JP, Mills R et al. (1997) Reductions in viral load and increases in T lymphocyte numbers in treatment-naïve patients with advanced HIV-1 infection treated with ritonavir, zidovudine and zalcitabine triple therapy. *Antiviral Ther* 2: 175–183
33 Carr A, Chuah J, Hudson J, French M, Hoy J, Law M, Sayer D, Emery S, Cooper DA (2000) A randomised, open-label comparison of three highly active antiretroviral therapy regimens including two nucleoside analogues and indinavir for previously untreated HIV-1 infection: the OzCombol study. *AIDS* 14: 1171–1180
34 Clendeninn N, Quart B, Anderson R, Knowles M, Chang Y (1988) Analysis of long-term virologic data from the Viracept® (Nelfinavir, NFV) 511 protocol using 3 HIV-RNA assays. *5th Conference on Retroviruses and Opportunistic Infections,* 1–5 February 1998, Chicago, Illinois, USA, Abstract 372
35 Bart PA, Gizzard GP, Gallant S, Welbon C, Graziosi C, Chave JP, Meylan P, Martin G, Steel H, Spreen W et al. (1998) Combination abacavir/amprenavir therapy in HIV-1-infected antiretroviral-naïve subjects with CD4+ counts >400 cells/µl and viral load >5,000 copies/ml. *5th Conference on Retroviruses and Opportunistic Infections,* 1–5 February 1998, Chicago, Illinois, USA, Abstract 365
36 Ledergerber B, Egger M, Opravil M, Telenti A, Hirschel B, Battegay M, Vernazza P, Sudre P, Flepp M, Furrer H et al. (1999) Clinical progression and virological failure on highly active antiretroviral therapy in HIV-1 patients: a prospective cohort study. Swiss HIV Cohort Study. *Lancet* 353: 863–868
37 Lucas GM, Chaisson RE, Moore RD (1999) Highly active antiretroviral therapy in a large urban clinic: risk factors for virologic failure and adverse drug reactions. *Ann Intern Med* 131: 81–87

38 Sanne I, Piliero P, Squires K, Thiry A, Schnittman S (2003) Results of a phase 2 clinical trial at 48 weeks (AI424-007): a dose-ranging, safety, and efficacy comparative trial of atazanavir at three doses in combination with didanosine and stavudine in antiretroviral-naïve subjects. *J Acquir Immune Defic Syndr* 32: 18–29
39 Cameron DW, Japour AJ, Xu Y, Hsu A, Mellors J, Farthing C, Cohen C, Poretz D, Markowitz M, Follansbee S et al. (1999) Ritonavir and saquinavir combination therapy for the treatment of HIV infection. *AIDS* 13: 213–224
40 Voigt E, Wickesberg A, Wasmuth JC, Gute P, Locher L, Salzberger B, Wohrmann A, Adam A, Weitner L, Rockstroh JK (2002) First-line ritonavir/indinavir 100/800 mg twice daily plus nucleoside reverse transcriptase inhibitors in a German multicentre study: 48-week results. *HIV Med* 3: 277–282
41 Montaner JS, Reiss P, Cooper D, Vella S, Harris M, Conway B, Wainberg MA, Smith D, Robinson P, Hall D et al. (1998) A randomized, double-blind trial comparing combinations of nevirapine, didanosine, and zidovudine for HIV-infected patients. *JAMA* 279: 930–937
42 Staszewski S, Morales-Ramirez J, Tashima KT, Rachlis A, Skiest D, Stanford J, Stryker R, Johnson P, Labriola DF, Farina D et al. (2000) Efavirenz plus zidovudine and lamivudine, efavirenz plus indinavir, and indinavir plus zidovudine and lamivudine in the treatment of HIV-1 infection in adults. *N Engl J Med*; 341: 1865–1873
43 van Leth F, Hassink E, Phanuphak P, Miller S, Gazzard B, Cahn P, Wood R, Squires K, Katlama C, Santos B et al. (2003) Results of the 2NN study: a randomised comparative trial of first-line antiretroviral therapy with regimens containing either nevirapine alone, efavirenz alone or both drugs combined, together with stavudine and lamivudine. *10th Conference on Retroviruses and Opportunistic Infections,* 10–14 February 2003, Boston, Massachusetts, USA, Oral Abstract 176
44 Barreiro P, Soriano V, Blanco F, Casimiro C, de la Cruz JJ, Gonzalez-Lahoz J (2000) Risks and benefits of replacing protease inhibitors by nevirapine in HIV-infected subjects under long-term successful triple combination therapy. *AIDS* 14: 807–812
45 Rachlis A, Becker S, Gill J, Dejesus E, Pierone G, Ruiz N, Aznar E (2001) Successful substitution of protease inhibitors with efavirenz in patients with undetectable viral loads. *1st International AIDS Society Conference on HIV Pathogenesis and Treatment,* 8–11 July 2001, Buenos Aires, Argentina, Abstract 418
46 Martinez E, Garcia-Viejo MA, Blanco JL, Bianchi L, Buira E, Conget I, Casamitjana R, Mallolas J, Gatell JM (2000) Impact of switching from human immunodeficiency virus type 1 protease inhibitor to efavirenz in successfully treated adults with lipodystrophy. *Clin Infect Dis* 31: 1266–1273
47 Ruiz L, Negredo E, Domingo P, Paredes R, Francia E, Balague M, Gel S, Bonjoch A, Fumaz CR, Johnston S et al. (2001) Antiretroviral treatment simplification with nevirapine in protease inhibitor-experienced patients with HIV-associated lipodystrophy: 1-year prospective follow-up of a multicenter, randomized, controlled study. *J Acquir Immune Def Syndr* 27: 229–236
48 Clumeck N, Goebel F, Rozenbaum W, Gerstoft J, Staszewski S, Montaner J, Johnson M, Gazzard B, Stone C, Athisegaran R et al. (2001) Simplification with abacavir-based triple nucleoside therapy *versus* continued protease inhibitor-based highly active antiretroviral therapy in HIV-1-infected patients with undetectable plasma HIV-1 RNA. *AIDS* 15: 1517–1526
49 Opravil M, Hirschel B, Lazzarin A, Furrer H, Chave JP, Yerly S, Bisset LR, Fischer M, Vernazza P, Bernasconi E et al. (2002) A randomized trial of simplified maintenance therapy with abacavir, lamivudine, and zidovudine in human immunodeficiency virus infection. *J Infect Dis* 185: 1251–1260
50 Carr A, Hudson J, Chuan J, Mallal S, Law M, Hoy J, Doong N, French M, Smith D, Cooper DA (2001) HIV protease inhibitor substitution in patients with lipodystrophy: a randomized, controlled, open-label, multicentre study. *AIDS* 15: 1811–1822
51 Staszewski S, Keiser P, Montaner J, Raffi F, Gathe J, Brotas V, Hicks C, Hammer SM, Cooper D, Johnson M et al. (2001) Abacavir-lamuvidine-zidovudine vs indinavir-lamivudine-zidovudine in antiretroviral naïve HIV-infected adults: a randomized equivalence trial. *JAMA* 285: 1155–1163
52 Martinez E, Mocroft A, Garcia-Viejo M, Perez-Cuevas JB, Blanco JL, Mallolas J, Bianchi L, Conget I, Blanch J, Phillips A et al. (2001) Risk of lipodystrophy in HIV-1 infected patients treated with protease-inhibitors: a prospective cohort study. *Lancet* 357: 592–598
53 Carr A, Samaras K, Chisholm J Cooper DA (1998) Pathogenesis of HIV-1-protease inhibitor-associated peripheral lipodystrophy, hyperlipidemia, and insulin resistance. *Lancet* 351: 1881–1883
54 Mallal SA, John M, Moore CB, James JR, McKinnon EJ (2000) Contribution of nucleoside ana-

logue reverse transcriptase inhibitors to subcutaneous fat wasting in patients with HIV infection. *AIDS* 14: 1309–1316
55 van der Valk M, Gisolf EH, Reiss P, Wit FW, Japour A, Weverling GJ, Danner SA (2001) Increased risk of lipodystrophy when nucleoside analogue reverse transcriptase inhibitors are included in the treatment of HIV-1 infection. *AIDS* 15: 847–855
56 Guidelines for using antiretroviral agents among HIV-infected adults and adolescents. Recommendations of the Panel on Clinical Practices for Treatment of HIV (2002) *MMWR Recomm Rep* May 17: 1–55. www.hivatis.org
57 The EACS Euroguidelines Group (2003) European guidelines for the clinical management and treatment of HIV infected adults in Europe. *AIDS* supplement to volume 17, 02 June 2003
58 Ruiz L, Carcelain G, Martinez-Picado J, Frost S, Marfil S, Paredes R, Romeu J, Ferrer E, Morales-Lopetegi K, Autran B et al. (2001) HIV dynamics and T-cell immunity after three structured treatment interruptions in chronic HIV-1 infection. *AIDS* 15: F19–F27
59 Dybul M, Chun TW, Yoder C, Hidalgo B, Belson M, Hertogs K, Larder B, Dewar RL, Fox CH, Hallahan CW et al. (2001) Short-cycle structured intermittent treatment of chronic HIV infection with highly active antiretroviral therapy: effects on virologic, immunologic, and toxicity parameters. *Proc Natl Acad Sci USA* 98: 15161–15166
60 Hirschel B, Fagard C, Oxenius A, Gunthard H, Garcia F. (2002) A prospective trial of treatment interruptions in HIV infection. *9th Conference on Retroviruses and Opportunistic Infections,* 24–28 February 2002, Seattle, Washington, USA, Abstract 528-M
61 Staszewski S, Gallant J, Pozniak A, Suleiman JMAH, Dejesus E, Koenig E, Coleman S, Lu B, Cheng AK, Coakley DF (2002) Efficacy and safety of tenofovir disoproxil fumarate *versus* stavudine when used in combination with lamivudine and efavirenz in HIV-1 infected naïve patients to antiretroviral treatment: 48 weeks interim results. *XIVth International AIDS Conference*, 7–12 July 2002, Barcelona, Spain, Abstract Or17
62 Little SJ, Holte S, Routy JP, Daar ES, Markowitz M, Collier AC, Koup RA, Mellors JW, Conninck E, Conway B et al. (2002) Antiretroviral-drug resistance among patients recently infected with HIV. *N Engl J Med* 347: 385–394
63 Tural C, Ruiz L, Holtzer C, Schapiro J, Viciana P, Gonzalez J, Domingo P, Boucher C, Rey-Joly C, Clotet B (2002) Clinical utility of HIV-1 genotyping and expert advice: the Havana trial. *AIDS* 16: 209–218
64 Katlama C, Dominguez S, Duvivier C, Delaugerre C, Peytavin G, Legrand M, Calvez V, Gourlain K, Costagliola D (2003) Long-term benefit of treatment interruption in salvage therapy (GIGHAART ARNS 097). *10th Conference on Retroviruses and Opportunistic Infections,* 10–14 February 2003, Boston, Massachusetts, USA, Oral Abstract 68
65 Lalezari JP, Henry K, O'Hearn M, Lefebvre E, Monaner J, Piliero P, Walmsley S, Chung J, Fang L, Delehanty J et al. (2002) Enfuvirtide (T-20) in combination with an optimized background (OB) regimen *versus* OB alone: Week 24 response among categories of treatment experience and baseline (BL) HIV antiretroviral (AVR) resistance. *42nd Interscience Conference on Antimicrobial Agents and Chemotherapy*, 27–30 September 2002, San Diego, California, USA, Abstract H-1074
66 Ho DD (1998) Toward HIV eradication or remission: the tasks ahead. *Science* 280: 1866–1867
67 Li TS, Tubiana R, Katlama C, Calvez V, Ait Mohand H, Autran B (1998) Long-lasting recovery in CD4 T-cell function and viral-load reduction after highly active antiretroviral therapy in advanced HIV-1 disease. *Lancet* 351: 1682–1686
68 Kovacs JA, Vogel S, Albert JM, Falloon J, Davey RT Jr, Walker RE, Polis MA, Spooner K, Metcalf JA, Baseler M et al. (1996) Controlled trial of interleukin-2 infusions in patients infected with the human immunodeficiency virus. *N Engl J Med* 335: 1350–1356
69 Hengge UR, Goos M, Esser S, Exner V, Dotterer, Wiehler H, Borchard C, Muller K, Beckmann A, Eppner MT et al. (1998) Randomized, controlled phase II trial of subcutaneous interleukin-2 in combination with highly active antiretroviral therapy (HAART) in HIV patients. *AIDS* 12: F225–F234
70 Losso MH, Belloso WH, Emery S, Benetucci JA, Cahn PE, Lasala MC, Lopardo G, Salomon H, Saracco M, Nelson E et al. (2000) A randomized, controlled, phase II trial comparing escalating doses of subcutaneous interleukin-2 plus antiretrovirals *versus* antiretrovirals alone in human immunodeficiency virus-infected patients with CD4+ cell counts ≥350 cells/mm^3. *J Infect Dis* 181: 1614–1621
71 Davey RT Jr, Murphy RL, Graziano FM, Boswell SL, Pavia AT, Cancio M, Nadler JP, Chaitt DG,

Dewar RL, Sahner DK et al. (2000) Immunologic and virologic effects of subcutaneous interleukin 2 in combination with antiretroviral therapy: a randomized controlled trial. *JAMA* 284: 183–189
72 Abrams DI, Bebchuk JD, Denning ET, Davey RT, Fox L, Lane HC, Sampson J, Verheggen R, Zeh D, Markowitz NP (2002) Randomized, open-label study of the impact of two doses of subcutaneous recombinant interleukin-2 on viral burden in patients with HIV-1 infection and CD4+ cell counts of ≥300 cells/mm^3. *J Acquir Immune Defic Syndr* 29: 221–231
73 Levy Y, Durier C, Krzysiek R, Rabian C, Capitant C, Lascaux AS, Michon C, Oksenhendler E, Weiss L, Gastaut JA et al. (2003) Effects of interleukin-2 therapy combined with highly active antiretroviral therapy on immune restoration in HIV-1 infection: a randomized controlled trial. *AIDS* 17: 343–351
74 Katlama C, Carcelain G, Duvivier C, Chouquet C, Tubiana R, De Sa M, Zagury L, Calvez V, Autran B, Costagliola D (2002) Interleukin-2 accelerates CD4 cell reconstitution in HIV-infected patients with severe immunosuppression despite highly active antiretroviral therapy: the ILSTIM study-ANRS 082. *AIDS* 16: 2027–2034
75 Frumkin L (1997) Role of granulocyte colony-stimulating factor and granulocyte-macrophage colony-stimulating factor in the treatment of patients with HIV infection. *Curr Opin Hematol* 4: 200–206
76 Hardy GA, Imami N, Gotch FM (2002) Improving HIV-specific immune responses in HIV-infected patients. *J HIV Ther* 7: 40–45
77 Altfeld M, Allen TM, Yu XG, Johnston MN, Agrawal D, Korber BT, Montefiori DC, O'Connor DH, Davis BT, Lee PK et al. (2002) HIV-1 superinfection despite broad CD8+ T -cell responses containing replication of the primary virus. *Nature* 420: 434–439

Combination Therapy of AIDS
Edited by E.D.A. De Clercq and A.-M.I. Vandamme

The basic principles for combination therapy

Cécile L. Tremblay[1] and Martin S. Hirsch[2]

[1] *C.H.U.M., Hôpital Hôtel-Dieu de Montréal, Pavillon Jeanne-Mance, Bureau 7-355, 3840, rue Saint-Urbain, Montréal Qc, H2W 1 T8, Canada*
[2] *Massachusetts General Hospital, Harvard Medical School, 65 Landsdowne St, Room 419, Cambridge, MA 02139, USA*

Introduction

The use of antiretroviral combinations in HIV-1 infection has led to a dramatic improvement in the morbidity and mortality associated with HIV/AIDS [1]. The concept of combining therapeutic agents is not unique to HIV infection and has proven successful in a variety of medical areas, most notably cancer. For example, the most significant successes in the fight against childhood leukemias were seen when anticancer drugs were used in combination. The use of combination therapies has also proven instrumental in advances made against infectious diseases such as tuberculosis and bacterial endocarditis.

The goals of combining therapeutic agents in HIV infection may be several: 1. To achieve greater efficacy secondary to additive or synergistic interactions among components of the combination. 2. To prevent or reduce the emergence of drug resistance. 3. To reduce individual drug toxicity by reducing doses of drugs that show synergistic antiviral effects. 4. To increase penetration in difficult to reach cell types or tissues. To achieve these goals, one may want to simultaneously target the same or different steps in the virus replication cycle, different cell activation states, different tissues or different cell reservoirs, or take advantage of varied pharmacokinetics among individual drugs and drug combinations.

Taking advantage of additive or synergistic drug interactions

By increasing the number of drugs in a combination, one would expect to increase the amount of viral suppression achieved, assuming favorable interactions among the individual drugs combined. This has been shown to be true both *in vitro* and *in vivo* comparing regimens containing 4, 3, 2 drugs and monotherapy [2–8]. However understanding potential interactions between individual drugs is crucial when choosing which combinations to study in patients. Not all combinations of antiretrovirals result in synergistic interactions against HIV-1. Drugs may compete with each other for binding sites, for intracellular enzymes, and for substrate availability. Antagonistic interactions may result from such competition. A good example is the antagonism seen *in vitro* and *in vivo* between zidovudine and stavudine [7, 9–11]. Zidovudine inhibits phosphorylation of stavudine because both nucleoside analogs compete for the same cellular thymidine kinase and zidovudine is preferentially phorphorylated [10, 11, 11a]. Clinical trials of this combination have also demonstrated apparent antagonism [9]. In contrast, many other agents, e.g., zidovudine and lamivudine, show strong synergy *in vitro* [7, 12–14], as well as in clinical trials [15–17]. Given synergistic interactions among individual drugs in multi-drug combinations, the more drugs in a combination, the greater the antiviral effect [6].

Preventing emergence of drug resistance

Genetic variants of HIV-1 conferring resistance to individual drugs can be selected by exposure to a particular drug. This may occur rapidly when drugs are used that have low genetic barriers to resistance, such as lamivudine or nevirapine [18–20]. Mutations will continue to accumulate in the viral genome unless the virus replication rate can be reduced below the level of detection. Antiretroviral combinations that prevent virus replication will decrease the selection pressure for the emergence of resistant HIV-1 variants [21, 22]. Furthermore, some mutations that confer resistance to one drug may increase the sensitivity of the virus to other drugs. For example, the M184V mutation that confers resistance to lamivudine, restores sensitivity to zidovudine-resistant virus that possess the M41L or 215Y mutations [12]. M184V may also increase susceptibility to the nucleotide analogs adefovir and tenofovir [23, 24]. The L74V mutation which confers reduced sensitivity to didanosine restores susceptibility to zidovudine of viruses with the zidovudine-induced T215Y mutation [25]. Enhanced susceptibility or "hypersusceptibility" to non-nucleoside reverse transcriptase inhibitors (NNRTIs) has been associated with multiple mutations conferring broad crossresistance to nucleoside reverse transcriptase inhibitors (NRTIs), and this phenomenon may result in an enhanced clinical response to NNRTIs [26]. Hypersusceptibility may also occur with protease inhibitors (PIs) where the D30N and N88S mutations induced by nelfinavir or indinavir confer increased sensitivity to amprenavir [27].

Reducing toxicity

All currently available antiretroviral drugs have toxicities that may affect quality of life and compromise adherence to therapy. By combining agents with favorable pharmacokinetic (PK) or synergistic antiviral properties, it may be possible to decrease the dose or dosing frequency of a drug and still achieve blood concentrations of the combination sufficient to achieve viral suppression. This approach is now widely utilized with ritonavir-boosted protease inhibitor (PI) regimens where ritonavir can be used at lower concentrations, thereby reducing its unpleasant gastrointestinal side-effects [28–31]. Ritonavir is a potent inhibitor of P450CYP3A4, thereby inhibiting the metabolism of other PIs (or the active metabolite, M8, in the case of nelfinavir). For this reason, ritonavir has been the drug of choice for PK enhancement of other PIs. Ritonavir increases the trough drug concentration (Cmin) and the area under the curve (AUC). It has a variable impact on the maximum drug concentration (Cmax) of the co-administered PI. In most cases Cmax is decreased or unchanged with the exception of indinavir where Cmax is increased [32]. Ritonavir also inhibits the P-glycoprotein (P-gp) system which serves as a cellular efflux pump responsible for removal of drug from the intracellular compartment [33]. Certain PIs such as amprenavir, lopinavir, ritonavir and nelfinavir also have CYP3A4 induction capabilities. Although these effects have been less studied, they can account for the reduced PI levels observed when combinations of some PI combinations are used [33a]. As we learn more about the pharmacokinetic interactions between different drugs, their use will become more rational. The use of therapeutic drug monitoring may aid substantially in this process.

Targeting the same or various steps of HIV replication cycle

One strategy to increase synergy between drugs is to target different steps of the HIV-1 replication cycle. The first class of inhibitors to be developed targeted the enzyme reverse transcriptase, preventing the conversion of HIV-1 genomic RNA into DNA, thereby decreasing the possibility of integrating viral DNA into host cell DNA. When the protease inhibitors were introduced, they offered the possibility of attacking the virus at another site, by reducing the production of infectious particles expressed from integrated viral genomes. The attachment/entry inhibitors have recently provided a new target for a possible multi-front attack. Many HIV attachment/entry inhibitors are synergistic when used with each other and with drugs from other classes [34, 35]. As more classes of inhibitors enter the clinic, such as integrase inhibitors or Tat inhibitors, we would expect to increase the possibilities of synergistic interactions among subsequent drug regimens.

It still remains an open question, however, whether it is better to attack the same enzyme, e.g., reverse transcriptase, simultaneously with multiple drugs, or to attack multiple targets. Three reverse transcriptase inhibitors (e.g., zidovudine, lamivudine, and efavirenz) in combination appear at least as good or better than other regimens that attack both reverse transcriptase and protease (ACTG 384, Dupont 006) [36, 36a].

Targeting various tissue or cellular reservoirs

The penetration of different drugs into various tissue or cellular reservoirs should also help guide the rational design of effective combination regimens. Antiretroviral agents penetrate different body compartments to varying degrees. The viral dynamics observed in semen [37–40], and cerebrospinal fluid (CSF) [41, 42] often differ with what is observed in plasma during treatment with potent antiretroviral regimens. The levels of protein binding and hydrophobicity are factors known to affect drug penetration into tissues. In the case of HIV-1 infection, the degree of cell activation may also be important in determining the activity of a drug [43, 44]. Resting memory CD4 lymphocytes have been found to harbor replication-competent viruses even when viral load is undetectable by measurement of plasma HIV-1 RNA [45–48]. Many currently available antiretroviral drugs are not active in resting cells that do not express markers indicative of HIV-1 infection. Therefore, development of new drugs or new strategies to target these sanctuaries is an important area for research into the improvement of combination therapy.

How to evaluate *in vitro* drug interactions

Several *in vitro* systems have been used to measure drug-drug interactions. In these assays, HIV-infected cells are cultured in the presence of various concentrations of drugs, alone and in combination, to generate dose-response curves. The interrelations between these curves are then analyzed. More than 20 different analytical methods have been developed to characterize such interactions [49]. Mathematical models integrate the dose-response curves of individual drugs into two- or three-dimensional graphs reflecting the dose-response surface of the combination studied. The results are then compared with the expected effects to determine whether synergistic, antagonistic or additive interactions occur. There is great variability in the systems used in terms of the analytical method (fractional product, multiple dose-response curves, isobologram, combination index, differential surface analysis, parametric surface fitting), the calculation of expected effects (interdependence or additivity), and the interpretation of results (qualitative, quantitative). The most widely used method for analysis is the median-effect principle developed by Chou and Talalay [50, 51]. This method is based on the Loewe additivity principle and defines synergy as the activity of the drugs in combination being greater than the sum of independent activities of individual drugs, and antagonism as the activity of the drugs in combination being smaller than the sum of independent activities. A combination index is derived that takes into account not only the potency of the drugs but also the shapes of the dose-effect curves. Other methods such as the independent effect method developed by Prichard and Shipman use Bliss independence as the null reference [52]. All the data points generated from a checkerboard assay are used to allow a three-dimensional analysis of drug interactions. Confidence bounds are determined from replicate data. If the 95% confidence intervals do not overlap the theoretic additive surface, then the interaction between drugs differs significantly from additive. Most drug combinations analyzed by both methods have shown similar results.

Can we extrapolate these results to *in vivo* settings?

The assays used to evaluate *in vitro* drug interactions have several limitations. They cannot reproduce the complex pharmacokinetics occurring in the human body, particularly the absorption, protein binding, different degrees of tissue localization or intracellular penetration, or the rate of drug elimination. However, *in vitro* analyses can help predict which drug interactions will occur at the therapeutic doses and help identify drug combinations with favourable profiles worthy of evaluation in clinical trials. If drug interactions are not favorable (i.e., are antagonistic), the benefit of combining drugs might be lost and one might be left with the antiviral effects of one drug and the toxicity of two. For example, as previously mentioned, zidovudine and stavudine are antagonistic against several HIV-1 isolates when combined *in vitro* [7, 10, 11]. When studied in two clinical trials, these *in vitro* results were validated *in vivo*. In ACTG 290, four antiretroviral regimens, zidovudine-stavudine, stavudine alone, zidovudine-didanosine and didanosine alone were compared in 144 zidovudine-experienced subjects. After 48 weeks of treatment, the proportion of subjects achieving plasma HIV-1 RNA levels below 500 copies/ml was 0% in the zidovudine-stavudine group *versus* 6% in the stavudine monotherapy group [9]. Similar results were observed in ACTG 298 which compared stavudine alone *versus* stavudine-zidovudine *versus* lamivudine-zidovudine in 128 treatment-naïve patients. In this population, the proportion of subjects with undetectable plasma HIV-1 RNA levels was only 8% in the zidovudine-stavudine group *versus* 31% in the stavudine monotherapy group [52a]. In both populations (drug experienced or naïve), the combination of zidovudine and stavudine was no more effective than stavudine alone.

In contrast, *in vitro* studies showing *in vitro* synergistic effects of zidovudine combined with agents such as alpha-interferon [53, 54], didanosine [55], lamivudine [7], zalcitabine [7] and abacavir [56], have led to pivotal clinical trials demonstrating the superiority of combination therapy over monotherapy in HIV-1 infection. Among those trials, ACTG 175 showed a 36% reduction in risk of disease progression or death for the zidovudine-didanosine arm compared to the zidovudine monotherapy arm [3]. In the Delta trial, in treatment-naïve individuals, there was a relative reduction in mortality of >42% for the zidovudine-didanosine arm and 32% for the zidovudine-zalcitabine arm, as compared to zidovudine-monotherapy [57].

Other combination regimens showing favourable *in vitro* drug interactions have proven to be extremely useful clinically. For example, the three-drug combination of indinavir, zidovudine, and lamivudine has synergistic interactions *in vitro* [14]. The clinical efficacy of this regimen has been amply validated in trials evaluating indinavir monotherapy and dual therapy with zidovudine and lamivudine compared to a combination of zidovudine, lamivudine and indinavir [14, 15, 17]. Clinical validation of *in vitro* results has also been demonstrated when nucleoside analogs are combined with non-nucleoside reverse transcriptase inhibitors. Zidovudine, lamivudine and efavirenz, as well as zidovudine, lamivudine and nevirapine [7], have synergistic drug interactions *in vitro*. These regimens have proven highly effective *in vivo* in a number of trial including DMP 266-005 006, the INCAS study, and ACTG 384 [2, 36, 36a, 58].

Composition of combination regimens

When designing an antiretroviral combination regimen, several questions need to be addressed: the number of drugs to employ, which classes of antiretroviral to use, which drugs within a class to use, and the dosing regimen (once daily, twice daily). It is not clear what is the optimal number of drugs for maximal suppression of HIV-1 replication, and this will likely depend on the potency of the individual drugs and their interactions with one another. *In vitro* studies have shown that when drug interactions are favorable, more drugs provide greater suppression than fewer [6, 7]. Preliminary clinical trials studying four- or five-drug regimens have shown a significant decrease in the first phase of decay of the viral load when compared with regimens employing fewer drugs. Whether the advantage is sustained over a long period of time and whether these regimens can be tolerated for long periods remain to be determined [5, 8]. Moreover, certain trials, e.g., ACTG 384, have shown that whereas certain four-drug regimens may be better than three-drug regimens, other three-drug regimens perform as well as those employing four drugs [58a].

As more drug resistant HIV-1 isolates emerge, new classes of potent antiretrovirals targeting different steps of the HIV replicative cycle are a welcome addition to the HIV arsenal. As these new drugs are being developed, *in vitro* drug interaction studies will be useful to determine how best to combine them with existing drugs against resistant viruses. An informative example are the interactions among HIV-1 attachment/entry inhibitors. Agents have been developed which target each step in the attachment/entry process including attachment to the CD4 receptor, binding to the co-receptor CCR5 or CXCR4 and fusion. Several of these agents have entered clinical trials, and one, T-20 or enfuvirtide, has shown considerable activity in individuals who have failed other regimens [59–62]. There have been concerns that the use of a CCR5 inhibitor could induce the virus to switch its co-receptor usage to the alternate co-receptor CXCR4 and might hasten disease progression [63]. Even if this is only a theoretical concern for the moment, it supports the notion of using these

CCR5 inhibitors in regimens containing drugs active against both R5 and X4 viruses. In this context, CCR5 inhibitor interactions have been evaluated with agents of the same class as well as reverse transcriptase and protease inhibitors. One such agent, SCH-C, showed highly synergistic interactions with T-20 which is active against both X4 and R5 viruses, as well as with other antiretrovirals from different classes [34]. High levels of synergy have also been observed when combining T-20 with the CXCR4 inhibitor AMD3100 [35].

Conclusions

It has proven difficult to accurately quantify drug interactions in an *in vivo* setting. This would require the use of several doses of various drugs, large sample sizes to allow statistical analysis and the occasional use of sub-optimal regimens that could lead to emergence of drug-virus resistance. Thus, *in vitro* drug combination studies will continue to serve as valuable surrogates to screen for unfavourable drug interactions and to help predict *in vivo* responses. There are other factors that will affect a drug combination's efficacy in a clinical setting, including the patient's adherence to the regimen, intercurrent illnesses that may affect viral replication, drug intolerance or toxicity, and drug pharmacokinetics. All these parameters have to be considered when designing combination regimens, in order to optimise efficacy and minimise toxicity.

References

1 Palella FJ Jr, Delaney KM, Moorman AC, Loveless MO, Fuhrer J, Satten GA, Aschman DJ, Holmberg SD, and The HIV Outpatient Study Investigators (1998) Declining morbidity and mortality among patients with advanced human immunodefiency virus infection. *N Engl J Med* 338: 853–860
2 D'Aquila RT, Hughes MD, Johnson VA, Fischl MA, Sommadossi JP, Liou SH, Timpone J, Myers M, Basgoz N, Niu M, Hirsch MS, and The National Institute of Allergy and Infectious Diseases AIDS Clinical Trials Group Protocol 241 Investigators (1996) Nevirapine, zidovudine, and didanosine compared with zidovudine and didanosine in patients with HIV-1 infection. A randomized, double-blind, placebo-controlled trial. *Ann Intern Med* 124: 1019–1030
3 Hammer SM, Katzenstein DA, Hughes MD, Gundacker H, Schooley RT, Haubrich RH, Henry WK, Lederman MM, Phair JP, Niu M et al. (1996) A trial comparing nucleoside monotherapy with combination therapy in HIV-infected adults with CD4 cell counts from 200 to 500 per cubic millimeter. *N Engl J Med* 335: 1081–1090
4 Havlir DV, Marschner JC, Hirsch MS, Collier AC, Tebas P, Bassett RL, Ioannidis JP, Holohan MK, Leavitt R, Boone G, Richman DD (1998) Maintenance antiretroviral therapies in HIV-infected subjects with undetectable plasma HIV RNA after triple-drug therapy. *N Engl J Med* 339: 1261–1268
5 Hoetelmans RMW, Reijers MHE, Weverling GJ, ten Kate RW, Wit FWNM, Mulder JW, Weigel HM, Jos Frissen PH, Roos M, Jurriaans S et al. (1998) The effect of plasma drug concentrations on HIV-1 clearance rate during quadruple therapy. *AIDS* 12: F111–F115
6 Mazzulli T, Rusconi S, Merrill DP, D'Aquila RT, Moonis M, Chou TC, Hirsch MS (1994) Alternating *versus* continuous drug regimens in combination chemotherapy of human immunodeficiency virus type 1 infection *in vitro*. *Antimicrob Agents Chemother* 38: 656–661

7 Merrill DP, Moonis M, Chou TC, Hirsch MS (1996) Lamivudine or stavudine in two- and three-drug combinations against human immunodeficiency virus type 1 replication *in vitro*. *J Infect Dis* 173: 355–364

8 Weverling GJ, Lange JMA, Jurriaans S, Prins JM, Lukashov VV, Notermans DW, Roos M, Schuitemaker H, Hoetelmans RMW, Danner SA et al. (1998) Alternative multidrug regimen provides improved suppression of HIV-1 replication over triple therapy. *AIDS* 12: F117–F122

9 Havlir DV, Tierney C, Friedland GH, Pollard RB, Smeaton L, Fox L, Sommadossi JP, Kessler H, Fife KH, Richman DD (2000) *In vivo* antagonism with zidovudine plus stavudine combination therapy. *J Infect Dis* 182: 321–325

10 Hoggard, PG, Sales SD, Kewn S, Sunderland D, Khoo SH, Hart CA, Back DJ (2000) Correlation between intracellular pharmacological activation of nucleoside analogues and HIV suppression *in vitro*. *Antiviral Chem Chemother* 11: 353–358

11 Hoggard PG, Kewn S, Barry MG, Khoo SH, Back DJ (1997) Effects of drugs on 2',3'-dideoxy-2',3'-didehydrothymidine phosphorylation *in vitro*. *Antimicrob Agents Chemother* 41: 1231–1236

11a Sommadossi JP, Zhou XJ, Moore J, Havlir DV, Friedland G, Tierney C, Smeaton L, Fox L, Richman DD, Pollard R and The ACTG 290 Team (1998) Impairment of stavudine (d4T) phosphorylation in patients receiving a combination of zidovudine (ZDV) and d4T (ACTG 290). 5th Conference on Retroviruses and Opportunistic Infections Chicago, Ill. USA, 1–5 February 1998, Abstract 3.

12 Larder BA, Kemp SD, Harrigan PR (1995) Potential mechanism for sustained anti-retroviral efficacy of AZT-3TC combination therapy. *Science* 269: 696–699

13 Eron JJ Jr, Johnson VA, Merrill DP, Chou TC, Hirsch MS (1992) Synergistic inhibition of replication of human immunodeficiency virus type 1, including that of a zidovudine-resistant isolate, by zidovudine and 2',3'-dideoxycytidine *in vitro*. *Antimicrob Agents Chemother* 36: 1559–1562

14 Snyder S, D'Argenio DZ, Weislow O, Bilello JA, Drusano GL (2000) The triple combination indinavir-zidovudine-lamivudine is highly synergistic. *Antimicrob Agents Chemother* 44: 1051–1058

15 Gulick RM, Mellors JW, Havlir DV, Eron JJ, Gonzalez C, McMahon D, Richman DD, Valentine FT, Jonas L, Meibohm A et al. (1997) Treatment with indinavir, zidovudine, and lamivudine in adults with human immunodeficiency virus infection and prior antiretroviral therapy. *N Engl J Med* 337: 734–739

16 Hammer SM, Squires KE, Hughes MD, Grimes JM, Demeter LM, Currier JS, Eron JJ Jr, Feinberg JE, Balfour HH Jr, et al. (1997) A controlled trial of two nucleoside analogues plus indinavir in persons with human immunodeficiency virus infection and CD4 cell counts of 200 per cubic millimeter or less. *N Engl J Med* 337: 725–733

17 Hirsch MS, Steigbigel R, Staszewski S, Mellors J, Scerpella E, Hirschel B, Lange J, Squires K, Rawlins S, Meibohm A, Leavitt R for the Protocol 039 Study Group (1999) A randomized, controlled trial of indinavir, zidovudine, and lamivudine in adults with advanced human immunodeficiency virus type 1 infection and prior antiretroviral therapy. *J Infect Dis* 180: 659–665

18 Boucher CA, Cammack N, Schipper P, Schuurman R, Rouse P, Wainberg MA, Cameron JM (1993) High-level resistance to (–) enantiomeric 2'-deoxy-3'-thiacytidine *in vitro* is due to one amino acid substitution in the catalytic site of human immunodeficiency virus type 1 reverse transcriptase. *Antimicrob Agents Chemother* 37: 2231–2234

19 Tisdale M, Kemp SD, Parry NR, Larder BA (1993) Rapid *in vitro* selection of human immunodeficiency type 1 resistant to 3'-thiacytidine inhibitors due to a mutation in the YMDD region of reverse transcriptase. *Proc Natl Acad Sci USA* 90: 5653–5656

20 Richman DD, Havlir D, Corbeil J, Looney D, Ignacio C, Spector SA, Sullivan J, Cheeseman S, Barringer K, Pauletti D et al. (1994) Nevirapine resistance mutations of human immunodeficiency virus type 1 selected during therapy. *J Virol* 68: 1660–1666

21 Coffin, JM (1995) HIV population dynamics *in vivo*: implications for genetic variation, pathogenesis, and therapy. *Science* 267: 483–489

22 Hirsch MS, Brun-Vezinet F, Clotet B, Conway B, Kuritzkes DR, D'Aquila RT, Demeter LM, Hammer SM, Johnson VA, Loveday C, et al. (2003) Antiretroviral drug resistance testing in adults infected with human immunodeficiency virus type I: 2003 recommendations of an International AIDS Society-USA panel. *Clin Infect Dis* 37: 113–128

23 Miller MD, Anton KE, Mulato AS, Lamy PD, Cherrington JM (1999) Human immunodeficiency virus type 1 expressing the lamivudine-associated M184V mutation in reverse transcriptase shows increased susceptibility to adefovir and decreased replication capability *in vitro*. *J Infect Dis* 179: 92–100

24 Naeger LK, Margot NA, Miller MD (2001) Increased drug susceptibility of HIV-1 reverse transcriptase mutants containing M184V and zidovudine-associated mutations: analysis of enzyme processivity, chain-terminator removal and viral replication. *Antiviral Ther* 6: 115–126
25 St Clair MH, Martin JL, Tudor-Williams G, Bach MC, Vavro CL, King DM, Kellam P, Kemp SD, Larder BA (1991) Resistance to ddI and sensitivity to AZT induced by a mutation in HIV-1 reverse transcriptase. *Science* 253: 1557–1559
26 Schulman N, Zolopa AR, Passaro D, Shafer RW, Huang W, Katzenstein D, Israelski DM, Hellmann N, Petropoulos C, Whitcomb J (2001) Phenotypic hypersusceptibility to non-nucleoside reverse transcriptase inhibitors in treatment-experienced HIV-infected patients: impact on virological response to efavirenz-based therapy. *AIDS* 15: 1125–1132
27 Ziermann R, Limoli K, Das K, Arnold E, Petropoulos CJ, Parkin NT (2000) A mutation in human immunodeficiency virus type 1 protease, N88S, that causes *in vitro* hypersensitivity to amprenavir. *J Virol* 74: 4414–4419
28 Cameron WD, Japour AJ, Xu Y, Hsu A, Mellors J, Farthing C, Cohen C, Poretz D, Markowitz M, Follansbee S et al. (1999) Ritonavir and saquinavir combination therapy for the treatment of HIV infection. *AIDS* 13: 213–224
29 Hsu A, Granneman GR, Cao G, Carothers L, Japour A, El-Shourbagy T, Dennis S, Berg J, Erdman K, Leonard JM, Sun E (1998) Pharmacokinetic interaction between ritonavir and indinavir in healthy volunteers. *Antimicrob Agents Chemother* 42: 2784–2791
30 Sham HL, Kempf DJ, Molla A, Marsh KC, Kumar GN, Chen CM, Kati W, Stewart K, Lal R, Hsu A et al. (1998) ABT-378, a highly potent inhibitor of the human immunodeficiency virus protease. *Antimicrob Agents Chemother* 42: 3218–3224
31 Van Heeswijk RPG, Veldkamp AI, Hoetelmans RMW, Mulder JW, Schreij G, Hsu A, Lange JMA, Beijnen JH, Meenhorst PL (1999) The steady-state plasma pharmacokinetics of indinavir alone and in combination with a low dose of ritonavir in twice daily dosing regiments in HIV-1 infected individuals. *AIDS* 13: F95–F99
32 Becker S, Fisher A, Flexner C, Gerber JG, Haubrich R, Kashuba AD, Luber AD, Piscitelli SC (2001) Pharmacokinetic parameters of protease inhibitors and the Cmin/IC50 ratio: call for consensus. *J Acquir Immune Defic Syndr* 2001;27: 210–211
33 Profit L, Eagling VA, Back DJ (1999) Modulation of P-glycoprotein function in human lymphocytes and Caco-2 cell monolayers by HIV-1 protease inhibitors. *AIDS* 13: 1623–1627
33a Peytavin G, Lamotte C, Duval X et al. (2001) Amprenavir plasma concentrations are dramatically decreased when combined with lopinavir/ritonavir in HIV-infected patients. Program and abstracts of the Second International Workshop on Clinical Pharmacology of HIV Therapy, April 2–4, 2001, Noordwijk, The Netherlands, Abstract 1, 14
34 Tremblay CL, Giguel F, Kollmann C, Guan Y, Chou TC, Baroudy BM, Hirsch MS (2002) Anti-HIV interactions of SCH-C (SCH 351125), a CCR5 antagonist, with other antiretroviral agents *in vitro*. *Antimicrob Agents Chemother* 46: 1336–1339
35 Tremblay CL, Kollmann C, Giguel F, Chou TC, Hirsch MS (2000) Strong *in vitro* synergy between the fusion inhibitor T-20 and the CXCR4 blocker AMD-3100. *J Acquir Immune Defic Syndr* 25: 99–102
36 Staszewski S, Morales-Ramirez J, Tashima KT, Rachlis A, Skiest D, Stanford J, Stryker R, Johnson P, Labriola DF, Farina D et al. (1999) Efavirenz plus zidovudine and lamivudine, efavirenz plus indinavir, and indinavir plus zidovudine and lamivudine in the treatment of HIV-1 infection in adults. *N Engl J Med* 341: 1865–1873
36a Robbins GK, De Gruttola V, Shafer RW, Smeaton LM, Snyder SW, Pettinelli C, Dube MP, Fischl MA, Pollard RB, Delapenha R et al. for the ACTG384 Team (2003) Comparison of sequential three-drug regimens for initial therapy. *N Engl J Med*; *in press*
37 Barroso PF, Schechter M, Gupta P, Melo MF, Vieira M, Murta FC, Souza Y, Harrison LH (2000) Effect of antiretroviral therapy on HIV shedding in semen. *Ann Int Med* 133: 280–284
38 Coombs RW, Speck CE, Hughes JP, Lee W, Sampoleo R, Ross SO, Dragavon J, Peterson G, Hooton TM, Collier AC et al. (1998) Association between culturable human immunodeficiency virus type 1 in semen and HIV-1 RNA levels in semen and blood: evidence for compartmentalization of HIV-1 between semen and blood. *J Infect Dis* 177: 320–330
39 Kiessling AA, Fitzgerald LM, Zhang D, Chhay H, Brettler D, Eyre RC, Steinberg J, McGowan K, Byrn RA (1998) Human immunodeficiency virus in semen arises from a geneticallly distinct virus reservoir. *AIDS Res Hum Retroviruses* 14 (suppl. 1): S33–S41
40 Vernazza PL, Eron JJ, Cohen MS, van der Horst CM, Troiani L, Fiscus SA (1994) Detection and biologic characterization of infectious HIV-1 in semen of seropositive men. *AIDS* 8: 1325–1329

41 Cheeseman SH, Hattox SE, McLaughlin MM, Koup RA, Andrews C, Bova CA, Pav JW, Roy T, Sullivan JL, Keirns JJ (1993) Pharmacokinetics of nevirapine: initial single-rising-dose study in humans. *J Infect Dis* 37: 178–182
42 Portegies, P (1998) HIV-1, the brain, and combination therapy. *Lancet* 346: 1244–1245
43 Gao WY, Shirasaka T, Johns DG, Broder S, Mitsuya H (1993) Differential phosphorylation of azidothymidine, dideoxycytidine, and dideoxyinosine in resting and activated peripheral blood mononuclear cells. *J Clin Invest* 91: 2326–2333
44 Lori F, Malykh AG, Foli A, Maserati R, De Antoni A, Minoli L, Padrini D, Degli Antoni A, Barchi E, Jessen H et al. (1997) Combination of a drug targeting the cell with a drug targeting the virus controls human immunodeficiency virus type 1 resistance. *AIDS Res Hum Retroviruses* 13: 1403–1409
45 Chun TW, Carruth L, Finzi D, Shen X, DiGiuseppe JA, Taylor H, Hermankova M, Chadwick K, Margolick J, Quinn TC et al. (1997) Quantification of latent tissue reservoirs and total body viral load in HIV-1 infection. *Nature* 387: 183–188
46 Chun, TW, Stuyver L, Mizell SB, Ehler LA, Mican JA, Baseler M, Lloyd AL, Nowak MA, Fauci AS (1997) Presence of an inducible HIV-1 latent reservoir during highly active antiretroviral therapy. *Proc Natl Acad Sci USA* 94: 13193–13197
47 Finzi D, Hermankova M, Pierson T, Carruth LM, Buck C, Chaisson RE, Quinn TC, Chadwick K, Margolick J, Brookmeyer R et al. (1997) Identification of a reservoir for HIV-1 in patients on highly active antiretroviral therapy. *Science* 278: 1295–1300
48 Wong, JK, Hezareh M, Günthard HF, Havlir DV, Ignacio CC, Spina CA, Richman DD (1997) Recovery of replication-competent HIV despite prolonged suppression of plasma viremia. *Science* 278: 1291–1295
49 Greco WR, Bravo G, Parsons JC (1995) The search for synergy: a critical review from a response surface perspective. *Pharmacol Rev* 47: 331–385
50 Chou T, Hayball MK (1996) *CalcuSyn for Windows, Multiple-drug dose-effect analyzer and manual.* Biosoft, Cambridge Place, Cambridge, UK
51 Chou T, Talalay P (1984) Quantitative analysis of dose effect relationships: The combined effects of multiple drugs or enzyme inhibitors. *Adv Enzyme Regul* 22: 27–55
52 Prichard MN, Shipman C Jr, (1996) Analysis of combinations of antiviral drugs and design of effective multidrug therapies. *Antiviral Ther* 1: 9–20
52a Havlir DV, Friedland G, Pollard R, Tierney C, Smeaton L, Fox L, Richman D (1998) Combination zidovudine (ZDV) and stavudine (d4T) therapy *versus* other nucleosides: report of two randomized trials (ACTG 290 and 298). 5th Conference on Retroviruses and Opportunistic Infections Chicago, Ill. USA, 1–5 February 1998, Abstract 2
53 Hartshorn KL, Vogt MW, Chou TC, Blumberg RS, Byington R, Schooley RT, Hirsch MS (1987) Synergistic inhibition of human immunodeficiency virus *in vitro* by azidothymidine and recombinant alpha A interferon. *Antimicrob Agents Chemother* 31: 168–172
54 Johnson VA, Merrill DP, Chou TC, Hirsch MS (1992) Human immunodeficiency virus type 1 (HIV-1) inhibitory interactions between protease inhibitor Ro 31-8959 and zidovudine, 2'3'-dideoxycytidine, or recombinant interferon-alpha against a zidovudine-sensitive or -resistant HIV-1 *in vitro*. *J Infect Dis* 166: 1143–1146
55 Antonelli G, Dianzani F, Bellarosa D, Turriziani O, Riva E, Gentile A (1994) Drug combination of AZT and ddI: synergism of action and prevention of appearance of AZT-resistance. *Antiviral Chem Chemother* 5: 51–55
56 Daluge SM, Good SS, Faletto MB, Miller WH, St Clair MH, Boone LR, Tisdale M, Parry NR, Reardon JE, Dornsife RE et al. (1997) 1592U89, a novel carbocyclic nucleoside analog with potent, selective anti- ıuman immunodeficiency virus activity. *Antimicrob Agents Chemother* 41: 1082–1093
57 Delta Coordinating Committee (1996) Delta: a randomised double-blind controlled trial comparing combinations of zidovudine plus didanosine or zalcitabine with zidovudine alone in HIV-infected individuals. *Lancet* 348: 283–291
58 Montaner JSG, Reiss P, Cooper D, Vella S, Harris M, Conway B, Wainberg MA, Smith D, Robinson P, Hall D et al. (1998) A randomized, double-blind trial comparing combinations of nevirapine, didanosine, and zidovudine for HIV-infected patients: the INCAS trial. Italy, The Netherlands, Canada, and Australia Study. *JAMA* 279: 930–937
58a Shafer RW, Smeaton LM, Robbins GK, De Gruttola V, Snyder SW, D'Aquila RT, Johnson VA, Morse GB, Nokta MA, Martinez AI et al. for the ACTG384 Team (2003) Comparison of four-drug and sequential three-drug regimens. *N Engl J Med*; *in press*

59 Doms RW, Moore JP (2000) HIV-1 membrane fusion: targets of opportunity. *J Cell Biol* 151: F9–F14
60 Kilby JM, Hopkins S, Venetta TM, DiMassimo B, Cloud GA, Lee JY, Alldredge L, Hunter E, Lambert D, Bolognesi D et al. (1998) Potent suppression of HIV-1 replication in humans by T-20, a peptide inhibitor of gp41-mediated virus entry. *Nat Med* 4: 1302–1307
61 Pilcher CD, Eron JJ Jr, Ngo L, Dusek A, Sista P, Gleavy J, Brooks D, Venetta T, DiMassimo E, Hopkins S (1999) Prolonged therapy with the fusion inhibitor T-20 in combination with oral antiretroviral agents in an HIV-infected individual. *AIDS* 13: 2171–2173
62 Strizki JM, Xu S, Wagner NE, Wojcik L, Liu J, Hou Y, Endres M, Palani A, Shapiro S, Clader JW et al. (2001) SCH-C (SCH 351125), an orally bioavailable, small molecule antagonist of the chemokine receptor CCR5, is a potent inhibitor of HIV-1 infection *in vitro* and *in vivo*. *Proc Natl Acad Sci USA* 98: 12718–12723
63 Mosier DE, Picchio GR, Gulizia RJ, Sabbe R, Poignard P, Picard L, Offord RE, Thompson DA, Wilken J (1999) Highly potent RANTES analogues either prevent CCR5-using human immunodeficiency virus type 1 infection *in vivo* or rapidly select for CXCR4-using variants. *J Virol* 73: 3544–3550

Combination Therapy of AIDS
Edited by E.D.A. De Clercq and A.-M.I. Vandamme

Comparison of the efficacy of HAART: single, dual or triple-class antiretroviral therapy

Roger Paredes, Bonaventura Clotet and Lidia Ruiz

IrsiCaixa Retrovirology Laboratory and HIV Unit, Hospital Universitari Germans Trias i Pujol, 08916 Barcelona, Spain

Introduction

Current antiretroviral therapy guidelines [1] recommend the combination of at least two nucleoside retrotranscriptase inhibitors (NRTI) plus at least one protease inhibitor (PI) and/or at least one nonnucleoside retrotranscriptase inhibitor (NNRTI) for the treatment of HIV infection. Different specific drugs are chosen to comply with this general scheme considering the stage of the disease, the occurrence of AIDS-defining events, plasma viral load and patient's immune status, prior pharmacologic history, likeliness to adhere to treatment, as well as drug interactions, adverse events and comorbidity.

One of the basic principles of modern antiretroviral therapy is to inhibit HIV's life cycle in different stages. This principle derives from the possibility to block retrotranscription and protease-mediated protein splicing at once by combining NRTIs (or NNRTIs) and PIs. To date, this strategy is believed to be the most effective to control viral replication and to limit the viral escape through the development and evolution of drug resistance mutations. Hence, several drugs are under development to block other stages of HIV replicative life cycle such as: entry inhibitors (T-20, T-1249, D-peptides, etc), uncoating inhibitors, assembly inhibitors, zinc finger (DNA complex) inhibitors, capsid protein inhibitors, etc.

However, although initial responses to currently available therapy are favourable, treatment failure is expected for the majority of patients in light of the clinical results. Even worse, every treatment failure decreases the likeliness to respond to ensuing salvage strategies and substantially restricts the therapeutic arsenal. There are only two ways to overcome this situation: a) to develop new drugs with complementary resistance patterns and/or sites of action, and b) to implement the use of currently available drugs at its maximum. One way to perform this goal is to maximize the number of combinations of antiretroviral drugs.

Following this aim, several studies – many of them still ongoing – have investigated the following questions:

1. Do combinations of three NRTIs *versus* two NRTIs + one PI and *versus* two NRTIs + one NNRTI, have a similar antiviral efficacy? If yes, Is this efficacy equally durable in time?
2. Do they achieve equal immunological benefits?
3. Does any of them confer a benefit regarding future salvage regimens?
4. Is there a differential improvement in quality of life, adverse events and in adherence issues for any of these combinations?
5. Is there a specific combination that is more suitable for starting therapy?

This chapter will review the results of the most relevant studies with the aim to compare the efficacy of different drug family combinations constituting HAART.

Initial antiviral efficacy

Single class HAART

Although several theoretical combinations may be possible, the concept of "single class HAART" is currently restricted to triple drug combinations exclusively including NRTIs (ddI/d4T/hydroxyurea, ddI/d4T/3TC or ABC/ZDV/3TC). The idea of combining three NRTIs as a unique antiretroviral strategy certainly seems odd for the majority of NRTIs, and it is dubious whether such combinations may be strictly considered as "highly active." In practice, the only combinations of three NRTIs suitable to that concept are those including abacavir (ABC). This NRTI has been recently commercialised and, due to its unique properties, is currently considered to be similar to a PI in terms of antiviral potency.

The only prospective available data about single class HAART combinations not including ABC comes from the Atlantic Study [2] that compares three arms including d4T/ddI plus either IDV, NVP, or 3TC. Definitive results of this study have not been published yet. However, although 6- and 12-month data showed virological equipotency of the three arms, 48-week data demonstrated a clear tendency (although not significant) of the arm including d4T/ddI/3TC to achieve the worst virological outcome. Definitive results will clarify this issue, although there is not much confidence in this combination.

Conversely, three-NRTI regimens including ABC have achieved quite better results in prospective comparative studies. The CNAAB3005 International Study Team [3] recently demonstrated that in antiretroviral-naïve HIV-infected adults, the triple nucleoside regimen composed of ABC/3TC/ZDV was equivalent to the regimen composed of IDV/3TC/ZDV in achieving a plasma HIV RNA level <400 copies/mL at 48 weeks. This was a placebo-controlled, double-blind, double-dummy study. The proportion of patients who met the end point of having an HIV RNA level of 400 copies/mL or less at week 48 was equivalent between the ABC group (51% [133/262]) and the IDV group (51% [136/265]), with a treatment difference of –0.6% (95% confidence interval [CI],

–9% to 8%). However, in patients with baseline HIV RNA levels greater than 100,000 copies/mL, the proportion of patients achieving <50 copies/mL was greater in the IDV group than in the abacavir group with 45% (45/100) *versus* 31% (30/96) and a treatment difference of –14% (95% CI, –27% to 0%) was observed. The two treatments were comparable with respect to their effects on CD4 cell counts. There was no difference between the groups in the frequency of treatment-limiting adverse events or laboratory abnormalities.

A second study comparing ABC *versus* IDV (CNA3014) [4] was presented at the 1st IAS Conference in Buenos Aires. This was an open-label study in which the lack of placebos allowed a much-reduced pill burden. CNA3014 also compared the efficacy and tolerability of ABC (300 mg twice daily, $n = 169$) *vs* IDV (800 mg three times daily, $n = 173$), each with *Combivir*. Results presented were similar to CNAAB3005 Study. Briefly, at week 48, 59% of the 164 patients receiving ABC achieved a pVL <50 copies/mL, compared with 48% of the 165 patients receiving IDV (P = NS). Of note, no differences were observed in the stratum of patients with high baseline viral load, although abacavir was superior to IDV in the low viral load stratum. Both treatment regimens were generally well tolerated. As well, although adherence was higher for the ABC/*Combivir* regimen this combination was only equally effective as IDV/*Combivir* at suppressing viral load to <50 copies/mL.

In conclusion, single class regimens including ABC have demonstrated a comparable efficacy to PI-containing regimens at least during the first 2 years of follow-up in antiretroviral naïve HIV-1 infected patients. Other 3-NRTI regimens sparing ABC have not demonstrated such equipotency. Conversely, comparative studies available to date show a tendency towards worst virological results of ABC-sparing 3-NRTI regimens in comparison with dual-class regimens. Triple NRTI regimens are not routinely recommended. However, they may be considered as starting regimens for patients with low viral loads who are reductant or unable to commit taking more complex HAART regimens, or in whom virologic failure seems inevitable and long-term options need to be preserved.

Dual class HAART

We consider dual class HAART those combinations including at least two NRTIs plus either a NNRTI or a PI. Recent important studies have demonstrated that certain NNRTI-based triple-drug combinations achieve an equal or even higher virological suppression than certain PI-based HAART regimens, at least in the short-term.

Efavirenz (EFV)

Staszewski et al. [2] recently demonstrated that the combination of EFV, ZDV and 3TC had greater antiviral activity and was better tolerated than the combination of IDV, ZDV, and 3TC. In an open-label, three-arm randomised study (DuPont 006 Trial) recently published, 450 patients who had not previously been treated with 3TC or any NNRTI or PI were recruited. At baseline, all patients had a CD4 cell count of more than 50 cells per cubic millimeter and a plasma HIV-1 RNA level of more than 10,000 copies per milliliter. Patients were randomly assigned to one of three regimens: EFV (600 mg daily) plus ZDV (300 mg twice daily) and 3TC (150 mg twice daily); the protease inhibitor IDV (800 mg every 8 h) plus ZDV and 3TC; or EFV plus IDV (1000 mg every 8 h). They found that suppression of plasma HIV-1 RNA to undetectable levels at week 48 of follow-up was achieved in more patients in the group given EFV plus NRTIs than in the group given IDV plus NRTIs (70 percent *vs* 48 percent, $p < 0.001$). The efficacy of the regimen composed by EFV plus IDV was similar (53 percent) to that of the regimen of IDV/ZDV/3TC. Of note, response rates in the subgroup of patients with HIV-1 RNA levels of at least 100,000 copies per milliliter at base line were the same as or greater than those in the subgroup of patients with base-line HIV-1 RNA levels of less than 100,000 copies per milliliter for the group given EFV plus NRTIs. Indeed, more patients discontinued treatment because of adverse events in the group given IDV and 2 NRTIs than in the group given EFV and two NRTIs (43 percent *versus* 27 percent, $p = 0.005$). Nevertheless, CD4 cell counts increased significantly with all combinations (range of increases, 180 to 201 cells per cubic millimeter). Interestingly, no significant differences were found regarding CD4 cell count increase between the groups. Several other studies have confirmed the superiority of EFV- *versus* IDV-based regimens [6].

It is unlikely that the benefit achieved by EFV-based combinations over those IDV-based can be generalised to their respective families. In other words, we cannot conclude from this data that NNRTI-based triple drug combinations are more potent than those including a PI. First, because there seem to exist significant differences in terms of antiviral potency within the different NNRTIs and, second, because studies comparing NNRTIs other than EFV with PIs have not demonstrated such superiority.

For instance, recent data [6] from the EuroSIDA cohort observational database including more than 2000 HIV-1 patients starting NNRTI showed that EFV-based regimens achieved a more prolonged long-term suppression of plasma HIV-1 RNA than NVP given under similar circumstances. Differences between both regimens were clearly significant, even when both were used as an initial HAART regimen.

Nevirapine (NVP)

Although there is consistent data arguing in favour of a higher antiviral potency of EFV- *versus* PI-containing combinations, results comparing NVP- with PI-containing regimens are slightly more controversial.

The first data demonstrating the antiviral and immunological efficacy of NVP came from the INCAS study [12]. In this study, 151 treatment-naïve patients with 200–600 CD4 cells/mm^3 were randomized to receive ZDV/ddI, ZDV/NVP, or ZDV/ddI/NVP. The triple-therapy arm was associated with the highest rise in CD4+ cell counts (120 cells/mm^3), the greatest decrease in HIV RNA from baseline (1.7 logs) and the greatest proportion of patients with undetectable virus after 52 weeks (67%).

One of the most important clinical trials comparing the efficacy of NVP-containing three-drug regimens is the Atlantic Study [2]. This is an ongoing open-label randomised international study comparing three treatment strategies: PI-based, NNRTI-based, and triple NRTI regimens. Patients included are antiretroviral treatment-naïve with HIV RNA >500 copies/mL and CD4 >200 cells/mm^3. Subjects are randomized to receive d4T/ddI plus either IDV, NVP, or 3TC. Definitive results have not yet been published, but preliminary data at 48 weeks of follow-up have been presented in international meetings. Of note, mean baseline viral load levels were low in all three groups. In the as-treated analysis, the percentage of patients with HIV RNA below 50 copies/mL in the IDV, NVP, and 3TC arms was 90%, 82%, and 78%, respectively at 48 weeks, whereas in the intent-to-treat (ITT) analysis, the percentages were 57%, 51%, and 49%, respectively. In conclusion, NVP- and IDV-based regimens performed similarly, at least in patients with low baseline viral load.

Another relevant comparative study has been the COMBINE Study [8]. Final data were presented in the 1st International AIDS Society Conference on HIV Pathogenesis and Treatment. This randomized, open-label, multicenter trial evaluated the efficacy and safety of NFV (1250 mg twice daily) or NVP (200 mg twice daily) both combined with coformulated ZDV/3TC (*Combivir*: ZDV 300 mg plus 3TC 150 mg twice daily). Up to 142 treatment-naïve patients were included. At baseline patients had a mean CD4 cell count of 359 cells/mm^3 (range, 10–908) and a mean viral load of 5.15 $\log_{10}$ copies/mL (range, 3.2–6.2). No differences were demonstrated between the two regimens at 12 weeks of follow-up regarding virological and immunological outcome. This conclusion has to be highlighted because prior interim analyses of this study showed a higher efficacy of NVP. According to final results, then, at 12 months (and by an ITT analysis), 60% of patients receiving NFV (CI_{90}, 49%–70%) *versus* 75% of those receiving NVP (CI_{90}, 65%–83%) achieved a plasma viral load <200 copies/mL ($p = .056$). Regarding ultrasensitive pVL analysis (<20 c/mL), 50% of NFV (CI_{90}, 40%–60%) *versus* 65.2% of NVP-treated patients (CI_{90}, 55%–75%) achieved undetectable pVL levels ($p = .065$). In addition, no differences were found in the subset of patients with baseline pVL > 100,000 copies/mL. Interestingly, again no differences were found in terms of CD4+ increase between the groups.

Conclusion

In conclusion, whereas three-drug combinations including EFV seem to achieve better virological results than those containing a PI for patients initiating HAART for the first time with baseline CD4 > 50 cells/mm^3, combinations including NVP perform as good as (but no better than) PI-containing regimens, at least during the first 1 or 2 years of treatment. However, it has to be highlighted that the only regimens that have demonstrated a clear antiviral efficacy for patients with advanced stages of HIV infection (CD4 < 50 cells/mm^3) are PI-based combinations.

Triple class HAART

Recent studies have tested antiretroviral drug combinations including the three available drug families in HIV-1-infected patients naïve for NNRTIs and/or PIs. These combinations have been frequently used as salvage regimens in antiretroviral experienced patients. In one hand, this is an attractive idea because it seeks the simultaneous suppression of the viral life cycle by three different mechanisms. Such combinations have demonstrated higher rates of viral suppression than standard regimens but, as well, studies presented have compared four-drug against three-drug regimens.

Albrecht et al. [9] have demonstrated that in HIV-infected patients previously treated with nucleoside analogues, treatment with nelfinavir (NFV) plus EFV and at least one new nucleoside analogue achieved a higher rate of viral suppression than did regimens with nucleoside analogues and NFV or EFV alone. Up to 195 patients who had been treated with NRTIs only, and had a plasma VL > 500 copies/ml, were randomly assigned to receive, in addition to 2 NRTIs, NFV, EFV or NFV+EFV. At week 48, the proportion of patients with a pVL < 500 copies/mL were, respectively, 74%, 60% and 35% in the EFV+NFV, EFV and NFV groups ($p = 0.001$). As well, triple therapy with EFV conferred a higher rate of pVL suppression at 48 weeks than triple therapy with NFV ($p = 0.004$). Quadruple therapy also achieved a higher rate of virologic suppression than triple therapy with EFV ($p = 0.008$). Once again, no differences were found between the groups regarding the increase in CD4 counts.

Although results from this study demonstrate that a regimen including 2 NRTIs+ EFV+NFV achieves better virological outcome than standard three-drug regimens, it is unknown to what extent this excess of benefit is due to the fact of including four drugs instead of three. As is well-known, four-drug regimens including two PIs plus two NRTIs have also demonstrated a higher virological suppression than single PI-based triple regimens [20]. Further studies comparing both kinds of four-drug combinations (two NRTIs + one NNRTI + one PI *versus* two NRTIs + 2 PIs) are needed to clarify this issue.

Immunological benefits

Despite the virological results presented before, to date, no PI-sparing combination has improved the immunological results previously attained with standard two NRTI + one PI strategies. Certainly, absolute and percentual CD4 counts are a slower marker of response than viral load reductions, and, of course, results regarding immune recovery are more difficult to interpret. However, none of the studies presented before (including CNAAB3005, CNA3014, DuPont 006, Atlantic, COMBINE, or the more recent by Albrecht et al) [2–5, 8] has demonstrated significant differences between the different study arms in terms of CD4 increase during their respective follow-ups. Furthermore, none has detailed modifications in naïve *versus* memory subpopulations or in other markers of immune reconstitution between groups.

Only recently, a substudy [9] of the Atlantic Study [2] that examined the effects of the three study arms (ddI/d4T plus either 3TC, NVP or IDV) on lymph node architecture and CD4 cell quantity in lymph nodes, reported that there was a trend towards fewer CD4 cells and higher levels of HIV p24 antigen in the single classs arm compared with the dual class arms. As well, the majority (6/10) of the single-class subjects had hyperplastic lymph nodes whereas most (8/11) of the subjects on PI- or NNRTI-based therapy had lymphoid hypoplasia or absence of follicle formation. Such findings suggest a better impact of dual class regimens on lymph nodes. However, duration of therapy prior to the lymph node biopsy was substantially shorter in the single-class group than in the dual-class group. Hence, it remains unclear whether these findings reflect differences in duration of treatment, incomplete penetration of all anatomic compartments by the mono-class regimen, or both.

In conclusion, presently available studies have not demonstrated a differential impact of single *versus* dual or triple class regimens in the improvement of the immune system of patients initiating these combinations for the first time. Data from the AVANTI and INCAS [12] groups comparing PI-based triple therapies and dual NRTI regimens, suggested that CD4 count improvement was not unique to HAART regimens containing PIs, but that it was related to the degree of viral load suppression accomplished, regardless of the modality of the treatment chosen. This would explain the similarity of short-term CD4 improvements seen in comparative studies between single, dual and triple class regimens, as such regimens achieved a similar antiviral efficacy. Hopefully, further studies will analyse this matter in depth, which is of capital relevance for clinicians due to its important clinical prognostic implications.

Durability of benefits

One of the main concerns of the new PI-sparing regimens is the durability of the favourable virological responses attained after their first initiation. In other words, are PI-sparing single and dual class regimens more prone to treatment failure than are PI-based strategies? This concern arises from the lower genetic barrier of NNRTIs and the possible lower antiviral potency of NRTIs.

Theoretical considerations

Drug failure is a multifactorial phenomenon occurring due to multiple reasons that gather simultaneously in a single patient. Below there is a diagram that shows that drug failure depends on favouring factors (in the numerator) and limiting factors (in the denominator). Drug failure is more likely if any of the following factors exist: a longer time on antiretroviral therapy, a longer prior antiretroviral experience, a more advanced stage of HIV/AIDS disease, a higher baseline pVL, a higher number or the appearance of certain specific resistance mutations before the onset of a new treatment, a higher pill burden of the new regimen, a higher co-morbidity and an older age at the time of initiating a new regimen. Conversely, factors that prevent treatment failure include: the higher genetic barrier of a drug or a combination of drugs, the better tolerability, the degree of adherence to treatment, the number of active drugs included in the new regimen, higher CD4 counts at baseline, patients's social and economical support and, finally, the degree of suppression of viral replication achieved with the new treatment combination. The likeliness of drug failure has to be considered in a comprehensive manner, taking all these factors into consideration simultaneously.

$$\text{Drug failure} = f\left(\frac{\text{Time on antivirals, number of prior antivirals, pill burden, stage of HIV/AIDS, baseline pVL, baseline resistance mutations, co-morbidity, age}}{\text{Drug genetic barrier, tolerability, adherence, number of new drugs, baseline CD4 counts, degree of viral suppression, social and economic support}}\right)$$

However, there are a number of factors that are **unmodifiable** when planning a new antiretroviral regimen (Tab. 1). These factors include: prior time on antiretroviral therapy, prior antiretroviral experience, stage of HIV/AIDS disease, baseline pVL, prior existing resistance mutations, co-morbidity, age, baseline CD4 counts. Such factors must be assumed and considered when designing a new regimen, but the clinician is not able to modify them.

In contrast, other factors influencing the likeliness of drug failure are certainly **controllable**. Therefore it is the task of the clinician to design a drug combination seeking the lowest pill burden, the higher genetic barrier, the higher adherence to treatment, the higher number of new drugs included and the higher degree of viral suppression attainable. Although it is more difficult, we believe that the physician should also cooperate with his/her community to improve patients's social and economical support.

Table 1. Factors influencing drug failure

Unmodifiable	Prior time on antiretroviral therapy Prior antiretroviral experience Stage of HIV/AIDS disease Baseline pVL Baseline CD4 counts Prior existing resistance mutations Co-morbidity Age
Controllable	Pill burden Genetic barrier Tolerability Adherence Number of new drugs Degree of viral suppression

Single, dual and triple class regimens show different characteristics regarding controllable factors influencing drug failure (Tab. 2).

Single class agents, and more specifically ABC/ZDV/3TC, imply the lowest pill burden of all regimens. *Trizivir*, a combination of ABC, ZDV and 3TC in a single capsule, is administered twice a day and has less food restictions than other combinations. Such low pill burden and other factors favour a high degree of adherence, as was recently demonstrated in the CNA3014 Study [4]. In this study, at 44–48 weeks of follow-up, 72% of patients receiving ABC/Combivir *versus* 45% of IDV/Combivir recipients maintained an adherence level above 95% ($p < 0.001$). However, despite a better adherence, ABC/*Combivir* was only equally effective as IDV/*Combivir* at suppressing viral load to <50 copies/mL. This result highlights that antiviral efficacy (as

Table 2. Characteristics of single, dual and triple class combinations regarding controllable factors influencing drug failure.

	Single class	Dual class		Triple class
		NNRTI-based	*PI-based*	
Pill burden	Lowest	Lower	Higher	Highest
Genetic barrier	Mid	Low	High	High
Adherence	Easiest	Easier	Difficult	Most difficult
Number of new drugs	[1]	[1]	[1]	[1]
Antiviral potency	Lower[2]/High[3]	High	High	High
Tolerability	Better	Better	Not good	Less

[1] depends on prior history
[2] for ddI/d4T/3TC and ddI/d4T/Hydroxyurea
[3] for ABC/ZDV/3TC

well as drug failure) is a multifactorial function to be considered comprehensively. In addition, the genetic barrier of ABC/ZDV/3TC is higher than that of NNRTI-based combinations as different sequential mutations are needed to acquire high-level resistance against such combination. However, there is a considerable degree of cross-resistance between ABC and ZDV, and previously existing mutations against ZDV confer resistance to ABC as well. Furthermore, 3TC is a drug with a very-low genetic barrier. Overall, single class combinations may be considered to have a lower genetic barrier than PI-based combinations. Finally, 3 NRTI combinations not including ABC are considered to have a lower antiviral potency than other combinations including NNRTIs and/or PIs. The observation that triple NRTI regimens were less effective than standard PI-based regimens among patients with a high baseline viral load suggests that these regimens are less potent. Nevertheless, direct comparative studies have shown antiviral equipotency of ABC/ZDV/3TC and IDV/ZDV/3TC for patients with a pVL < 100.000 copies/mL at baseline, and similar studies are ongoing comparing ABC/ZDV/3TC and EFV + two NRTIs. However, PI-containing dual class regimens are more effective for patients with high baseline replicative levels.

Regarding dual class therapies, both NNRTI- and PI-based combinations achieve a high antiviral potency. Protease inhibitor-based combinations have a higher genetic barrier. Conversely, NNRTI-including therapies usually have a lower pill burden and easier administration schedules that enable a better adherence to medication. However, only PI-based therapies have demonstrated a high antiviral potency in patients with low baseline CD4 counts (<50 cells/mm^3).

Finally, although triple class regimens are more potent than dual and single class regimens (with the exception of 2 PI+2NRTI combinations), act at two different points of HIV's life cycle but by three different mechanisms and have a high genetic barrier, they imply the highest pill burden and the highest difficulty for drug adherence. As well, there is the concern of lack of salvage regimens once they fail.

Taking into consideration all these facts, *a priori* one would expect triple class regimens to achieve the most durable response – in case of appropriate adherence – closely followed by PI-based strategies, since both kinds of regimens have a high antiviral potency combined with a high genetic barrier, although they are poorly tolerated as compared to other regimens. Overall, the duration of antiviral response to dual class combinations including NNRTIs may be shorter because, although such regimens have a comparable antiviral efficacy and a better tolerability, they have the lowest genetic barrier. Finally, single class agents may show the shortest duration of virological response (with ABC–containing regimens being a possible exception) due to the lower antiviral potency and a mid-level genetic barrier. These speculations have been partially tested in several studies.

Clinical studies on durability of response

Available prospective data about long-term ABC/ZDV/3TC effectivity is restricted to CNAAB3005 and CNA3014 Studies that showed that this effectivity was maintained during 48 weeks in patients with baseline viral load < 100.000 copies/mL. Further follow-up is needed to determine eventual differences with PI-based combinations as well as comparative data with EFV/ZDV/3TC are ongoing. Definitive data of the Atlantic Study are expected to be published although preliminary results (not confirmed) point to a higher rate of virological failure for ddI/d4T/3TC in comparison with ddI/d4T/IDV and ddI/d4T/NVP arms.

Two- and 3-year follow-ups of the DuPont 006 trial were presented in the 8th CROI held in Chicago in February 2001 [13]. A total of 40% of subjects experienced virologic failure or discontinued treatment on ZDV/3TC/EFV at 2 years, compared with 60% of subjects on ZDV/3TC/IDV or EFV/IDV. However, in all arms, the rates of virologic failure alone were substantially less; only 20% of subjects had virologic failure in the ZDV/3TC/EFV arm. The rates of virologic rebound in subjects who had achieved a plasma HIV-1 RNA level of less than 50 copies/mL were 8% during the first year, 7% during the second year, and 3.5% during the third year of follow-up. In addition, while the rate of sustained virologic suppression was significantly lower in individuals who entered the study with CD4 cell counts less than 200 cells/mm^3 in the ZDV/3TC/IDV and EFV/IDV arms; differences in suppression rates by baseline CD4 cell counts did not reach statistical significance in the ZDV/3TC/EFV arm. Hence the antiviral efficacy of EFV-containing dual class regimens was maintained at least during the first 3 years.

In an observational database-based study recently published [14], the authors concluded that PI-containing regimens had a lower risk of treatment failure than NNRTI-containing regimens. In this study they compared the effectiveness of first protease inhibitor (PI)-containing and non-nucleoside reverse transcriptase inhibitor (NNRTI)-containing regimens, analysing data from three large HIV patient databases: Apache HIV InsightTM (APACHE), Target Management Services (TMS) and Clinical Partners (CP). Briefly, they found the type of regimen was not associated with time to undetectable viral load in any of the three databases, all of which had high power to detect a difference. PI-containing regimens were significantly less likely to fail after reaching undetectable viral loads for APACHE and CP patients (relative hazard, 1.7; 95% confidence interval, 1.3–2.1 and relative hazard, 1.6; 95% confidence interval, 1.0–2.5 respectively). These results remained significant after allowing for an unmeasured confounder with moderate effect on risk. No significant association between time to failure and regimen was found for TMS patients, possibly due to low power (67% to detect a relative hazard of 1.5). No difference was found between regimens in the time taken for an increase of >100 cells/μl in CD4 T-cell counts. In the APACHE database, those on NNRTI-containing regimens were more likely to have a failing CD4 T-cell response.

Such results are somewhat discordant with the positive data obtained by NNRTIs in prospective studies, and probably present data that are more close to standard clinical practice than controlled prospective trials. However, we cannot definitively conclude that NNRTI-based combinations fail more frequently than those that are PI-based, until more data are available in that direction.

Unfortunately, data on durability of virological response regarding triple class combinations are limited to the study by Albrecht et al. [9], where the favourable virological outcome of 2NRTI + EFV + NFV was maintained during a 48-week follow-up. It would be worthwhile that further comparative studies of triple class combinations included 2 NRTI + 2 PI regimens, as well.

Salvage regimen options

Perhaps the most attractive aspect of single class combinations is the possibility to spare both NNRTIs and PIs for subsequent salvage regimens. Single class strategies would be the most favourable option in terms of sparing subsequent salvage strategies. However, these regimens are, at present, not reccomended for use as initial therapy in antiretroviral naïve patients due to the doubts on their antiviral potency mainly in patients with high viral load.

Dual-class regimens including PIs are the second most favourable regimen in order to spare salvage options because, overall, virological failure to a certain PI frequently permits the use of another PI and, secondly, spares the whole NNRTI family. Conversely, failure to a NNRTI-based regimen almost precludes the use of a second NNRTI, although spares PIs for salvage.

Finally, there is strong concern in using combinations of drugs from all three groups because, despite the probable excellent initial results, patients finally developing resistance to such regimens would have few good options for subsequent therapy. As well, drug interactions and adverse events must be monitored in patients receiving such combinations. Therefore, triple class combinations, at present, seem to be more suitable for salvage therapy regimens and in patients who initiate therapy at a very advanced stage.

Adverse events

Comparative data about adverse events are also relevant in terms of morbidity, mortality and have a significant impact on drug adherence and quality of life. Special interest exists regarding long-term adverse effects of medication, especially on lipodystrophy-associated body fat redistribution and metabolic disturbances, as well as drug-induced osteopenia. A detailed description of PI-related toxicity is beyond the scope of this review as this is a well-known issue and has been extensively discussed elsewhere.

Overall, PI-sparing regimens have shown a lower incidence of short-term mild (grade 1 or 2) adverse events, a lower rate of treatment interruptions and a better tolerability, although the incidence of severe (grade 3 or 4) adverse events is similar to PI-containing combinations. However, important drug adverse reactions have to be considered for these regimens, as well.

Single-class regimens

The CNAAB3005 Study [3] comparing ABC/ZDV/3TC *versus* IDV/ZDV/3TC, found that the study treatments were equally well tolerated for up to 48 weeks, with ABC-related hypersensitivity occurring as the most important adverse reaction. The most common (5%) drug-related adverse events that were of moderate to severe intensity (grades 2–4) included nausea (with or without vomiting), malaise and fatigue, headache, and renal signs and symptoms. The proportion of patients with severe laboratory abnormalities was similar between groups. Four deaths were reported during the study. In the abacavir-3TC-ZDV group, one death was attributed to hypersensitivity reaction that occurred following rechallenge with abacavir approximately 3 weeks after initiating study treatment, and two were attributed to cardiac arrhythmia and myocardial infarction occurring 30 to 35 weeks after initial study treatment. The latter two events were not considered to be related to abacavir. In the IDV/ZDV/3TC group, one death was attributed to drug overdose (heroin and cocaine), which occurred approximately 6 weeks after initiating treatment. Hypersensitivity reactions were seen in 7% of the patients in the ABC/ZDV/3TC group and in 2% of the patients in the IDV/ZDV/3TC group. In the ABC/ZDV/3TC group, symptoms generally occurred within 6 weeks of initiating abacavir, and included fever and rash accompanied by gastrointestinal tract–related symptoms, such as nausea, vomiting, and diarrhea. In the IDV/ZDV/3TC group, symptoms were less severe, were gastrointestinal in nature, and included rash or fever but not both concurrently.

Regarding ddI/d4T/3TC and ddI/d4T/HU regimens, there is concern for additive NRTI toxicity, specially regarding pancreatitis, neuropathy, anaemia and other forms of mitochondrial toxicity like lactic acidosis, that may eventually be severe and lead to the death of the patient. However, such additive toxicity has not been properly studied yet.

Therefore, ABC-related sensitivity should be promptly detected and patients experiencing this reaction should never be rechallenged to ABC. Other three NRTI regimens should not be considered for patients with high risk or prior history of mitochondrial toxicity.

NNRTI-based dual class regimens

EFV-containing regimens cause less systemic and gastrointestinal symptoms but are associated to a high incidence of urticariform rash and central nervous system symptoms. In the DuPont 006 trial [2], the rates of nausea (27%), vomiting (15%), pain (predominantly of the flank, 11%), and increased bilirubin levels (8%) were significantly higher (by no more than 12, 7, and 2 and less than one percentage points, respectively) in the group given two NRTIs + IDV than in the two groups treated with the EFV-containing regimens. The rates of maculopapular rash and insomnia were higher in the group given EFV + IDV than in the group given two NRTIs + IDV (16% *versus* 6% and 6% *versus* 1%, respectively). Dizziness and impaired concentration were more frequent in the group given two NRTIs + EFV than in the group treated with two NRTIs + IDV (9% *versus* 8% and 9% *versus* 8%, respectively). Both groups receiving the regimens containing EFV had higher rates of rash (34% for each group) than the group receiving two NRTIs + IDV (18%). None of the patients who received EFV had a severe rash, and the median duration of rash was 14 days. There were no significant differences among the groups with regard to discontinuation of treatment because of rash. Central nervous system symptoms reported in all treatment groups included, but were not limited to, dizziness, impaired concentration, insomnia, and abnormal dreaming. The incidence was 58% in the group given two NRTIs + EFV, 53% in the group given EFV+IDV, and 26% in the group given two NRTIs + IDV ($p < 0.001$). No severe adverse events were reported, and the median duration of symptoms was 19 to 22 days. There were no significant differences among the groups with regard to discontinuation of treatment as a result of these central nervous system symptoms. In another study [15], treatment failure and CNS side-effects were associated with low and high EFV plasma levels, respectively. According to the authors' conclusions, the important inter-individual variability in EFV levels strongly argued for dose adjustment on the basis of therapeutic drug monitoring to optimize treatment.

Similar rates of urticariform rash have been reported for NVP-based regimens, although these regimens do not imply a significant incidence of CNS symptoms. However, there is important concern regarding NVP-induced hepatotoxicity. In a study by Barlett et al. [8] about 9% of NVP-treated subjects (but no EFV-recipients) had grade 4 hepatic toxicity. Grade 3/4 liver toxicity on NVP was twice as common in women as in men, which was the only real predictor of these elevations in this study. Most of this toxicity was evident by the fourth study week and in many cases investigators were able to treat through grade 3 toxicity. However, two women died secondarily to complications of hepatic failure that occurred early in the course of therapy. As a result of this study, the package insert for NVP has been changed to recommend monitoring of liver function tests every 2 weeks when NVP therapy is initiated. In a multicenter, retrospective cohort study of all patients who received NVP over a 5-year period [17], severe rash was noted in nine of 95 women and three of 263 men (risk ratio [RR], 8.31; 95% confidence interval [CI], 2.3–30.0; $p = .005$). Women were more likely to discontinue NVP therapy because of rash (RR, 4.5; 95% CI, 1. 9–10.5; $p = .0005$). After adjusting for age and baseline CD4 cell count in multivariate analysis, women had a seven-fold increase in risk for severe rash and were 3.5 times more likely to discontinue NVP therapy. In a retrospective review study [18] including naïve patients beginning HAART between January 1997 and January 2000, severe hepatic injury developed in 21 of 222 patients (9%) receiving either PI or NNRTI-based terapies (10% of patients on PIs, 9% of patients on NNRTIs, and 9% of patients on PIs + NNRTIs. Predictor variables in the final multivariate model were: alcohol abuse (risk ratio [RR], 5.87; 95% confidence interval [CI], 1.49–23.15; $p = .01$], positive HCV serology (RR, 3.99; 95% CI, 1.32–12.10; $p = .01$], and older age (RR, 1.11; 95% CI, 1.04–1.18; $p = 0.001$).

The majority of adverse events related to NNRTI-based regimens occur during the first month of therapy and are often predictable and manageable without treatment interruption. EFV is not recomended for patients with baseline phychiatric disturbances and should be administered preferably at night unless vivid dreams predominate. Caution must be given when administering NVP to women, HCV-positive patients, people abusing drugs and the elderly. Of interest, a common practice intended to prevent NVP-related rash like the administration of corticosteroids during the first month of treatment has been demonstrated to be detrimental and to increase the severity of the rash [19]. Antihistaminics may become a useful alternative to limit NNRTI-related rash.

Conversely, NNRTI-based strategies might be especially useful for patients experiencing lipodystrophy-related dislipidemia. Data from the Atlantic Trial [20] (as well as data of our group currently in press) have shown that a regimen of d4T/ddI plus NVP, but not IDV or 3TC, resulted in a lipid profile which is associated with a sharp reduction in the atherogenic index, what has led to a reduction of the coronary artery disease risk in other settings. More specifically, the initiation of NVP in patients suffering lipodystrophy-related dislipidaemia implied a decrease in total and LDL-cholesterol, as well as an increase in HDL-cholesterol values and size. Although total triglyceride levels remained unchanged, a significant reduction in the VLDL1 fraction was also seen in patients receiving NVP. EFV-related results are more controversial. Therefore, NNRTI-based regimens may be especially indicated in patients suffering dislipidemia. Of note, no effect of NNRTI-including regimens on body fat redistribution has been observed.

Triple class agents

Scarce data on adverse events are available regarding triple-class agents. In the study conducted by Albrecht et al. [9], there were no significant differences among the three treatment groups with respect to the time to the development of the first grade 3 (severe) or grade 4 (life-threatening) signs or symptoms, which occurred in a total of 36 patients (18%), 12 patients in the NFV group, 10 in the EFV group, and 14 in the NFV-plus-EFV group ($p > 0.3$ for all pairwise comparisons). The most frequent grade 3 or 4 signs or symptoms reported were aches, pain, or discomfort (in 13 patients [7%]) and diarrhea or loose stools (in seven patients [4%]). A grade 3 or 4 rash developed in only three patients (2%). Grade 2 (moderate) peripheral neuropathy was observed in six patients (3%).also, there were no significant differences among the three treatment groups in the time to development of grade 3 or 4 laboratory abnormalities, which occurred in a total of 58 patients (30%), 23 patients in the NFV group, 17 in the EFV group, and 18 in the NFV-plus-EFV group ($p > 0.3$ for all pairwise comparisons). The most common laboratory abnormalities were creatine kinase elevations (in 23 patients [12%]) and elevated triglyceride levels (in 21 patients [11%]). However, further data are required to assess the exact incidence of triple class related combinations.

Starting therapy

Taking into consideration the above discussed facts, it is clear that neither single class nor triple class combinations should be considered, at present, as suitable for an initial antiviral strategy. Single class combinations are currently not recommended, basically due to the doubts still existing on their antiviral potency. Perhaps the only combination suitable for antiretroviral naïve patients with less than 100,000 copies of plasma HIV-1 RNA at baseline is ABC/ZDV/3TC, which would have the additional benefit of sparing NNRTIs and PIs for subsequent salvage regimens. However, further follow-up and in-depth comparative studies will be needed to permit this inclusion into international HIV treatment guidelines.

Triple class combinations raise important concerns about toxicity and limit future salvage regimens. To date, such combinations should be reserved for rescue therapies.

Hence, the most suitable combinations for starting treatment at present are dual class NNRTI- and PI-based strategies. There has been much debate on whether PIs or NNRTIs should be used first as both strategies present important advantages and limitations. Factors in favour and against starting with either PIs or NNRTIs are summarized in Table 3. The final decision remains the responsibility of clinician, and must be individualised to every patient considering together all the factors involved, whether controllable or not.

Conclusions

Although PI-containing regimens were considered the standard of care for HIV infection until recently, new antiretroviral drug combinations, especially those based on NNRTIs, are currently considered at least comparable in terms of antiviral efficacy to PI-based regimens. However, there are multiple inquiries that must be solved in the immediate following years. PI-sparing combinations will have to be followed further to determine the durability of the response. More in-depth insights on the potential of these regimens for the immune recovery will have to be assessed. Also, comparative studies of the

Table 3. Factors in favour and against starting with either PI or NNRTI-based dual class regimens in antiretroviral-naïve patients (adapted from [21])

PI-based regimens	NNRTI-based regimens
In favour	
1. Data are extensive.	1. In one large study (DuPont 006), viral suppression was greater in the EFV/ZDV/3TC arm than in the IDV/ZDV/ETC arm.
2. Long-term viral suppression (>72 weeks) only demonstrated with PI-based regimens.	2. Convenient dosing schedules (NVP, EFV).
3. Prolong disease-free survival.	3. Fewer pills per day are required.
4. Quality of life is enhanced.	4. Use of NNRTI-based regimens may allow for subsequent use of PI-based salvage regimens (although this argument also applies in the opposite direction).
5. Viral suppression is commonly observed in various tissue reservoirs, including the genital tract and CSF.	5. Drug exposure is greater with NNRTIs than with the PIs.
6. High genetic threshold	6. There is no association between the use of NNRTIs and the development of long-term toxicity.
7. Only PI effective in patients with advanced disease (CD4+ < 50 cells/mm^3)	
8. Early virologic failure with a PI-based regimen is commonly associated with little or no detectable resistance.	
9. Dual PI-based regimens permit twice daily doses without food restrictions, and high protein-corrected trough concentration/IC50 ratio.	
Against	
1. Complex food requirements and/or strict dosing schedules.	1. Less long-term experience
2. Long-term adherence difficult.	2. Long-term safety has not been determined.
3. Drug-drug interactions are common.	3. Unclear whether NNRTI-based triple therapy regimens are associated with a survival benefit.
4. Cross-resistance is common among the PIs, making sequential use of these agents difficult.	4. Virtually complete cross-resistance among the NNRTIs.
5. Long-term side-effects have been associated with PI therapy. e.g., lipodystrophy and other metabolic abnormalities.	5. NNRTIs can have significant side-effects.
	6. Small genetic barrier.

various NNRTIs, particularly regarding NVP and EFV, will have to be presented. Simpler dosage schedules and further data on treatment adherence and quality of life will be required. In conclusion, there is still a long way to go!

References

1 US Dept of Health and Human Services (2001) Guidelines for the use of antiretroviral agents in HIV-infected adults and adolescents. Washington, DC: US Dept of Health and Human Services;

February 2001. Published at: http://hivatis.org/trtgdlns.html. Available at: *http://hiv.medscape.com/guidelines/adults*

2 Murphy RL, Katlama C, Johnson V, Squires K, Horban A, Gatell JM, Clotet B, Staszewski S, van Leeuwen R, Clumeck N et al. (1999) The Atlantic Study: A randomized, open-label trial comparing two protease inhibitor (PI)-sparing antiretroviral strategies *versus* a standard PI-containing regimen. Program and abstracts of the 39th Interscience Conference on Antimicrobial Agents and Chemotherapy (ICAAC), 26–29 September 1999, San Francisco, California, USA, Abstract LB-22

3 Staszewski S, Keiser P, Montaner J, Raffi F, Gathe J, Brotas V, Hicks C, Hammer SM, Cooper D, Johnson M et al. (2001) Abacavir-lamivudine-zidovudine vs IDV-lamivudine-zidovudine in antiretroviral-naïve HIV-infected adults: a randomized equivalence trial. *JAMA* 285: 1155–1163

4 Vibhagool A, Cahn P, Schechter M, Soto-Ramirez L, Montroni M, Smaill F, Thomas N (2001) Abacavir/combivir (ABC/COM) is comparable to indinavir/combivir in HIV-1 infected antiretroviral therapy naïve adults: preliminary results of a 48-week open label study (CNA3014). Program and abstracts of the 1st IAS Conference on HIV Pathogenesis and Treatment, 8–11 July 2001, Buenos Aires, Argentina, Abstract 63

5 Staszewski S, Morales-Ramirez J, Tashima K, Rachlis A, Skiest D, Stanford J, Stryker R, Johnson P, Labriola DF, Farina D et al. (1999) Efavirenz plus zidovudine and lamivudine, efavirenz plus indinavir, and indinavir plus zidovudine and lamivudine in the treatment of HIV-1 infection in adults. Study 006 Team. *N Engl J Med* 341: 1865–1872

6 Fischl M, Castro J, Monroig R, Scerpella E, Thompson L, Rechtine D, Thomas D (2001) Impact of directly observed therapy on long-term outcomes in HIV clinical trials. Program and abstracts of the 8th Conference on Retroviruses and Opportunistic Infections, 4–8 February 2001, Chicago, Illinois, USA, Abstract 528

7 Phillips AN, Pradier C, Lazzarin A, Clotet B, Goebel F, Hermans P, Antunes F, Ledergerber B, Kirk O, Lundgren J (2001) Virological and clinical outcome of NNRTI-containing regimens for 1932 patients in EuroSIDA. Program and abstracts of the 8th Conference on Retroviruses and Opportunistic Infections, 4–8 February 2001; Chicago, Illinois, USA, Abstract 324

8 Podzamczer D, Ferrer E, Consiglio E, Gatell J, Perez P, Perez J, Luna E, Gonzalez A, Pedrol E, Lozano L (2001) Final 12-month results from the COMBINE study: a randomized, open, multicenter trial comparing combivir plus nelfinavir or nevirapine in naïve patients. Program and abstracts of the 1st IAS Conference on HIV Pathogenesis and Treatment, 8–11 July 2001, Buenos Aires, Argentina, Abstract 7

9 Albrecht MA, Bosch RJ, Hammer SM, Liou SH, Kessler H, Para MF, Eron J, Valdez H, Dehlinger M, Katzenstein DA (2001) Nelfinavir, efavirenz, or both after the failure of nucleoside treatment of HIV infection. *N Engl J Med* 345: 398–407

10 Kirk O, Katzenstein TL, Gerstoft J, Mathiesen L, Nielsen H, Pedersen C, Lundgren JD (1999) Combination therapy containing ritonavir plus saquinavir has superior short-term antiretroviral efficacy: a randomized trial. *AIDS* 13: F9–F16

11 Schacker T, Nguyen P, Gatell J, Horban A, Berzins B, Lange J, Murphy R (2001) Persistent depletion of CD4+ T cells in lymphoid tissues among patients randomized to triple nucleoside *versus* protease or nonnucleoside containing regimens in the Atlantic study. Program and abstracts of the 1st IAS Conference on HIV Pathogenesis and Treatment, 8–11 July 2001, Buenos Aires, Argentina, Abstract 26

12 The AVANTI and INCAS Study Groups (2000) Highly active antiretroviral therapy including protease inhibitors does not confer a unique CD4 cell benefit. *AIDS* 14: 1383–1388

13 Levy R, Labriola D, Ruiz N (2001) Low two-year risk of virologic failure with first regimen HAART. Program and abstracts of the 8th Conference on Retroviruses and Opportunistic Infections, 4–8 February 2001, Chicago, Illiniois, USA, Abstract 325

14 Ghani AC, Henley WE, Donnelly CA, Mayer S, Anderson RM (2001) Comparison of the effectiveness of non-nucleoside reverse transcriptase inhibitor-containing and protease inhibitor-containing regimens using observational databases. *AIDS* 15: 1133–1142

15 Marzolini C, Telenti A, Decosterd LA, Greub G, Biollaz J, Buclin T (2001) Efavirenz plasma levels can predict treatment failure and central nervous system side-effects in HIV-1-infected patients. *AIDS* 15: 71–75

16 Bartlett J (2001) Severe liver toxicity in patients receiving two nucleoside analogues and a non-nucleoside reverse transcriptase inhibitor. Program and abstracts of the 8th Conference on Retroviruses and Opportunistic Infections, 4–8 February 2001, Chicago, Illinois, USA, Abstract 19

17 Bersoff-Matcha SJ, Miller WC, Aberg JA, van Der Horst C, Hamrick Jr, HJ, Powderly WG, Mundy LM (2001) Sex differences in nevirapine rash. *Clin Infect Dis* 32: 124–129
18 Nunez M, Lana R, Mendoza JL, Martin-Carbonero L, Soriano V (2001) Risk factors for severe hepatic injury after introduction of highly active antiretroviral therapy. *J Acquir Immune Defic Syndr* 27: 426–431
19 Montaner JSG, Gigliotti M, Cahn P et al. (2000) The effects of a short course of prednisone on the incidence of rash associated with nevirapine. Abstract and poster presentation WePpB1378 at the *XIIIth International AIDS Conference*, 9–14 July 2000, Durban, South Africa
20 Van Der Valk M, Reiss P, Molhuizen H, Stellato R, Katlama C, Horban A, Glesby M, Behrens G, Clotet B, Kastelein J et al. (2001) Nevirapine containing potent antiretroviral therapy results in ananti- atherogenic plasma lipid profile: Results from the Atlantic Trial. Program and abstracts of the 8th Conference on Retroviruses and Opportunistic Infections, 4–8 February 2001, Chicago, Illinois, USA, Abstract 436
21 Steven Deeks (2001) Optimizing Antiretroviral Therapy: Strategies and Regimens. Medscape HIV/AIDS Clinical Management. Available at: *http://hiv.medscape.com/Medscape/HIV/ClinicalMgmt/CM.v03/public/index-CM.v03.html*

[illegible]

Combination Therapy of AIDS
Edited by E.D.A. De Clercq and A.-M.I. Vandamme

Pharmacokinetics and pharmacodynamics of HAART

Michael Kurowski

Therapia GmbH, Rubenstrasse 125, D-12157 Berlin, Germany

The presently employed components of HAART are derived from a variety of chemically and pharmacologically different drug classes. The first compounds, which possessed antiretroviral activity were designed as nucleoside analogues in order to inhibit the reverse transcriptase. The first drug zidovudine (AZT) became available in 1987 and was followed soon by several compounds, which were directed at the same pharmacological target and derived from the same chemical class. The early experiences with nucleoside analogues (NRTI) in the course of antiviral treatment revealed a number of limitations: The medication had to be administered frequently at high doses with a considerable number of adverse effects. The most significant clinical restriction, however represented the emergence of mutant virus strains, which led to a partial or complete loss of activity. Consequently NRTI were combined at full doses, e.g., AZT with 3TC (lamivudine) or ddI (didanosine) with d4T (stavudine), which led to a higher antiviral potency and a prolonged period of time until mutations occurred in treated patients.

A new therapeutic era started with the introduction of protease inhibitors (PI), saquinavir being the first drug launched in 1997. Until today, nine different protease inhibitors have been introduced to antiretroviral therapy. These drugs were derived from either structure-based or substrate based design targeted at the HIV-protease. Chemically their common feature is characterized by a peptidomimetic structure with the exception of atazanavir. This class of drugs allowed to target the HI-virus at a different enzyme in the cycle of viral replication. Combining PIs with nucleoside analogues allowed a pronounced reduction of viral load and a sustained suppression until the occurrence of mutations. The galenical preparation and pharmacokinetics of PIs proved to be rather complicated as these drugs interact with various human enzyme systems involved in absorption, distribution and elimination. This property also led to significant interactions with other antiretroviral drugs and components of frequently administered comedications. Initially PIs were administered at high doses thrice daily due to their short to intermediate plasma half-lives, until it was learned that a combination with small doses of ritonavir enhanced absorption and prolonged the elimination of other protease inhibitors given at full doses. Combination therapy with two PIs at regular doses with or without a "backbone" of nucleosides is still under clinical investigation.

Nonnucleoside inhibitors of the reverse transcriptase (NNRTI) were introduced as a new class of drugs with a higher inhibitory potency and longer elimination half-lives as compared to NRTIs. Presently three different drugs – delarvidine, efavirenz and nevirapine – are used in combination with NRTIs and PIs. They are characterized by a small molecular weight, high lipophilicity and an intermediate to long elimination half-life allowing b.i.d. or q.d. dosage regimens.

Finally, the first inhibitor of viral entry became available for a distinct population of patients. Enfuvirtide (T-20) is a synthetic peptide of 36 amino acids, which has to be administered subcutaneously. It introduces a new facet in the pharmacological armamentarium against HIV-infection.

These different pharmacological classes will be detailed with respect to their pharmacokinetic and pharmacodynamic properties. As the number of different combinations and strategies has grown vast and each combination affects the pharmacokinetic properties of its components a selection of frequently applied therapeutic regimen will be made.

Nucleoside reverse transcriptase inhibitors (NRTI)

Principle of action

NRTI represent prodrugs, which need to be activated by phosphorylation in lymphocytes. The phosphorylation occurs sequentially at the 5'-position by different kinases forming the mono- di and triphosphate. The rate-limiting process may be the synthesis of the active triphospate from the diphosphate, with a saturable enzymatic step or one of the preceeding enzymatic steps. In resting peripheral blood lymphocytes the activation rate has been found to be lower for lamivudine [1]. In contrast, with some nucleosides activation rates are higher in the resting state. The triphosphates competitively inhibit the incorporation of thymidin into viral DNA by reverse transcriptase. Following incorporation into viral DNA the triphosphates terminate the elongation of viral DNA by preventing additional phosphodiester linkages to the unreactive 3'-groups. Thus the synthesized DNA remains incomplete [2]. In certain combinations two nucleosides (e.g., AZT and 3TC) may synergistically inhibit reverse transcriptase and HIV-replication, which however is not a general rule. NRTI with the exception of lamivudin possess considerable activity against mammalian DNA polymerase γ and may be incorporated into mammalian mitochondrial DNA [3, 4]. Thus, nucleosides have the ability to induce clinically significant hepatic and haematological adverse events, neuropathy and myopathy.

Absorption, distribution and elimination

Generally, NRTI are hydrophilic compounds, which resolve quickly in the gastrointestinal milieu. The bioavailabilities range between 40 and 90%. The time to reach Cmax occurs between 0.5 and 1.5 h. With the exception of abacavir and lamivudin NRTI are poorly bound to plasma proteins (<5%). Due to their low lipophilicity they have limited access to adipose tissues and hepatocytes. Thus, the volume of distribution of NRTI ranges around 1 L/kg, which indicates their presence in the plasma water. From these pharmacological characteristics it can be expected, that NRTI do have a limited potential for pharmacokinetic interactions.

Nucleosides can be regarded as produgs as they are phosphorylated intracellularly in three steps. In lymphocytes triphosphates are formed by a group of kinases that have different activities for the administered nucleosides. The intracellular kinetics and elimination half-lives vary considerably among nucleosides, individuals, and from the pharmacokinetics in plasma. As the triphosphates persist much longer in lymphocytes clinical trials are presently being conducted to test q.d. dosage regimen for certain nucleosides as 3TC and AZT [5].

The elimination of nucleosides occurs predominantly via filtration in the kidneys as unchanged drug or in its conjugated form.

Tenofovir disoproxil fumarate (tdf) is administered as a prodrug ester, which requires enzymatic cleavage before the phosphonomethyl ether can enter the target cells. Its intracellular activation includes a two-step phosphorylation to the triphopsphate (tenofovir diphosphate). The diphosphate inhibits competitively the HIV-1 revese transcriptase that terminates the growing DNA chain. The prodrug principle with a sequential formation of the phosphonomethyl ether allows single daily administrations of tdf. Its pharmacokinetics is linear and the terminal elimination half-life ranges between 11.7 and 15.4 h depending on the treatment duration [6]. This dosage regimen has been demonstrated to provide sufficient amounts of the activated form to inhibited viral replication continuously.

In the combination with didanosin tdf augments the Cmax of ddI by 28% and the AUC by 40% [7]. Thus, it has been recommended to reduce the ddI dose from 400 to 250 mg in this combination. The pharmacokinetic characteristics of the NRTIs are summarized in Table 1.

Protease inhibitors

Principle of action

Aspartyl protease was the second enzyme identified as a target of antiretroviral therapy. As the first approved drug saquinavir became available by the end

Table 1. Nucleoside reverse transcriptase inhibitors (NRTI)

Drug	Abacavir	Didanosine	Lamivudine	Stavudine	Zalcitabine	Zidovudine	Tenofovir DF*
T1/2 (h)	1.5	1.5	5.7	1.3–1.4	2.0	1.1	21.6 ± 18.5
Cmax (mg/ml)	3.0 ± 0.9	1.5–2.3	1.5 ± 0.5	0.7–2.0	0.025	1.9	0.296 ± 0.09
Cmin (mg/ml)	<0.1	<0.01	0.33 ± 0.22	0.02	<0.005	0.1	
AUC (0–12 h) (ng*h/ml)	6.02 ± 1.73	1.20–2.36	17.1 ± 6.46	1.3	0.072	1.7	2.29 ± 0.69
Bioavailability (%)	83	~42	80–85	~86	>80	60–70	25–40
Protein Binding (%)	~49	<5	<36	<5	<4	34–38	0.7
Vd (L/kg)	0.8	1.08	1.3	0.66	0.53	1.6	1.3 ± 0.6
pKa	0.4, 5.1	9.1	4.3	10	4.4	9.7	*
Lipid Solubility (Octanol/Phosphate buffer)	Lipophilic (N/A)	Slightly lipophilic (N/A)	Hydrophilic (N/A)	Hydrophilic (N/A)	Hydrophilic (0.04)	Slightly lipophilic (1.15)	
CSF:plasma ratio	0.42	0.19	0.12	0.2	0.14	0.5	
Semen:plasma ratio			9.1 (2.3–16.1)	>1		5.9 (0.95–13.5)	
CLH (%)			5–10		~15	50–80	
CLR of parent drug (%)	2	40–50	70	40	75	>20	
Renal impairment	No dosage adjustment required, but should be avoided in end-stage renal disease	Dose modification recommended	Dose modification recommended	Dose modification recommended	Dose modification recommended	Should not be necessary for those with CLCR ≥ 5 ml/min, but recommended during haemodialysis or peritoneal dialysis.	(N/A)
Hepatic impairment	Close monitoring required, data limited	Not known, close monitoring recommended	No dose modification required	PK not altered in 5 non-HIV+ patients	(N/A) Limited data	Data limited; dose modification may be required.	(N/A)

*Tenofovir DF is a prodrug of Tenofovir (PMPA). (Compiled and modified according to data published at the internet page: *http://www.hivpharmacology.com*)

of 1995, soon followed by a number of other protease inhibitors, indinavir, ritonavir and nelfinavir.

The HIV-protease consists of two identical subunits of 99 amino acid residues arranged in a symmetric fashion. The active site of the enzyme contains two catalytic aspartic acid residues. Substrates are bound by the active site aspartic acids, Asp 25 and Asp 125 and in addition by a pair of hairpin structures, frequently related to as "flaps" of the enzyme.

The antiviral effect of protease inhibitors on infected cells occurs at a late point in the infectious cycle in contrast to reverse transcriptase inhibitors. By electron microscopy it was shown that PI prevent maturation of new viral particles, leading to a loss of infectivity [8]. Thus, PIs prevent copies of the HIV virus from being successfully assembled and released from the infected CD4+ T-cells.

Absorption, distribution and elimination

The protease inhibitors that are presently in clinical use share a number of common biopharmaceutical features. They resemble petidomimetic compounds containing stabilized amide bonds in order to prevent hydrolytic cleavage during their gastrointestinal passage.

The molecular weights range between 500 and 800 and they are poorly soluble in aqueous media. Therefore some manufacturers chose to formulate them as liquids in gelatin capsules. The bioavailability of most PIs is limited by these properties. In addition, PIs are substrates of drug transporting enzymes, such as p-glycoprotein (PgP), which reduce their uptake from the gastrointestinal tract. Most PIs are also subject to a profound first pass metabolism, before they enter the systemic circulation.

Following absorption, drugs are rapidly distributed throughout the body, with some of the drug free in the plasma and the major fraction reversibly bound to plasma proteins; an equilibrium is achieved between free and bound drug. With the exception of indinavir protease inhibitors are highly bound to plasma proteins, especially to α1-acid glycoprotein. Plasma protein binding interactions have been described for some highly bound drugs, including phenytoin and warfarin. One highly bound drug can displace another, leading to a transitory increase in plasma concentration. This effect is short-term, as the displaced drug is available for metabolism and excretion [9]. In practice such reactions are rare, particularly for antiretroviral agents. The high protein binding of protease inhibitors can be viewed as the major cause for extremely high apparent volumes of distribution, which can reach as much as 700 l/kg for saquinavir.

Elimination occurs by metabolism and renal or hepatic excretion of the parent compound or the formed metabolites.

Several hepatic enzymes of the cytochrome family are involved in the metabolism, CYP 3A4 being the dominant species among them. Ritonavir has been shown to be one of the most potent inhibitors of CYP 3A4. During a therapy including ritonavir (RTV) it can be observed that initially RTV blocks its own metabolism leading to a prolongation of its half-life and an accumulation in the systemic circulation. As a consequence of this blockade additional CYP 3A4 is formed in the liver. The induction process results in an acceleration of the RTV-metabolism and a reduction of steady-state plasma levels. A complete steady-state with PI-containing regimen can be assumed after 1 week of treatment, but it cannot be ruled out that genetically determined induction processes may require in certain cases 3 weeks or longer.

Predominant elimination half-lives of the protease inhibitors range between 2 and 10 h, which determines a t.i.d. or b.i.d. dosage regimen. Mostly the predominant half-lives are superponed by much longer terminal half-lives, which do not become apparent in regular dosage intervals. Under certain conditions a q.d. dosage schedule may be applied. Further details of the PK-parameters are listed in Table 2.

The boosted PIs

The described effect of combining PIs with RTV can be therapeutically used in order to prolong half-lives and dosage intervals and to reduce PI doses. Adding a small dose of RTV without the objection of reaching HIV-inhibitory RTV-concentrations is referred to as "boosting." In contrast, the combination of two or more PIs at full doses has been recently focused on in several clinical studies. The aim of double PI-treatment is to prevent early appearance of viral mutations and to reach different tissues or compartments.

The "boosting" effect of RTV on other PIs can be differentiated into "Cmax-boosting," e.g., with saquinavir, or "half-life boosting", e.g., with amprenavir.

RTV at small doses inhibits PgP and other drug transporters as well as Cytochrom P450 3A4. In the combination with saquinavir the amount of drug absorbed and the overall exposure is raised, resulting in higher AUC and Cmax concentrations and the metabolism is blocked, which leads to a longer half-life. In the combination with RTV both available galenic formulations produce comparable plasma concentrations [10].

When combining amprenavir with RTV an increase of APV maximum concentration cannot be observed, but a reduction of hepatic clearance with an increase of the APV half-life leads to a pronounced elevation of trough concentrations under "steady- state" conditions. In most combinations with RTV both components "Cmax-boosting" and "half-life boosting" are involved at different degrees. The measurable plasma concentrations attained with boosted regimens vary with the dose taken and correlate with the clinical activity of the drug. However, a sub-therapeutic plasma level is less likely to occur with a boosted PI than with a single PI. For example, Kaletra® (lopinavir coformulated with low-dose ritonavir) has been developed to achieve high and sus-

Table 2. Protease inhibitors (PI)

Drug	Amprenavir	Indinavir	Lopinavir	Nelfinavir	Ritonavir	Saquinavir/ soft gel	Saquinavir/ hard gel
T1/2 (h)	7–10	2	5–6	3–5	3–5	7–12	7–12
Cmax (mg/ml)	5.36	8.97	9.6 ± 4.4	4.0	11.2 ± 3.6	1.0–2.2 (1200 mg tds)	0.25 (600 mg tds)
Cmin (mg/ml)	0.28	0.146	5.5 ± 4.0	0.7–2.2	3.7 ± 2.6	0.10–0.22 (1200 mg tds)	0.038 (600 mg tds)
AUC (0–12 h) (ng*h/ml)	18.46	21.9	82.8 ± 44.5	15.5	121.7 ± 53.8	3.5–7.2 (1200 mg tds)	0.76 (600 mg tds)
Bioavailability (%)	Not determined	~65	Not determined	70–80 (with food)	Not determined	Not determined	4
Protein Binding (%)	90	60	98–99	>98	97	97	98
Vd (L/kg)	6.1	0.4–1.74		2–7	~0.3–0.6	700	700
pKa	1.9	6.2		6.0, 11.1	2.8	1.1, 7.1	1.1, 7.1
Lipid Solubility (Octanol/Phosphate buffer)	Lipophilic (N/A)	Hydrophilic (N/A)		Lipophilic (N/A)	Lipophilic (4)	Lipophilic (N/A)	Lipophilic (N/A)
CSF:plasma ratio	0.03	0.14	Consistently undetectable	Consistently undetectable	0.0–0.52	Negligible	Negligible
Semen:plasma ratio	0.2	1.9	0.07	0.07	<0.04	<0.04	<0.04
CLH (%)	~90	~80	~92	78	~86	88	88
CLR of parent drug (%)	<3	<20	<3	1.2	3.5	1.0	1.0
Renal impairment	Not investigated, but no dose modification should be necessary	Not investigated	Not investigated	Not investigated	Not investigated	Not investigated	Not investigated
Hepatic impairment	Dose modification recommended	Not investigated	Not investigated Lopinavir concentrations may be increased	Not investigated	Not investigated, but use with caution	Not investigated	Not investigated

(Compiled and modified according to data published at the internet page: *http://www.hivpharmacology.com*)

tained plasma levels of lopinavir. Without the addition of ritonavir the pharmacokinetics of lopinavir would not persist at inhibitory concentrations.

In contrast, the incidence of drug-related toxicity in patients taking a combination of indinavir and ritonavir correlates with the plasma levels of indinavir [11]. In such patients, drug level monitoring can be useful to establish an appropriate dose.

Drug interactions and therapeutic drug monitoring (TDM)

Due to frequent drug interactions seen with PIs, the high interindividual variability of plasma levels and the almost unpredictable individual concentrations achieved with various dosages and combinations a Therapeutic Drug Monitoring (TDM) has been proposed and offered by several groups [12–14]. TDM may assist in finding dosages for individual patients and to correct too low or high plasma concentrations.

Usually, the determined plasma concentration value includes both the fraction of free drug in the plasma and the fraction that is plasma protein bound. For must drugs, total plasma drug concentration can be used for TDM because the plasma-bound fraction of the drug is generally constant and predictable. However, when comparing plasma concentrations in *in vitro* measures, such as the IC50 for a virus strain, a correction needs to be made for plasma protein binding, since only the free fraction can diffuse into the intracellular compartment and have a pharmacologic action. This correction can be done using mathematically or physically determined correction factors (using methods such as ultracentrifugation, ultrafiltration, or equilibrium dialysis). However, there is a lack of standardization on plasma protein binding correction methods, so care should be taken when comparing results derived from different centers [13].

RTV and other PI can also interfere in a similar manner with other drugs, which are metabolised through the same pathway. The magnitude of the interaction can hardly be predicted in an individual, as the variability of the effect is high and the interaction is often obscured by a large number of concomitantly administered drugs. Therefore, it has been proposed to employ therapeutic drug monitoring (TDM) for PI and NNRTI as well as for interacting comedications. A growing use for TDM is to identify and manage drug interactions. In addition to many other medications taken by HIV positive patients, the metabolism of the PIs and the NNRTIs involves several members of the cytochrome P450 enzyme group. Therefore, any drug that alters the activity of these enzymes can affect the plasma levels of the PIs or the NNRTIs and *vice versa*.

Drugs such as ketoconazole and clarithromycin act as cytochrome P450 enzyme inhibitors (increasing the plasma levels of drugs metabolized by cytochrome P450), while drugs such as rifampicin are inducers of Cytochrome P3A4 and PgP (decreasing the plasma levels of such drugs). All PIs act to some extent as enzyme inhibitors, with ritonavir having the greatest effect. This mechanism has been clinically applied in order to boost PIs, whereas a low dose of ritonavir is used to inhibit the metabolism of other PIs, such as lopinavir, indinavir, amprenavir, and saquinavir.

Additionally, in some circumstances, some of the PIs (e.g., nelfinavir, amprenavir, and ritonavir) and NNRTIs (e.g., efavirenz and nevirapine) also induce metabolic enzymes.

When a number of enzyme inducers or enzyme inhibitors are taken (as is the case in many antiretroviral regimes), the resulting plasma levels can be difficult to predict, and so TDM may play a role in designing and monitoring such combinations.

Although ritonavir boosts the plasma levels of amprenavir, when lopinavir is added to a regimen containing amprenavir/ritonavir, the plasma levels of amprenavir are reduced. Similarly, when amprenavir is added to a regimen containing lopinavir/ritonavir, the levels of lopinavir are reduced [15]. This complex relationship involves both, inhibition and induction of metabolic enzymes. The application of TDM may be beneficial, especially when the combination is used as part of a salvage regimen (where the required inhibitory concentration has to be higher).

Interactions may also occur when a PI is combined with an NNRTI. For example, when lopinavir/ritonavir is combined with efavirenz, the plasma level of lopinavir is lowered. A dose of 533 mg/133 mg is now recommended for adults taking lopinavir/ritonavir with efavirenz or nevirapine [16]. Efavirenz also lowers the plasma levels of the other PIs. Similarly, this effect can be overcome by adding small doses of ritonavir and, eventually raising the PI-doses. The interaction between nevirapine and PI seem to be less pronounced, but few data are available from clinical trials.

TDM may help to determine the most appropriate dose of a PI to be used in combination with efavirenz or nevirapine – the dose that would achieve sustained viral suppression while avoiding toxicity.

The whole subject of drug interactions during antiretroviral therapy is very complex. A useful resource to guide clinical management is the website: http://www.hiv-druginteractions.org.

Interindividual variation

Although some population reference values have been defined for the PIs and NNRTIs, significant variation does occur in the plasma drug levels achieved in different individuals. The majority of published reference values have been derived from studies investigating previously antiretroviral-naïve individuals and often single-drug regimes. The variability of plasma trough levels in HIV-infected patients is described by a normal distribution and, for the most PIs (except indinavir) by sub-normal trough levels in the lower tail of that distribution [17].

The observed median and 95% confidence intervals of Pis in frequently applied dosage regimens are:

PI-regimen (doses in mg)	Median (ng/ml)	Normal range, 95% ci (ng/ml)
APV/r 600/100 b.i.d.	1618	581–4505
IDV/r 800/100 b.i.d.	710	70–7155
LPV/r 400/100 b.i.d.	5010	77–8414
NFV 1250 b.i.d. (incl. M8 metabolite)	1859	495–6975
SQV/r 1000/100 b.i.d.	804	1249–19398

Reference values for antiretroviral-experienced individuals and for those taking multidrug therapy, particularly salvage regimes, are likely to be different for a variety of reasons, including increases of viral resistance, and the consequences of drug interaction (involving not only the antiretroviral agents, but other concomitant drugs). This could mean that a range of reference values has to be defined for varying levels of experience with therapy (treatment-naïve *versus* treatment-experienced *versus* salvage) and for different therapeutic combinations. Such concerns can be partially overcome using a measurement value that adjusts for viral susceptibility and which accounts for differences between viral strains.

Intraindividual variation

Plasma concentrations of a drug can vary between each dose and from one dose to the next day. This variation is reduced in the setting of a controlled study. In reality, the conditions under which patients take their medications can vary widely, leading to potentially large variations on plasma concentration. Factors that can influence these variations include the time when the dose is taken, whether is taken before, with, or after food, or even the composition of the food itself and the quantity of fluid consumed during the day. These, and some of the other factors can influence plasma concentrations. The interaction between these factors can lead to very complex dosing schedules for patients taking combinations of antiretroviral agents.

The pharmacokinetics of some drugs display a circadian rhythm. This means that a plasm ι concentration profile reached after an afternoon dose is different from the profile after a morning dose [18]. This difference may be due to diurnal variations of several physiological and pharmacological parameters.

It has also been shown, that namely with protease inhibitors pharmacokinetic properties can change during long-term treatment, probably due to complex induction or adaptation processes [19].

Non-nucleoside reverse transcriptase inhibitors (NNRTI)

NNRTIs block the reproduction of HIV earlier in its cycle than either the nucleoside/nucleotide analogues or the protease inhibitors. By binding tightly in a non-competitive manner to tyrosin, located at the positions 181 and 188 of the p66-subunit of the reverse transcriptase reverse transcriptase, NNRTIs prevent viral RNA from being converted to DNA.

Presently three different drugs, delarvidine, efavirenz and nevirapine, with quite heterogeneous chemical properties are in clinical use. They share moderate to high lipophilicity and relatively small molecular weights. Delarvidine and efavirenz are highly bound to plasma proteins (98–99%). NNRTIs are predominantly eliminated by hepatic metabolism. Efavirenz and nevirapine can induce the metabolism of other drugs leading to lower plasma concentrations, e.g., nevirapine accelerates the elimination of methadone or rifabutin. These interactions invoke the need for plasma level monitoring and adaptation of dosages.

Due to their long elimination half-life efavirenz (45 h) can be administered once daily and nevirapine (25–30 h) twice daily (the option of a qd regimen is presently under investigation). From these pharmacokinetic properties it can be concluded that a "steady-state" is reached after 2 weeks following the initiation of a treatment or a change of the dosage regimen (Tab. 3).

Some clinical studies attempted to correlate pharmacokinetic parameters and pharmacodynamics of NNRTI. In a group it has been observed, that 50% of patients with low efavirenz levels (<1000 microg/l, 14 h after administration) experienced virological failure [20]. This was contrasted by a proportion of 22 and 18% in patients with 1000–4000 microg/l or more than 4000 microg/l, respectively. CNS toxicity was approximately three times more frequent in patients with high efavirenz levels (>4000 microg/l) compared with patients with 1000–4000 microg/l. The authors concluded that treatment failure and CNS side-effects are associated with low and high efavirenz plasma levels, respectively. Moreover, they suggested that the marked inter-individual variability in efavirenz levels strongly argues for dose adjustment on the basis of therapeutic drug monitoring to optimize treatment.

Entry inhibitors, fusion inhibitors

Enfuvirtide (T-20)

Enfuvirtide is the first of two peptidic inhibitors of viral entry into mammalian cells which has found its way into clinical treatment of HIV-infection. It is intended for administration in patients failing multiple combinations as an ultimate therapeutic option. It has been derived from the HIV-1 transmembrane glycoprotein gp41 as a partial structure of the C-terminal heptad repeat 2

Table 3. Non-nucleoside reverse transcriptase inhibitors (NNRTI)

Drug	Delavirdine	Efavirenz	Nevirapine
T1/2 (h)	6	45	25–30
Cmax (mg/ml)	19.3	4.07	5.74–6.69
Cmin (mg/ml)	8.3	1.77	2.88–3.73
AUC (0–12 h) (ng*h/ml)	99.5	58.1	56.2
Bioavailability (%)	Not determined	Not determined	~93
Protein Binding (%)	~98	>99	~60
Vd (L/kg)	0.7	2.4	1.21
pKa	4.3–4.6	10.2	2.8
Lipid Solubility (Octanol.Phosphate buffer)	Slightly lipophilic (2.98)	Lipophilic (N/A)	Slightly lipophilic (1.8)
CSF:plasma ratio	0.4	0.61	0.63
Semen:plasma ratio	~2.0	0.09 (0.03–0.43)	0.6–1.0
CLH (%)	95	>70	90
CLR of parent drug (%)	<5	<1	<3
Renal impairment	Not investigated	Not investigated	Dosage modification required. Patients with CLCR ≥ 20 ml/min do not require dose adjustment
Hepatic impairment	Not investigated	Not adequately investigated	Dosage modification not required in mild to moderate hepatic dysfunction

(Compiled and modified according to data published at the internet page: *http://www.hivpharmacology.com*)

(HR2) domain. T-20 has been shown to bind gp41 after the interaction of gp120 with its cellular receptors [21].

As a synthetic peptide containing 36 amino acids it has to be applied parenterally. Following subcutaneous doses of 2 mg/kg b.i.d. in children and adolescents, pharmacokinetics have been investigated at "steady-state" [22]. Peak concentrations of 5.8 μg/ml were observed after 4.7 h similarly in both groups. Trough concentrations of 2.8 μg/ml were found after 12 h. 89% of the dose was absorbed from the injection site, 83% entered the systemic circulation. In a previous study with single sc doses of 45, 90 or 180 mg, enfuvirtide was eliminated from plasma in a dose-dependent manner with half-lives ranging from 3.46 to 4.35 h [23].

Conclusion

Antiretroviral combinations and dosage regimen should be based on rational considerations, such as clinical patient status, pre-treatment, existing mutations of the virus, tolerability and other factors. As the durability of the effect is limited in time, an increasing number of patients encounter limited therapeutic options with complex dosage regimen including a growing number of new drugs.

The options of possible combinations including comedications lead to almost unpredictable pharmacokinetics and interactions. In this situation TDM will play an increasing part allowing at least to control plasma concentrations and to verify the probability of a clinical success.

TDM will gain more acceptance in antiretroviral therapy if more research, including large randomized trials to assess its clinical utility and routine applicability, is achieved. An international consensus must be reached on targets and tests used. Also, standardization between centers is needed on units of measurement, values for virtual phenotypes, and the way results are reported. Another option in the future may arise from the determination of intracellular drug concentrations, especially for certain classes of antiretroviral drugs such as the NRTIs.

References

1 Kewn S, Veal GJ, Hoggard PG, Barry MG, Back DJ (1997) Lamivudine (3TC) phosphorylation and drug interactions *in vitro*. *Biochem Pharmacol* 54(5): 589–595

2 Langtry HD, Campoli-Richards DM (1989) Zidovudine. A review of its pharmacodynamic and pharmacokinetic properties, and therapeutic efficacy. *Drugs* 37(4): 408–450

3 Hart GJ, Orr DC, Penn CR, Figueiredo HT, Gray NM, Boehme RE, Cameron JM (1992) Effects of (–)-2'deoxy-3'-thiacytidine (3TC9 5'-triphosphate) on human immunodefency virus reverse transcriptase and mammalian DNS polymerases α, β, and γ. *Antimicrob Agents Chemother* 36: 1688–1694

4 Perry CM, Faulds D (1997) Lamivudine: a review of its antiviral activity, pharmacokinetic properties and therapeutic efficacy in the management of HIV infection. *Drugs* 53 (4): 657–680

5 Rodman JH, Robbins BL, Martinez J, Lindsey JC, Fridland A, Rodriguez JF, Flynn PM (2003) Intracellular phosphorylation of zidovudin and lamivudin in peripheral blood mononuclear cells in HIV-infected adolescents and young adults on once- vs twice-daily regimens. Abstract 147, 10th Conference on Retroviruses and Opportunistic Infections, Boston, US
6 Barditch-Crovo P, Deeks SG, Collier A, Safrin S, Coakley DF, Miller M, Kearney BP, Coleman RL, Lamy PD, Kahn JO et al. (2001) Phase I/II trial of the pharmacokinetics, safety, and antiretroviral activity of tenofovir disoproxil fumarate in human immunodeficiency virus-infected adults. *Antimicrob Agents Chemother* 45(10): 2733–2739
7 Roberts NA, Craig JC, Duncan IB (1992) HIV proteinase inhibitors. *Biochem Soc Trans* 20: 513–516
8 Kearney BP, Flaherty J, Wolf J, Barriere S, Wong S, Chen SS, Wulfsohn M, Sayre J, Coakley D (2001) Coadministration of tenofovir DF and didanosine: pharmacokinetic drug-drug interaction and safety evaluations. Abstract 172, 8th European Conference on Clinical Aspects and Treatment of HIV Infection, Athens, Greece
9 Stockley IH (1992) General considerations and an outline survey of some basic interaction mechanism. In: IH Stockley (ed.): *Drug interactions*, 3rd ed. Blackwell Scientific Publications, Oxford, 1–14
10 Kurowski M, Sternfeld T, Sawyer A, Hill A, Mocklinghoff C (2003) Pharmacokinetic and tolerability profile of twice-daily saquinavir hard gelatin capsules and saquinavir soft gelatin capsules boosted with ritonavir in healthy volunteers. *HIV Med* 2: 94–100
11 Solas C, Basso S, Poizot-Martin I, Ravaux I, Gallais H, Gastaut JA, Durand A, Lacarelle B (2002) High indinavir Cmin is associated with higher toxicity in patients on indinavir-ritonavir 800/100 mg twice-daily regimen. *J Acquir Immune Defic Syndr* 29(4): 374–377
12 Khoo Sh, Gibbons SE, Back DJ (2001) Therapeutic drug monitoring as a tool in treating HIV infection. *AIDS* 15(Suppl. 5): 171–181
13 Back D, Gatti G, Fletcher C, Garaffo R, Haubrich R, Hoetelmans R, Kurowski M, Luber A, Merry C, Perno C-F (2002) Therapeutic drug monitoring in HIV infection: current status and future directions. *AIDS* 16 (suppl 1): S5–S37
14 Back DJ. Khoo SH, Gibbons SE, Merry C (2001) The role of therapeutic drug monitoring in treatment of HIV infection. *Br J Clin Pharmacol* 51: 301–308
15 Basso S, Solas C, Quinson AM, Ravaux I, Poizot-Martin I, Bacconier J, Durand A, Lacarelle B (2002) Pharmacokinetic interaction between lopinavir/r and amprenavir in salvage therapy. *J Acquir Immune Defic Syndr* 31(1): 115–117
16 Bertz R, Lam W, Hsu A, Rynkiewicz K, Foit C, Bryan P, Sylte J, Brun S, Granneman GR, Sun E (2000) Assessment of the pharmacokinetic interaction between ABT-378/ritonavir (ABT-378/r) and efavirenz (EFV) in healthy volunteers and in HIV+ subjects. abstract 424, 40th Interscience Conference on Antimicrobial Agents and Chemotherapy, Toronto, Canada,
17 Stocker H, Kruse G, Moecklinghoff C, Hill A, Sawyer A, Kurowski M (2003) Identification of sub-groups of PI treated patients with abnormally low plasma PI drug levels. abstract 5.4, *4th International Workshop on Clinical Pharmacology of HIV Therapy*, Cannes, France
18 Justesen US, Pedersen C (2002) Diurnal variation of plasma protease inhibitor concentrations. *AIDS* 16(18): 2487–2489
19 Gisolf EH, van Heeswijk RP, Hoetelmans RW, Danner SA (2000) Decreased exposure to saquinavir in HIV-1-infected patients after long-term antiretroviral therapy including ritonavir and saquinavir *AIDS* 14(7): 801–805
20 Joshi AS, Barrett JS, Fiske WD, Pieniaszek HJ, Ludden TM, Bacheler LT, Ruiz NM (1999) Population Pharmacokinetics of Efavirenz in Phase II Studies and Relationship with Efficacy. Abstract 1201, 39th Interscience Conference on Antimicrobial Agents and Chemotherapy, San Francisco, US
21 Furuta RA, Wild CT, Weng Y, Weiss CD (1998) Capture of an early fusion-active conformation of HIV-1 gp41. *Nat Struct Biol* 5(4): 276–279
22 Bellibas SE, Pepper T, Dorr A, Bertasso A, Sista P, Duff F, Kolis SJ DeLora P (2002) Pharmacokinetics and Safety of 2 mg/kg Enfuvirtide (T-20) in Combination therapy in HIV-infected children and adolescents. [abstract TuBeB4632], XIV World AIDS Conference, Barcelona, Spain, July 7–12, 2002
23 Zhang X, Nieforth K, Lang JM, Rouzier-Panis R, Reynes J, Dorr A, Kolis S, Stiles MR, Kinchelow T, Patel ICH (2002) Pharmacokinetics of plasma enfuvirtide after subcutaneous administration to patients with human immunodeficiency virus: Inverse Gaussian density absorption and 2-compartment disposition. *Clin Pharmacol Ther* 72(1): 10–19

Combination Therapy of AIDS
Edited by E.D.A. De Clercq and A.-M.I. Vandamme

Primary HIV infection: from diagnosis to treatment

Luc Perrin and Marie-Charlotte Bernard

Division of Infectious Diseases and Laboratory of Virology, Geneva University Hospital, 1211 Geneva 14, Switzerland

Clinical presentation

Signs and symptoms associated with the acute illness, also referred to as acute retroviral syndrome (ARS), are observed in approximately two-thirds of HIV-infected patients [1, 2]. Similar symptoms have been reported in patients infected through sexual intercourse and through the intravenous route except for mucosal lesions in the genital tract which are more frequently observed in patients contaminated through sexual intercourse. The ARS usually spontaneously resolves within 2 to 3 weeks but may last for months in rare cases. The ARS, originally described as an influenza or mononucleosis like illness is characterized by fever (the most common symptom), sore throat, pharyngitis, arthralgia, myalgia, weight loss and lymphadenopathy [3, 4]. Some signs and symptoms should araise a high degree of suspicion such as generalized lymphadenopathy accompanied by a mononucleosis-like illness and/or a cutaneous rash associated with mucocutaneous ulcers involving the oral cavity (buccal mucosa, gingival, palate) or the genital (penis, vagina) or anal mucosa [1–5]. The cutaneous rash presenting as a maculo-papular rash affecting predominantly the trunk is particularly suggestive and is present in 40 to 80% of persons with acute infection [1, 5, 6]. Some serious but rare complications including hepatitis, respiratory failure, rhabdomyolysis and HIV-1 encephalitis are also observed [6]. Other infections indicative of profound immunosuppression such as oral candidasis, oesophagitis due to candida albicans or cytomegalovirus and multidermal zoster are not uncommon during ARS [7]. The acute phase of HIV-1 infection is often undiagnosed or misdiagnosed [8, 9] due to the poor specificity of most symptoms and because anti-HIV-1 antibodies are not detected during the first week of ARS. This underscores the need to obtain an accurate history of exposure in the context of unclear symptomatology.

Diagnosis

Early diagnosis requires the selection of assays able to detect either viral antigens and/or HIV-1 RNA in the blood [10, 11]. In the absence of anti HIV antibodies, detection in the serum of p24 antigen (p24 antigen, a product of the gag gene is the HIV-1 antigen present in the highest concentration in serum) is characteristic of very early PHI. New assays recently developed over the last 3 years, but not available in the US, allow for the systematic co-detection of anti-HIV antibodies and p24 antigen; this obviously presents advantages in terms of the systematic identification of patients with PHI [12]. Acute HIV-1 infection can also be diagnosed through the detection of plasma viral RNA using assays based on the amplification of nucleic acid through the polymerase chain reaction (PCR) or though signal amplification (branched DNA) in conjunction with a characterization of antibodies by ELISA or Western blot.

Retrospective studies have indicated that specific antibody tests become positive, on average, 3 to 4 weeks after the infectious contact and that p24 antigen test and polymerase chain reaction (PCR) assays for the detection of plasma viral RNA become positive around 5 to 7 and 10 days earlier, respectively [13]. As co-infection with other viruses may modify the course of HIV infection, screening tests for hepatitis B, C, GBV C and cytomegalovirus should be performed, especially in intravenous drug users more likely to have been previously exposed to these pathogens or to be co-infected. Successive Western blots with identification of an increased number of positive HIV-1 specific bands over time illustrate the progressive emergence of specific antibodies. There are no other specific laboratory findings in PHI but CD4+ cell counts are usually decreased. The fall of CD4 counts mainly affects the CCR5-negative subset of CD4 and may rarely reach $<200/mm^3$ and allow for the development of opportunistic infections [7]. Over the weeks following PHI, the ratio of CD4+ over CD8+ T-cells is inverted. A moderate thrombocytopenia is found in approximately half of the PHI patients [1]. Elevated hepatic enzyme levels are detected in around 20% of the patients, and, in our experience, in the absence of active liver disease linked to chronic or acute hepatitis B or C, a moderate increase of hepatic enzymes (less than three-fold the normal range) did not preclude antiviral treatment based on protease inhibitor (PI)-containing regimens.

Early changes in viremia and CD4 cell counts

After migration of infected dendritic cells into lymphoid tissues, viral replication is extensive around 6 to 15 days after infection [14, 15]. This results in very high viremia in the range of 1 to 10 millions copies/ml [11, 16]. High viremia however rapidly decreases with the amendment of symptoms and viremia of 5,000 copies/ml should not exclude the diagnosis of PHI in patients in whom ARS has regressed or disappeared in the weeks preceding testing. During this early phase of the infection, the virus rapidly disseminates in the lymphoid tissues which are the main site of viral replication and in whom the viral burden is much higher than in peripheral blood. Within weeks viremia decreases by several logs and CD4 cell counts increase in parallel with a massive CD8 oligoclonal expansion with activation markers such as CD38 and

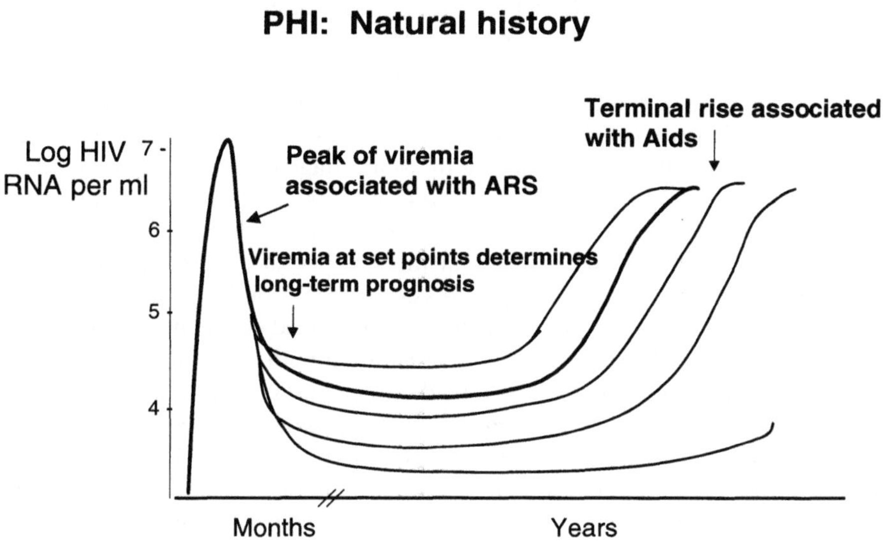

Figure 1. Natural history and viremia predict long-term disease outcome.
- Following acute HIV infection, viremia falls even without drugs, the immune responses are antivirally effective.
- The set points are established within a few months and predict long-term disease prognosis.

HLA-DR [17]. Six months after the acute infection, viremia and CD4 stabilize to a set point which is predictive of the long-term course of the disease in the absence of treatment [18, 19]. At this time point, viremia varies among individuals from a few copies to more than 100,000 copies/ml. It is thought that the control of HIV replication is due to cell-mediated responses. Indeed, neutralising antibodies are either rarely detected at this stage, or, when detected are directed to rapidly changing epitopes [20].

Factors affecting treatment decision

Over the last 5 years, there has been a trend towards a less aggressive approach for treatment initiation. In patients with established infection this has translated into guidelines based more on CD4 cell counts (<350/mm^3) than on viremia for initiating of treatment [21]. The new guidelines are the consequence of a better appraisal of two main factors: the failure to achieve viral eradication with the presently available drugs and the impact of short- and long-term toxicities associated with anti-retroviral treatment. Despite this new trend, most guidelines recommend to treat patients with acute HIV infection within the context of clinical trials. We will here focus on three mains topics: factors supporting or against treatment in PHI, results of clinical trials and potential strategic moves for PHI patients on HAART for years.

Viremia level at set point, that is 3 to 6 months after infection, can be considered as the best predictor of long term HIV disease outcome. A number of factors determine and/or modulate viremia at set point including immune activation at the time of infection, host and viral genetic factors [22] and quantitative and qualitative differences in the immune response [17, 23]. However, at the time of treatment decision, during the acute stage of the infection, it is not possible to accurately predict viremia levels at set point, thus some of the arguments in favor of therapeutic intervention are based on clinical observations.

It has been shown in particular that the severity of symptoms during the ARS are predictive of the time of progression to AIDS or death [24]. The same correlation has been observed for the incubation period between the time of the infection and the ARS [25]. As patients identified with acute infection tend to be more symptomatic than those not identified, treatment is generally proposed to those patients with a potentially worse prognosis. Other arguments in favor of treatment refer to the pathogenesis of the infection. Early treatment results in the preservation of a relatively intact immune system including HIV specific helper T-cell responses which are rapidly lost in the absence of treatment [26]. These specific CD4+ helper responses are only partially restored in patients treated at later stages of the disease. Indeed, in the patients treated at later stages of HIV-1 infection, immune responses to environmental antigens such as cytomegalovirus and candidasis are rapidly restored after initiation of potent antiviral therapy, while HIV-1 specific responses are not or only partially restored [27]. These HIV-1 specific responses might be crucial in the long term, i.e., in the context of strategic treatment interruption or in patients who opt for treament discontinuation. Other data derived from experimental infections in macaques suggest that a short course of anti-retroviral treatment at the time of acute infection results in lower viremia in treated animals as compared to non-treated controls [28]. Early treatment has also been shown to prevent the viral differentiation into multiple viral quasispecies, some of which might carry mutations associated with resistance to antiviral drugs [17, 29]. Other arguments refer to a limitation of viral spreading in sanctuaries such as the central nervous system where antiviral might not be fully effective and to a limitation of the size of the pool of long-lived latently infected cells.

Follow up data collected on patients on highly active anti-retroviral therapy (HAART) have also shown that residual viremia levels are lower and that the decrease of proviral DNA is more pronounced in PHI patients as compared to patients treated at the time of established infection [29–32]. In a controlled study, the mean viremia levels over time and nadir were significantly lower in PHI patients as compared with patients starting HAART at the time of established infection. In addition half of the PHI patients showed persistently undetectable viremia level (less than 2.5 copies/ml) during a more than 2-year follow-up, whereas in the chronic patients' group only occasional samples had undetectable viremia with most values fluctuating beween 10 and 200 copies/ml [30]. These data indicate that very early treatment either reduces the size of the pool of infected cells which release viruses or that the spreading of virus to sanctuaries, where antiviral drugs might not reach optimal concentrations, has been limited by early therapeutic intervention.

An additional benefit of treatment of PHI in terms of public health is a limitation of the transmission of new infections since there is evidence that patients are highly infectious for their sexual partners at the time of acute HIV infection. Along this line, a recent study has shown that in countries where anti-retroviral drugs are largely available, up to 40% of new infections are transmitted by early HIV-infected patients [33].

There are, however, also arguments in favor of postponing treatment: the complete spontaneous regression of the ARS within 3 weeks in more than 95% of patients followed by an asymptomatic phase usually lasting years can argue against early treatment. Among other elements suggesting to postpone treatment are drug-induced toxicities, difficulties in ingesting multiple drugs at various times, thus causing reduced adherence and change in lifestyle. Finally, the main argument against treatment of PHI is the development of resistance to antiviral drugs in poorly adherent patients, as this may impair therapeutic options in the future. However, this option is incumbent on the patient, and, thus, the key element to consider for treatment is the patient's commitment for a treatment which has to be pursued for years. If the patient opts for treatment (in our AIDS Center 90% of the patients with PHI do so), there is a necessity for continuous information and support.

Clinical trials

Trials focusing on treatment

There is only one randomized placebo-controlled study of antiviral therapy started at the time of PHI that involved comparison between zidovudine, a drug with low antiviral efficacy in monotherapy *versus* placebo [34]. After 6 months the patients in the zidovudine arm had an increase of 173 CD4+ cells/mm^3 as compared to an increase of 6 CD4+ cells/mm^3 in the placebo arm. Moreover, zidovudine-treated patients developed fewer opportunistic infections than the untreated patients but, as expected, there was no statistical difference in viremia levels between both arms. All the subsequent clinical trials had no control arms and were based on combination of two nucleoside reverse transcriptase inhibitors [35, 36] or triple therapy consisting either of two nucleoside reverse transcriptase inhibitors and a protease inhibitor, or of three nucleoside analogs inhibitors of the reverse transcriptase [37–39]. As in trials including patients with established HIV infection on HAART, adherence is the key factor for virological and CD4 responses. In patients adherent to HAART within a median of 3 to 4 months a rapid fall of viremia below the level of detection was observed. This was associated with a CD4 increase and a CD8 decrease with a median CD4/CD8 ratio close to or above 1 at 6 months and a rapid decrease of activation markers at the surface of CD8 lymphocytes [40].

The results reported in PHI patients are similar to those reported in drug-naïve patients treated later in the course of the infection, although, as mentioned earlier, virological control and preservation of HIV-specific immune responses is better achieved by HAART in PHI. Whether this translates into clinical benefit has not been conclusively established. This is due to the lack of placebo-controlled trials in PHI and to the lack of prospective or retrospective large cohorts of untreated PHI patients.

In the context of HAART treatment, it has been shown that transmission of drug-resistant variants occurs at a frequency of approximately 10–15% in Western countries where anti-retroviral drugs have been largely available [32, 41, 42]. PHI patients infected with resistant variants have been shown to have a slower viremia decline on HAART than those infected with wild-type virus [42, 43]. It is therefore rational to initiate treatment as soon as possible, to perform (genotypic or phenotypic) resistance testing and later on to modify treatment accordingly. In any case, storage of plasma samples collected as soon as possible will allow to perform resistance testing in case of incomplete virological response.

Trials focusing on pathogenesis

Several trials, with small numbers of patients included, had focused on the pathogenesis of HIV infection. They have underlined that early treatment of PHI is associated with the preservation or development of strong CD4 helper HIV-specific responses [26], and also with an increase of neutralising antibodies [44]. Early treatment is associated with a breath and magnitude of virus-specific cytotoxic T-lymphocytes (CTL) responses, significantly lower in treated PHI patients than those observed in patients treated at the chronic stage of the infection [17]. There were also differences in the frequency of recognition of HIV-specific T-cell epitopes as shown in particular for the A2 restricted SLYNTVATL epitope recognised by most HLA A2 chronic patients and infrequently by HLA A2 PHI patients [17]. This demonstrates that there is variation in immunogenicity among CTL epitopes. Other studies in monkeys have shown that the quality of the CTL responses and in particular the affinity of the CTL may play a key role in terms of their antiviral efficacy. These findings may have implications for vaccine design. This also suggests that the degree of exposure to HIV antigens is a key factor driving CTL responses [45]. As expected, several studies also report a less diverse viral population in PHI patients as compared to patients treated at the time of established infection [17, 29].

Trials focusing on HAART plus immune interventions

As early antiviral treatment in PHI is associated with preservation of HIV-specific immune responses and a limitation of the size of the pool of HIV producing cells, adjunctive therapies should be more effective in PHI patients. Several trials have thus combined HAART and immune interventions in PHI patients, these include strategic treatment interruptions, therapeutic vaccination, cyclosporin A, and gamma-interferon. Gene therapy represents another attractive option still poorly explored in HIV infection.

The aim of strategic treatment interruption following HAART is to boost HIV-specific immune responses by re-exposure of the patients to their own virus in order to achieve a better control of viral replication after the final treatment interruption, thus allowing a lower exposure to the antiretroviral drug. Additional benefit can be envisaged in terms of treatment cost. This approach has shown limited benefit with respect to the viremia levels reached after treatment cessation in patients treated at the chronic stages, but in uncontrolled studies appears to be associated with a better control of viral replication in patients treated at the time of acute HIV infection [46, 47].

Other trials have explored the potential benefit of adding therapeutic vaccines to HAART in early PHI patients [48, 49]. One of these trials (QUEST) is a large ongoing international controlled study which evaluates in early PHI patients treated for more than 18 months by HAART the potential benefit of adding one or two therapeutic vaccines (Remune or Remune plus Alvac vCP1452) to HAART before treatment interruption [49]. As there is a control group immunized with placebo vaccine, it would be possible to evaluate the impact of vaccine(s) on the containment of viral replication, preservation of CD4 levels and changes in proviral DNA and immunologic parameters. The trial will end this year and analysis of results will be available early next year. Other therapeutic vaccination trials in conjunction with HAART in PHI patients are planned for next year.

Immune interventions are also considered using IL-2 or have been explored in PHI patients using cyclosporin A and interferon-alpha. Cyclosporin A has the potential to decrease immune activation which is thought to favor HIV dissemination and replication. It has been used for 2 weeks in conjunction with HAART in very early PHI patients and was associated with a rapid CD4 recovery to levels higher than those observed in historic controls on HAART alone [50].

Interferon-alpha acts at several steps of the viral life cycle; it inhibits the transcriptional activity of the long-term repeat, the translation of viral RNA and the assembly of viral particles. Potentially, these post-integration effects can potentiate the activity of protease inhibitors. In addition, interferon-alpha has immunomodulatory activities: it improves CD4 survival, favours Th 1 type response, activates CD8 CTL and increases the presentation of viral antigens by the major histocompatibility complex. All these activities can be expected to favour control of HIV replication. In the ANRS086 study the pegylated form of alpha-interferon was administered subcutaneously once a week for 12 weeks in conjunction with HAART in 12 patients with acute HIV infection [51]. This regimen was well tolerated and allowed an early control of HIV replication and a rapid decay of proviral DNA. Both the decays of viremia and that of proviral DNA were more rapid than those observed in trials where PHI patients were treated with HAART alone.

Strategic moves for PHI patients on HAART for years

As patients treated at the chronic stages, PHI patients develop metabolic abnormalities and changes in habitus (lipodystrophy) following HAART treatment, eventually after less than 2 years treatment [52]. As a number of these PHI patients would have spontaneously reached CD4 and viremia levels not requiring treatment according to the current treatment guidelines, and as viral eradication is no more a reasonable goal, there is a clear trend to interrupt antiretroviral treatment in PHI patients treated for years. This assumption should be modulated by the fact that a number of PHI patients, included in clinical trials, were highly symptomatic at the time of the acute infection and thus more likely to progress rapidly.

Other factors may be considered: they include levels of proviral DNA which are lower in those patients who will have lower or no viremia rebound. Host genetic factors are today an active field of investigations and might become in the future a key determinant to evaluate viremia at set point and more importantly the clinical course of the infection in an increased number of patients. Presently only few host genetic polymorphisms have been found to be associated with a substantial resistance to infection or progression of disease, they are mainly coding for virus entry second receptor (notably, but not exclusively, protection from infection or delay in disease was observed in homozygotes or heterozygotes for the delta 32 deletion in the gene coding for the major co-receptor for HIV entry in CD4 called CCR5) or HLA alleles (notably, but not exclusively, the HLA B 5701 class 1 allele has been shown to be strongly associated with long-term non-progression) [53, 54]. One can reasonably guess that with the identification of more polymorphisms in genetic factors potentially involved in either resistance to infection or to progression, these markers will be used increasingly in the context of treatment decision. Viral factors modulating disease progression such as deletions in the Nef gene have also been identified, but overall their frequency is low in patients with a slow disease progression [55]. There is evidence that the pathways for acquisition of mutations associated with drug resistance, in particular for protease inhibitors, differ with HIV subtypes [56], and the issue of a modulation of the clinical course of HIV disease in relation to HIV subtypes is the subject of several only partially conclusive investigations [57].

In practical terms there are two main possibilities to envisage strategic decisions in long term HAART-treated PHI patients who are becoming "intolerant" to treatment or present major drug-induced toxicities; each of them being better than progressive poor adherence or intermittent or abrupt non-monitored treatment cessation. The first consists in the inclusion of these patients in strategic interruption trials or in therapeutic vaccination trials since the design

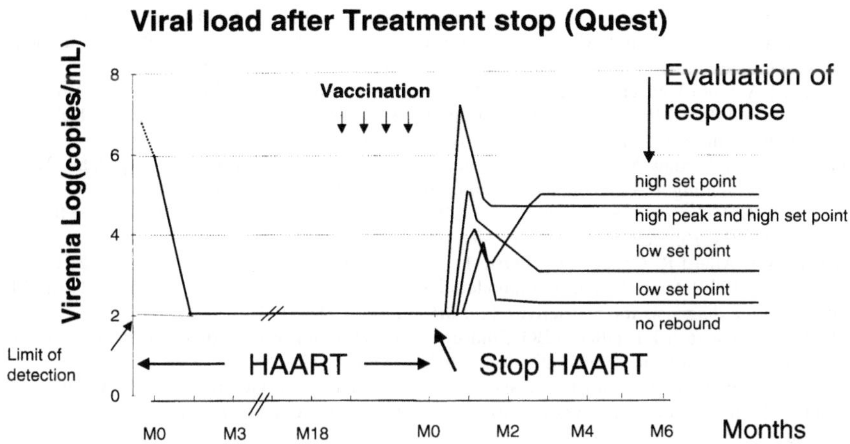

Figure 2. Schematic representation of the Quest study. The primary end point between the placebo and the "vaccinated" groups will be based on viremia levels 6 months after stopping all treatment.The study is still unblinded, viremia levels after treatment stop are based on viremia patterns observed in patients enrolled in Quest.

of both types of trials includes treatment cessation at the end of the trial in order to evaluate the efficacy of the intervention (see, as an example, the design of the Quest trial). The second is planned treatment interruption closely monitored by the treating physician. In both instances patients should be monitored at frequent intervals weekly, next every 15 days, and then monthly. Patients should be aware that usually symptoms reminiscent of the acute infection but weaker, mainly fever and fatigue, can occur during the month following interruption of treatment and that rarely a typical ARS will be observed. In most patients, viremia rebounds will be observed between 7 and 28 days after treatment cessation. Most patients will have a peak viremia rebound between 30,000 and 500,000 copies/ml with a decrease over the following months. Such rebounds have been well documented in large numbers of patients treated at the time of PHI or in chronic patients. However, results from large series are generally not published (example of large trials with treatment stop: the Quest study for PHI and the SSITT study for chronic patients [49, 58]). This means that a viremia rebound to more than 50,000 copies/ml is not an indication for the immediate re-initiation of treatment as long as CD4 cell counts are higher than 350/mm^3. As most patients who stop treatment have episodes of high viremia, they should be aware of their high infectivity and take adequate measures to avoid transmission of the infection to their partners.

References

1 Kinloch-de Loës S, de Saussure Ph, Saurat JH, Stalder H, Hirschel B, Perrin LH (1993) Symptomatic primary infection due to human immunodeficiency virus type 1: review of 31 cases. *Clin Infect Dis* 17: 59–65

2 Tindall B, Barker S, Donovan B, Barnes T, Roberts J, Kronenberg C, Gold J, Penny R, Cooper D, and the Sydney AIDS Study Group (1988) Characterization of the acute clinical illness associated with human immunodeficiency virus infection. *Arch Intern Med* 148: 945–949
3 Cooper DA, Gold J, Maclean P, Donovan B, Finlayson R, Barnes TG, Michelmore HM Brooke P, Penny R, and the Sydney AIDS Study Group (1985) Acute AIDS retrovirus infection. Definition of a clinical illness associated with seroconversion. *Lancet* i: 537–540
4 Gaines H, von Sydow M, Pehrson PO, Lundbergh P (1988) Clinical picture of primary HIV-1 infection presenting as a glandular-fever-like illness. *Br Med J* 297: 1363–1368
5 de Jong MD, Hulsebosch HJ, Lange JMA (1991) Clinical, virological and immunological features of primary HIV-1 infection. *Genitourin Med* 67: 367–373
6 Vanhems P, Allard R, Cooper DA, Perrin L, Vizzard J, Hirschel B, Kinloch-de Loës S, Carr A, Lambert J (1997) Acute human immunodeficiency virus type 1 disease as a mononucleosis-like illness: is the diagnosis too restrictive? *Clin Infect Dis* 24: 965–970
7 Pedersen C, Gerstoft J, Lindhardt BO, Sindrup J (1987) Candida esophagitis associated with acute human immunodeficiency virus infection. *J Infect Dis* 156: 529–530
8 Clark SJ, Kelen GD, Henrard DR, Daar ES, Craig S, Shaw GM, Quinn TC (1994) Unsuspected primary human immunodeficiency virus type 1 infection in seronegative emergency department patients. *J Infect Dis* 170: 194–197
9 Galetto-Lacour A, Kinloch-de Loës S, Hirschel B, Perrin L (1995) Primo-infection VIH: un diagnostic souvent évoqué mais posé tardivement. *Schweiz Med Wochenschr* 125: 341–346
10 Von Sydow M, Gaines H, Sonnerborg A, Forsgren M, Pehrson PO, Strannegard O (1988) Antigen detection in primary HIV infection. *Br Med J* 296: 238–240
11 Baumberger CH, Kinloch-de-Loës S, Yerly S, Hirschel N, Perrin LH (1993) High levels of circulating RNA in patients with symptomatic HIV-1 infection. *AIDS* 7 (Suppl 2): 59–64
12 Yerly S, Simon F, Perrin L (1999) Diagnostic précoce des primo-infections VIH: utilisation d'un test de dépistage combiné (antigène p24 et anticorps anti-VIH). *Schweiz Med Wochenschr* 129: 319–322
13 Busch MP, Satten GA (1997) Time course of viremia and antibody seroconversion following human immunodeficiency virus exposure. *Am J Med* 102 (Suppl 5B): 117–124
14 Pantaleo G, Graziosi C, Butini L, Pizzo PA, Schnittman SM, Kotler DP, Fauci AS (1991) Lymphoid organs function as major reservoirs for human immunodeficiency virus. *Proc Natl Acad Sci USA* 88: 9838–9842
15 Fox CH, Tenner-Racz K, Racz P, Firpo A, Pizzo PA, Fauci AS (1991) Lymphoid germinal centers are reservoirs of human immunodeficiency virus type 1 RNA. *J Infect Dis* 164: 1051–1057
16 Clark SJ, Saag MS, Decker WD, Campbell-Hill S, Roberson JL, Veldkamp PJ, Kappes JC, Hahn BH, Shaw GM (1991) High titers of cytopathic virus in plasma of patients with symptomatic primary HIV-1 infection. *N Engl J Med* 324: 954–960
17 Altfeld M, Rosenberg ES, Shankarappa R, Mukherjee JS, Hecht FM, Eldridge RL, Addo MM, Poon SH, Phillips MN, Robbins GK et al. (2001) Cellular immune responses and viral diversity in individuals treated during acute and early HIV-1 infection. *J Exp Med* 193: 169–180
18 Henrard DR, Philipps JF, Muenz LR, Blattner WA, Wiesner D, Eyster ME, Goedert JJ (1995) Natural history of HIV-1 cell-free viremia. *JAMA* 274: 554–558
19 Mellors JW, Kingsley LA, Rinaldo CR Jr, Todd JA, Hoo BS, Kokka RP, Gupta P (1995) Quantitation of HIV-1 RNA in plasma predicts outcome after seroconversion. *Ann Intern Med* 122: 573–579
20 Richmann DD, Wrin T, Little S, Petropoulos C (2002) Rapid evolution of the neutralizing antibody response following primary HIV infection. *9th Conference on Retroviruses and Opportunistic Infections*, Seattle, WA, USA, 24–28 February 2002, Abstract LB5
21 BHIVA Writing Committee on behalf of the BHIVA Executive Committee (2001) British HIV Association (BHIVA) guidelines for the treatment of HIV-infected adults with antiretroviral therapy. *HIV Med* 2: 276–313
22 Smith MW, Dean M, Carrington M, Winkler C, Huttley GA, Lomb DA, Goedert JJ, O'Brien TR, Jacobson LP, Kaslow R et al. (1997) Contrasting genetic influence of CCR2 and CCR5 variants on HIV-1 infection and disease progression. Hemophilia Growth and Development Study (HGDS), Multicenter AIDS Cohort Study (MACS), Multicenter Hemophilia Cohort Study (MHCS), San Francisco City Cohort (SFCC), ALIVE Study. *Science* 277: 959–965
23 Pantaleo G, Demarest JF, Schacker T, Vaccarezza M, Cohen OJ, Daucher M, Graziosi C, Schnittman SS, Quinn TC, Shaw GM et al. (1997) The qualitative nature of the primary immune response to HIV infection is a prognosticator of disease progression independent of the initial

level of plasma viremia. *Proc Natl Acad Sci USA* 94: 254–258
24 Lindbäck S, Broström C, Karlsson A, Gaines H (1994) Does symptomatic primary HIV-1 infection accelerate progression to CDC stage IV disease, CD4 count below 200 × 10^6/l, AIDS, and death from AIDS? *Br Med J* 309: 1535–1537
25 Vanhems P, Hirschel B, Phillips AN, Cooper DA, Vizzard J, Brassard J, Perrin L (2000) Incubation time of acute human immunodeficiency virus (HIV) infection and duration of acute HIV infection are independent prognostic factors of progression to AIDS. *J Infect Dis* 182: 334–337
26 Rosenberg ES, Billingsley JM, Caliendo AM, Boswell SL, Sax PE, Kalams SA, Walker BD (1997) Vigorous HIV-1-specific CD4+ T cell responses associated with control of viremia. *Science* 278: 1447–1450
27 Autran B, Carcelain G, Li TS, Blanc C, Mathez D, Tubiana R, Katlama C, Debre P, Leibowitch J (1997) Positive effects of combined antiretroviral therapy on CD4+ T cell homeostasis and function in advanced HIV disease. *Science* 277: 112–116
28 Wodarz D, Arnaout RA, Nowak MA, Lifson JD (2000) Transient antiretroviral treatment during acute simian immunodeficiency virus infection facilitates long-term control of the virus. *Philos Trans R Soc Lond B Biol Sci* 355: 1021–1029
29 Karlsson AC, Birk M, Lindback S, Gaines H, Mittler JE, Sonnerborg A (2001) Initiation of therapy during primary HIV type 1 infection results in a continuous decay of proviral DNA and a highly restricted viral evolution. *AIDS Res Hum Retroviruses* 17: 409–416
30 Yerly S, Kaiser L, Perneger TV, Cone R, Opravil M, Chave JP, Furrer H, Hirschel B, Perrin L, the Swiss Cohort Study (2000) Time of initiation of antiretroviral therapy: impact on HIV-1 viraemia. *AIDS* 14: 243–249
31 Yerly S, Perneger T, Vora S, Hirschel B, Perrin L (2000) Decay of cell-associated HIV-1 DNA correlates with residual replication in patients treated during acute HIV-1 infection. *AIDS* 14: 2805–2812
32 Ngo-Giang-Huong N, Deveau C, Da Silva I, Pellegrin I, Venet A, Harzic M, Sinet M, Delfraissy JF, Meyer L, Goujard C et al. (2001) Proviral HIV-1 DNA in subjects followed since primary HIV-1 infection who suppress plasma viral load after one year of highly active antiretroviral therapy. *AIDS* 15: 665–673
33 Yerly S, Kaiser L, Race E, Bru JP, Clavel F, Perrin L (1999) Transmission of antiretroviral-drug-resistant HIV-1 variants. *Lancet* 354: 729–733
34 Kinloch-de Loës S, Hirschel BJ, Hoen B, Cooper DA, Tindall B, Carr A, Saurat JH, Clumeck N, Lazzarin A, Mathiesen L et al. (1995) A controlled trial of zidovudine in primary human immunodeficiency virus infection. *N Engl J Med* 333: 408–413
35 Perrin L, Yerly S, Marchal F, Schockmel G A, Hirschel B, Fox CH, Pantaleo G (1998) Virus burden in lymph nodes and blood of subjects with primary human immunodeficiency virus type 1 infection on bitherapy. *J Infect Dis* 177: 1497–1501
36 Lafeuillade A, Poggi C, Tamalet C, Profizi N, Tourres C, Costes O (1997) Effects of a combination of zidovudine, didanosine, and lamivudine on primary human immunodeficiency virus type 1 infection. *J Infect Dis* 175: 1051–1055
37 Smith D, Berrey MM, Robertson M, Mehrotra D, Markowitz M, Perrin L, Clumeck N, Lazzarin A, Burckhardt B, Weber R et al. (2000) Virological and immunological effects of combination antiretroviral therapy with zidovudine, lamivudine, and indinavir during primary human immunodeficiency virus type 1 infection. *J Infect Dis* 182: 950–954
38 Hoen B, Dumon B, Harzic M, Venet A, Dubeaux B, Lascoux C, Bourezane Y, Ragnaud JM, Bicart-See A, Raffi F et al. (1999) Highly active antiretroviral treatment initiated early in the course of symptomatic primary HIV-1 infection: results of the ANRS 053 trial. *J Infect Dis* 180: 1342–1346
39 Markowitz M, Vesanen M, Tenner-Racz K, Cao Y, Binley JM, Talal A, Hurley A, Jin X, Chaudhry MR, Yaman M et al. (1999) The effect of commencing combination antiretroviral therapy soon after human immunodeficiency virus type 1 infection on viral replication and antiviral immune responses. *J Infect Dis* 179: 527–537
40 Tilling R, Kinloch S, Goh LE, Cooper D, Perrin L, Lampe F, Zaunders J, Hoen B, Tsoukas C, Andersson J et al. (2002) Parallel decline of CD8+/CD38+ T cells and viraemia in response to quadruple highly active antiretroviral therapy in primary HIV infection. *AIDS* 16: 589–596
41 Harzic M, Pellegrin I, Deveau C, Chaix ML, Dubeaux B, Garrigue I, Ngo N, Rouzioux C, Goujard C, Hoen B et al. (2002) Genotypic drug resistance during HIV-1 primary infection in France (1996–1999) frequency and response to treatment. *AIDS* 16: 793–796

42 Grant RM, Hecht FM, Warmerdam M, Liu L, Liegler T, Petropoulos CJ, Hellmann NS, Chesney M, Busch MP, Kahn JO (2002) Time trends in primary HIV-1 drug resistance among recently infected persons. *JAMA* 288: 181–188
43 Little SJ, Daar ES, D'Aquila RT, Keiser PH, Connick E, Whitcomb JM, Hellmann NS, Petropoulos CJ, Sutton L, Pitt JA et al. (1999) Reduced antiretroviral drug susceptibility among patients with primary HIV infection. *JAMA* 282: 1142–1149
44 Montefiori DC, Hill TS, Vo HT, Walker BD, Rosenberg ES (2001) Neutralizing antibodies associated with viremia control in a subset of individuals after treatment of acute human immunodeficiency virus type 1 infection. *J Virol* 75: 10200–10207
45 Jin X, Ogg G, Bonhoeffer S, Safrit J, Vesanen M, Bauer D, Chen D, Cao Y, Demoitie MA, Zhang L et al. (2000) An antigenic threshold for maintaining human immunodeficiency virus type 1-specific cytotoxic T lymphocytes. *Mol Med* 6: 803–809
46 Rosenberg ES, Altfeld M, Poon SH, Phillips MN, Wilkes BM, Eldridge RL, Robbins GK, D'Aquila RT, Goulder PJ, Walker BD (2000) Immune control of HIV-1 after early treatment of acute infection. *Nature* 407: 523–526
47 Hermans P (2001) Current review and clinical management of patients with primary HIV-1 infection: limits and perspectives. *Biomed Pharmacother* 55: 301–307
48 Jin X, Ramanathan M Jr, Barsoum S, Deschenes GR, Ba L, Binley J, Schiller D, Bauer DE, Chen DC, Hurley A et al. (2002) Safety and immunogenicity of ALVAC vCP1452 and recombinant gp160 in newly human immunodeficiency virus type 1-infected patients treated with prolonged highly active antiretroviral therapy. *J Virol* 76: 2206–2216
49 Goh LE, Perrin L, Hoen B, Cooper D, Phillips A, Janossy G, Sonnenborg A, Tsoukas C, Lampe F, Kinloch S (2001) Study protocol for the evaluation of the potential for durable viral suppression after quadruple HAART with or without HIV vaccination: the QUEST study. *HIV Clin Trials* 2: 438–444
50 Rizzardi GP, Harari A, Capiluppi B, Tambussi G, Ellefsen K, Ciuffreda D, Champagne P, Bart PA, Chave JP, Lazzarin A et al. (2002) Treatment of primary HIV-1 infection with cyclosporin A coupled with highly active antiretroviral therapy. *J Clin Invest* 109: 681–688
51 Emilie D, Burgard M, Lascoux-Combe C, Laughlin M, Krzysiek R, Pignon C, Rudent A, Molina JM, Livrozet JM, Souala F et al. (2001) Early control of HIV replication in primary HIV-1 infection treated with antiretroviral drugs and pegylated IFN alpha: results from the Primoferon A (ANRS 086) Study. *AIDS* 15: 1435–1437
52 Narciso P, Tozzi V, D'Offizi G, De Carli G, Orchi N, Galati V, Vincenzi L, Bellagamba R, Carvelli C, Puro V (2001) Metabolic and morphologic disorders in patients treated with highly active antiretroviral therapy since primary HIV infection. *Ann NY Acad Sci* 946: 214–222
53 Samson M, Libert F, Doranz BJ, Rucker J, Liesnard C, Farber CM, Saragosti S, Lapoumeroulie C, Cognaux J, Forceille C et al. (1996) Resistance to HIV-1 infection in caucasian individuals bearing mutant alleles of the CCR-5 chemokine receptor gene. *Nature* 382: 722–725
54 Migueles SA, Sabbaghian MS, Shupert WL, Bettinotti MP, Marincola FM, Martino L, Hallahan CW, Selig SM, Schwartz D, Sullivan J et al. (2000) HLA B*5701 is highly associated with restriction of virus replication in a subgroup of HIV-infected long term nonprogressors. *Proc Natl Acad Sci USA* 97: 2709–2714
55 Rhodes DI, Ashton L, Solomon A, Carr A, Cooper D, Kaldor J, Deacon N (2000) Characterization of three nef-defective human immunodeficiency virus type 1 strains associated with long-term nonprogression. Australian Long-Term Nonprogressor Study Group. *J Virol* 74: 10581–10588
56 Caride E, Hertogs K, Larder B, Dehertogh P, Brindeiro R, Machado E, de Sa CA, Eyer-Silva WA, Sion FS, Passioni LF et al. (2001) Genotypic and phenotypic evidence of different drug-resistance mutation patterns between B and non-B subtype isolates of human immunodeficiency virus type 1 found in Brazilian patients failing HAART. *Virus Genes* 23: 193–202
57 Alexander CS, Montessori V, Wynhoven B, Dong W, Chan K, O'Shaughnessy MV, Mo T, Piaseczny M, Montaner JS, Harrigan PR (2002) Prevalence and response to antiretroviral therapy of non-B subtypes of HIV in antiretroviral-naïve individuals in British Columbia. *Antiviral Ther* 7: 31–35
58 Hirschel B (2001) Planned interruptions of anti-HIV treatment. *Lancet Infect Dis* 1: 53–59

Combination Therapy of AIDS
Edited by E.D.A. De Clercq and A.-M.I. Vandamme

Salvage therapy

Marianne Harris and Julio S.G. Montaner

British Columbia Centre for Excellence in HIV/AIDS, 667-1081 Burrard Street, Vancouver, BC V6Z 1Y6, Canada

Introduction

Substantial declines in the rates of AIDS-related illness, hospitalization, and death in people with HIV infection have been achieved in the current era of highly active antiretroviral therapy (HAART) [1–3]. Sustained suppression of viral load below the level of detectability is required to achieve optimal impact on immunologic function and clinical outcomes [4, 5]. If plasma viral load is inadequately suppressed, either failing to become undetectable or rebounding to consistently detectable levels after initial undetectability, virologic treatment failure is said to have occurred. Failure to reestablish virologic control leads to immunologic deterioration and ultimately progression to AIDS.

Unfortunately, virologic failure occurs in a substantial proportion of patients taking HAART and failure rates increase with successive regimens [6, 7]. Factors associated with treatment failure include lower baseline CD4 cell count, higher viral load, prior AIDS diagnosis, previous suboptimal use of antiretroviral therapies, and comorbid conditions such as hepatitis C. Inadequate plasma drug concentrations (for example, due to drug interactions, impaired absorption, or suboptimal adherence) in the presence of ongoing viral replication will invariably promote the emergence of viral strains with reduced susceptibility to one or more antiretroviral agents. Given the substantial degree of cross-resistance between the available antiretrovirals within each drug class, this will compromise the efficacy of successive HAART regimens and therefore virologic failure may become more likely.

In order to prevent this sequence of events, and thereby minimize the chances of clinical disease progression, patients experiencing virologic failure on HAART are offered salvage or rescue therapy. Selection of the regimen is critical; while there are currently no standards of care in this area, some information is available pertaining to predictors of success or failure of salvage regimens. Pending availability of true second-line antiretroviral agents within existing drug classes or new agents aimed towards new targets (without cross-resistance with existing agents), the response to salvage therapy may be optimized by appropriate use of current technologies and novel treatment strategies.

Management of early HAART failure

Current guidelines recommend initiating therapy with two nucleosides (NRTIs) plus a nonnucleoside (NNRTI), a protease inhibitor (PI), or a third nucleoside [8]. When virologic failure is confirmed, it is recommended that at least two and preferably all three drugs in the regimen be changed, avoiding as much as possible cross-resistance between the old and the new regimen. Best results are generally observed when at least one of the new drugs belongs to a drug class that the patient has not previously used. The likelihood of virologic success with second-line therapies is typically lower than that seen in first-line therapy, and this tends to decrease sharply with subsequent regimens [7], due at least in part to the accumulation of mutations conferring reduced drug susceptibility. Thus, strategic selection of the initial as well as each subsequent salvage regimen is of critical importance to maximize virologic potency and sustain long-term benefit.

Drug sequencing within a class

For patients who have initially received an NNRTI-based regimen, another NNRTI-based regimen is unlikely to be a viable option, as there is broad, high-level cross-resistance within currently available members of this drug class. NNRTI cross-resistance is particularly conferred by the presence of a K103N mutation (the most frequent mutation selected when failing therapy with efavirenz [9] or nevirapine [10]). There is clinical evidence that, in the absence of K103N, some nevirapine failures may be salvaged with efavirenz [11]; however, this maneuvre may be unsuccessful [12] and result in further reduced susceptibility to the concomitant nucleosides. This approach is not generally recommended. Preliminary evidence suggests that second generation NNRTIs currently in development (e.g., capravirine; TMC-125) may have resistance profiles different from currently available agents and may retain antiviral potency even after NNRTI failure in the presence of the K103N mutant.

A number of salvage studies have examined treatment options for patients failing first or second PI-based regimens. Initial studies looked at salvage of a single PI failure, indinavir or ritonavir, with dual PI regimens (ritonavir plus saquinavir, or nelfinavir plus saquinavir) plus nucleosides. Rates of virologic success were low, <20% at 20–39 weeks [13–15], likely due to prolonged treatment with the initial PI and accumulation of mutations conferring cross-resistance to the PIs in the salvage regimen. Some evidence of better success has been observed following nelfinavir failure [16], apparently due to selection of a D30N mutation which reduces susceptibility to nelfinavir without conferring cross-resistance to other PIs. Tebas et al. [17] observed that 71% of 24 patients had viral load below the limit of quantitation after 24 weeks on a ritonavir plus saquinavir salvage regimen, despite prolonged nelfinavir pretreatment (median 24 weeks). However, the success of this PI-sequencing strategy has not been consistently reproduced and mutations conferring PI cross-resistance, notably L90M, can also be seen following nelfinavir failure [18, 19]. More recently, it has become clear that, guided by appropriate use of resistance testing results, a PI-based regimen may be more predictably salvaged with an alternative PI or a combination of PIs. Ritonavir-boosted PIs may be more effective in this context, largely due to higher overall drug exposure. In the future, the availability of new PIs with unique resistance profiles (e.g., tipranavir, TMC-114) may enhance the potential for PI-based regimens to be used to salvage PI failures.

Adding a new class

A common theme emerging from the available salvage studies is that inclusion of a new drug class in the new regimen substantially improves the likelihood of a successful outcome. This has been observed when an NNRTI was included in the new regimen following initial failure of a single PI-based regimen [14, 20–23] as well as following sequential failures on more than one PI [24]. In this context, it may be necessary to adjust the dosage of the PI (including a ritonavir-boosted PI) upward to compensate for the drug interaction with the NNRTI (nevirapine or efavirenz); however, this may not be necessary in the case of low-level PI resistance. The principle of adding a new class may also be applied to initial NNRTI-based regimens, which should be amenable to successful salvage with PI-based regimens, as there is no evidence of cross-resistance between the PI and NNRTI classes [8]. However, this strategy has not been extensively studied in clinical trials [25, 26].

Role of viral load in when to switch

Salvage studies also point clearly to better rates of virologic success when treatment is adjusted as soon as detectable viral load has been confirmed, when the viral load is still relatively low [7, 20, 27]. In earlier studies where this issue was not recognized, or before other options were available, success rates of salvage (especially of PIs with PIs) were poor due to sequential accumulation of PI-associated mutations. Early failure of two nucleosides plus indinavir is associated with resistance mutations to nucleosides (especially 3TC) but not to PIs [28, 29]; hence early intervention, when viral load detectability (>50 copies/mL) is confirmed, should preserve the PI class as an option for PI salvage. Pharmacologic enhancement, for example boosting PI levels with low-dose ritonavir, may be sufficient to overcome low- to mid-level PI resistance. If viral replication is allowed to proceed further in the presence of a partially-suppressive regimen, sequential accumulation of mutations and high-level resistance will be the result, significantly compromising the prospect of successful salvage.

However, the definition of virologic failure may not always be clear-cut. It is clear that maximal virologic suppression to below the level of quantitation of the most sensitive commercial assays currently available (<50 copies/mL) is necessary for sustained virologic response [4]; however, assay variability (generally three-fold) should be kept in mind. Isolated, low-level rebounds or blips above assay detectability have been observed in patients taking effective antiviral therapy [30, 31] and do not necessarily predict treatment failure. Such blips should trigger immediate re-evaluation of plasma viral load to ensure they are not the beginning of a true sustained rebound in viral load. There is some concern that such plasma viral load blips may reflect ongoing low-level viral replication. It is unclear whether these events could reliably be identified if more sensitive assays (<5 or <2 copies/mL) were available for routine use. Even if that were the case, there is no evidence at present that such assays would have better predictive value than the assays currently used (usually <50 copies/mL) or that monitoring with a more sensitive assay would result in better outcomes.

Role of CD4 in when to switch

Generally speaking, virologic rebound precedes CD4 decline, so therapy should be changed before CD4 deterioration has had a chance to occur. CD4 monitoring clearly has an important role in the decisions when to commence antiviral therapy and when to start opportunistic infection prophylaxis, but has a limited role in the decision when to change therapy. If the viral load is consistently undetectable but the CD4 remains very low, there may be a role for immune modulation or antiviral intensification; however, these approaches remain investigational until their overall safety, efficacy, risk-benefit, and cost-effectiveness are fully characterised. On the other hand, cases of apparently good CD4 response despite persistently detectable viral load are not infrequent [6, 32, 33]. These are generally referred to as “discordant responses”. A recent study indicates that most discordant responses are relative, transient and often related to transient previous decreases in plasma viral load, which may not have been detected in the absence of a rigorous and frequent viral load monitoring routine [34]. True sustained discordant responses are probably rare, and a decline in the CD4 is the inevitable outcome of sustained viral replication [35]. Thus, viral load should remain the primary determinant of therapy failure and, whenever possible, undetectable viral load should be the goal of antiretroviral therapy [5].

Triple class failure

Once patients have experienced failure on and developed resistance to all three available classes of drugs, successful salvage therapy becomes more difficult.

Clearly until agents from new antiretroviral classes become available there will be no options left for adding a new class of drugs. Generally dismal results of standard salvage therapy after three-class failure have led some clinicians to adopt alternative therapeutic goals in such patients, in whom undetectable viral loads may be impossible to achieve [36]. Partially suppressive regimens may provide clinical benefit to patients unable to tolerate more aggressive regimens, by maintaining selective pressure on the virus and favouring multi-drug resistant strains, which may have reduced replicative capacity relative to wild-type virus.

On the other hand, in selected patients with multi-drug resistant virus, it may be possible to achieve undetectable viral load by recycling drugs in multiple drug rescue therapy (MDRT), or mega-HAART regimens. This strategy was studied in an observational fashion among 106 patients with prior failure of at least two regimens [37]. Patients had a median prior exposure to antiretrovirals of 43 months and 38% had been exposed to NNRTIs. The rescue regimen consisted of five to nine drugs (Median 5): up to four nucleosides, up to two PIs, up to two NNRTIs, with or without hydroxyurea. Median baseline (pre-MDRT) viral load was 4.79 $\log_{10}$ copies/mL (interquartile range 4.18–5.34) and CD4 cell count was 180/mm^3 (interquartile range 90–260). Fifty-six patients (59%) of 95 with available baseline phenotypes had fourfold or greater decreased susceptibility to seven or more antiretroviral drugs before starting MDRT. Six patients discontinued MDRT due to severe adverse drug events or laboratory toxicity. After 1 year, 48% of patients as treated and 40% of patients by intent to treat analysis had plasma viral load < 400 copies/mL, and 49% (52/106) had plasma viral load < 400 copies/mL on at least two consecutive visits during the study. Plasma viral load < 50 copies/mL on at least two consecutive visits was achieved in 33% (35/106). Previous experience and evidence of resistance to individual drugs are not sufficient to rule out a potential response to recycled drugs in MDRT. Potential drawbacks of this strategy include toxicity, difficult adherence, drug interactions, impact on quality of life, and cost. Taking these into consideration, undetectable viral load remains an achievable goal for selected heavily pretreated patients, and should not be abandoned without a trial of a carefully selected MDRT regimen [37, 38].

Optimizing response to salvage therapy

Resistance testing

The benefit of genotypic resistance testing in guiding the selection of second-line therapy after failure of nucleoside and PI-based therapy has been demonstrated in two prospective, randomized controlled trials [39, 40]. With appropriate interpretation, resistance testing permits the avoidance of drugs which are less likely to be effective, thus reducing toxicity and cost of the salvage regimen. Absence of apparent resistance is not a guarantee that a drug will be effective in that individual. For example, when a drug is withdrawn, resistance mutations selected by that agent may revert to minority populations, which are not detected by current resistance tests. Genotypic tests will not detect archived mutations which arose as a result of a regimen used in the past or prior to a drug holiday.

Appropriate clinical use of genotypic data must be guided by expert interpretation. This can be obtained using publicly available databases or simply using "virtual phenotypes" derived from large correlational databases of genotypic mutational patterns and their respective phenotypes [41]. This method of interpretation is dependent on the quality of the database from which it is derived, which must be continually updated for new mutational patterns selected by new drugs or combinations.

It should be noted that a single resistance test can only reliably characterize the predominant circulating viral species. As such the test cannot reliably ascertain the degree of resistance that may be present in archival or integrated virus, particularly as it relates to regimens that the patient may have used in the distant past. Furthermore, the virus has been shown to rapidly shift towards the wild type genotype following even a few drug-free weeks or months. It is therefore critical to use all available information when selecting an appropriate salvage regimen. Careful documentation of the antiretroviral therapy history and plasma viral load response is of paramount importance in this setting. In addition, liberal use of resistance testing can be of substantial value. From a practical standpoint, it is useful to construct a "Virtual Virus" representing the worst documented level of resistance achieved for each drug over the duration of the patient's history of antiretroviral therapy exposure. This takes into account the effects of all previous drug exposure and the heterogeneity of viral populations, and corrects for the apparent loss of resistance to drugs which the patient is not currently taking. For patients previously exposed to more than one regimen, the "Virtual Virus" is likely to provide a more accurate prediction of future response than a single genotype or virtual phenotype.

Phenotypic testing, a more direct measure of reduced viral susceptibility to drugs, is more easily interpreted than genotypic resistance testing, but is less frequently used in the clinic setting due to comparatively high cost and longer turnaround time. Like genotypic testing, phenotype results are more likely to be useful in selecting an effective regimen in patients with less treatment experience [42] and have little impact on rates of long-term success in multidrug resistant patients [43, 44]. Phenotypic testing is unlikely to confer any additional benefit in a setting where genotypic testing with adequate interpretation is available for standard care [45].

Therapeutic drug monitoring and pharmacokinetic enhancement

Plasma drug levels have been shown to correlate with virologic effect and toxicity for most protease inhibitors and NNRTIs. For PI-based regimens, suboptimal drug concentrations, as well as resistance, are independent predictors of virologic response to therapy [46]. Adjusting drug doses on the basis of measured plasma drug levels (therapeutic drug monitoring or TDM), in addition to results of resistance testing, may provide a means to optimize antiretroviral regimens. The benefit of TDM has been demonstrated in a preliminary way in drug-naïve patients starting regimens including indinavir or nelfinavir [47]. Nelfinavir-treated patients whose physicians had access to TDM information had significantly better rates of response (viral load < 500 copies/mL) after 12 weeks on therapy than those who did not. Indinavir-treated patients whose physicians had access to TDM results had better outcomes due to lower rates of toxicity. In another study including previously antiretroviral-naïve patients, concentration-controlled therapy with zidovudine, lamivudine and indinavir resulted in a greater proportion of patients with viral load < 50 copies/mL at 52 weeks [48]. These results, while encouraging, highlight the likelihood that the impact of TDM on clinical management will be dependent on the regimen as well as the clinical setting. The role of TDM in salvage therapy remains to be evaluated in prospective clinical trials.

Another unresolved issue which impacts the use of TDM in salvage therapy is the definition of desirable target levels for plasma drug concentrations. Even in patients taking their first-line therapy, therapeutic targets for plasma drug levels have not been clearly defined. Moreover, in experienced patients the target levels are expected to vary between patients according to the degree of viral resistance. Where modest levels of reduced susceptibility are present, particularly in early failure with a few mutations, resistance might be overcome by pharmacologic enhancement to increase drug levels over the IC_{50} or IC_{90} in that individual [18]. For example, low to intermediate levels of resistance to protease inhibitors have been successfully overcome by pharmacologic enhancement with low-dose ritonavir, which increases plasma levels of saquinavir, indinavir, lopinavir, or amprenavir [49–51]. However, the majority of patients with extensive PI experience have high-level resistance, with IC_{50} or IC_{90} far beyond what could reasonably be achieved *in vivo* without undue toxicity. Thus, while the use of inhibitory quotients (Cmin/IC_{50}) in determining the optimal target drug levels in individuals with resistant virus is potentially attractive, its practical use may be limited.

TDM may have a critical role in optimizing responses to MDRT. When several PIs and NNRTIs are used together, multiple unpredictable drug interactions can occur and optimal drug doses are difficult to establish. TDM could be used to maximize levels of individual drugs while avoiding extremely high levels likely to be associated with drug toxicity, which is a critical determinant for the success of MDRT.

Adherence

As with other chronic diseases, non-adherence to medication regimens is a common cause of treatment failure and relapse in HIV. Unfortunately, particularly high levels of adherence (>90%) are required to avoid emergence of resistance and ensure sustained response to antiviral therapy [52]. Fortunately, substantial strides have been made in recent years towards making antiviral regimens more user-friendly. Ritonavir boosting of PIs has enabled twice daily dosing of saquinavir and indinavir and removed food restrictions for indinavir. Studies have demonstrated equivalent effect of nelfinavir in twice daily dosing compared to three times daily. Even multidrug salvage regimens can usually now be prescribed in a twice-daily regimen.

However, non-adherence continues to be a major cause of treatment failure and simplifying the schedule is not the only answer. Clearly, side-effects continue to be a major factor in patients' motivation to comply with their drug regimens. Toxicities to previous regimens need to be taken into account when selecting a salvage regimen, in order to avoid recycling drugs with previous serious toxicities or using other drugs with overlapping toxicity profiles (for example, peripheral neuropathy). While monitoring for serious toxicities (hepatotoxicity, pancreatitis, etc.) is routine, minor nuisance side-effects such as fatigue, diarrhea, and nausea may have a substantial impact on adherence and may not be routinely monitored by the treating physician. Early identification and symptomatic intervention for these minor side-effects may benefit adherence and long-term outcome.

While the pivotal role of adherence in treatment success is widely recognized, as yet there is no gold standard for assessing an individual's adherence to antiretroviral therapy in the clinical setting. Physicians are notoriously poor judges of an individual's ability to comply with therapy. Expected non-compliance should not be used as an excuse to deny patients necessary therapy, even with complicated salvage regimens. Adherence to salvage therapy might be better than that to earlier therapies, enhanced by the patient's awareness that there are limited future options available. Ideally, adherence might be enhanced by multidisciplinary strategies including counselling and tailoring the regimen to individual needs.

Immune-based therapies

Long-term non-progressors with HIV have evidence of strong HIV-specific immune function, which may not be reconstituted by HAART initiated after primary infection. Thus a number of modalities which have the potential to enhance the patient's immune response to HIV have been tested in conjunction with standard antiretroviral therapy. These include interleukins, interferons, vaccines and others. Interleukin-2 (IL-2) has been perhaps the most extensively studied thus far, since the recognition that deficient IL-2 production is one of the first immunologic defects in HIV disease. Unfortunately, immune activation of lymphocytes and macrophages by IL-2 or other cytokines is associated with enhanced HIV replication in the activated, infected cells [53, 54]. Thus IL-2 therapy should not be considered in salvage therapy without first establishing maximal viral suppression with HAART. Regarding interferon therapy, promising results have recently been reported using pegylated interferon alpha-2B in treatment-experienced patients failing their current HAART regimen [55]. Short-term viral load decreases were observed without significant CD4 cell count changes. However, 38% of patients receiving the highest (3.0 mcg) dose had Grade 3 or 4 adverse events (fatigue, flu-like symptoms, and hypertriglyceridemia).

It is likely that a better response to immune-based treatments will be achieved in early HIV disease, even during primary infection, before HIV-specific immune function is lost. These modalities are less likely to play a role in the salvage therapy of patients with advanced disease. One exception might be in the situation where a patient has good virologic suppression on antiviral therapy, with persistently low CD4 cell counts. Further research is required to determine whether immune enhancement might be of benefit in this situation.

Structured treatment interruptions

Treatment interruptions prior to salvage therapy in patients on a failing regimen have been shown to allow the predominant viral population to shift towards the wild-type species. Preliminary observational reports suggested that this shift could be associated with better virologic response to subsequent therapy [56]. However, initial virologic responses may not be maintained over the longer term, as resistance mutations re-emerge rapidly when therapy is reintroduced [57]. Furthermore, declines in CD4 cell counts during treatment interruption may not be reversed once therapy is resumed [56, 58]. Recent controlled studies of immediate *versus* deferred salvage therapy have produced conflicting results. One such study demonstrated greater 24-week viral load decreases in patients who started multidrug salvage therapy following an 8-week treatment interruption, with better responses seen in patients who had reversion of HIV resistance mutations during the treatment interruption [59]. Another similar study failed to show even a short-term virologic benefit of treatment interruption prior to salvage therapy [60]. Pending long-term evaluation, treatment interruption as a strategy to enhance response to subsequent salvage therapy should be approached with caution and cannot be generally recommended outside the context of research protocols.

Adjuvants

Hydroxyurea and mycophenolic acid have been found to have synergistic antiviral effects *in vitro* with didanosine and abacavir, respectively [61, 62]. Early clinical studies demonstrated that hydroxyurea conferred an additional benefit in reducing viral load when used as an adjuvant to didanosine [63, 64]; however, more recently it has become evident that hydroxyurea also increases the likelihood of nucleoside-related toxicities such as pancreatitis [65, 66] and peripheral neuropathy [67]. The risks of toxic effects with hydroxyurea probably outweigh its potential benefits in the treatment of HIV. For potentiation of abacavir effects with mycophenolic acid, clinical data are limited and this approach remains investigational.

New agents

A number of new agents are soon to be available which show promise for patients with previous treatment failure. These include drugs from existing classes which have unique resistance profiles (capravirine, TMC-125, tipranavir) and drugs from new classes, where cross-resistance with currently available agents is unlikely. Tenofovir DF, a nucleotide reverse transcriptase inhibitor with activity against nucleoside-resistant strains of HIV, has demonstrated durable 48-week suppression of HIV RNA in treatment-experienced patients, and has a toxicity profile indistinguishable from placebo [68]. In two large studies totalling nearly 1000 patients, the fusion inhibitor T-20 (or enfuvirtide) has demonstrated significant reductions in viral load and good tolerability over 6 months in highly treatment-experienced patients [69, 70]. Unfortunately, new agents often become available individually through expanded access programs and are incorporated singly into the treatment of patients with extensive previous failure and resistance to other drugs. The result often is sequential monotherapy and rapid development of resistance to the new agent. Ideally, unless a new agent is being added to intensify an already partly successful regimen, two or more new agents should be added together to avoid inevitable failure from sequential monotherapy [71].

Conclusions

Management of early treatment failure can usually be successfully accomplished by addressing reasons for failure (such as adherence problems or drug-resistant virus) and, if necessary, changing to new drugs, preferably from a class new to that patient. Sequencing drugs within a class may become a more viable option with availability of newer agents with unique resistance profiles. Once patients have experienced failure on and developed resistance to all three available classes of drugs, selecting a salvage therapy regimen with a reasonable chance of durable virologic success becomes more challenging. However, even in the face of extensive previous drug exposure and reduced viral susceptibility, undetectable viral load may be achievable for some patients using carefully selected multiple drug rescue therapy (MDRT). Expert use of resistance testing and therapeutic drug monitoring offer rational means to optimize salvage therapy, particularly in the context of complex multiple-drug regimens. Availability of experimental agents from new antiviral classes may also be of help in the management of previous treatment failure; however, adding these agents individually to failing regimens results in sequential monotherapy, which must be avoided if lasting benefit is to occur. In the setting of salvage therapy, the potential contributions of immune-based therapies, treatment interruptions, and adjunctive therapies such as mycophenylate remain investigational.

Portions of this chapter have been previously published in Montaner JSG, Harris M. Management of HIV-infected patients with multidrug resitant virus. Current Infectious Disease Reports (2002, 4:259–265).

References

1 Palella FJ Jr, Delaney KM, Moorman AC, Loveless MO, Fuhrer J, Satten GA, Aschman DJ, Holmberg SD (1998) Declining morbidity and mortality among patients with advanced human immunodeficiency virus infection. *N Engl J Med* 338: 853–860

2 Hogg RS, Heath KV, Yip B, Craib KJ, O'Shaughnessy MV, Schechter MT, Montaner JS (1998) Improved survival among HIV-infected individuals following initiation of antiretroviral therapy. *JAMA* 279: 450–454

3 Mocroft A, Katlama C, Johnson AM, Pradier C, Antunes F, Mulcahy F, Chiesi A, Phillips AN, Kirk O, Lundgren JD (2000) AIDS across Europe, 1994–98: the EuroSIDA study. *Lancet* 356: 291–296

4 Raboud JM, Montaner JSG, Conway B, Rae S, Reiss P, Vella S, Cooper D, Lange J, Harris M, Wainberg MA et al. (1998) Suppression of plasma viral load below 20 copies/ml is required to achieve a long-term response to therapy. *AIDS* 12: 1619–1624

5 Wood E, Yip B, Hogg RS, Sherlock CH, Jahnke N, Harrigan PR, O'Shaughnessy MV, Montaner JS (2000) Full suppression of viral load is needed to achieve an optimal CD4 cell count response among patients on triple drug antiretroviral therapy. *AIDS* 14: 1955–1960

6 Ledergerber B, Egger M, Opravil M, Telenti A, Hirschel B, Battegay M, Vernazza P, Sudre P, Flepp M, Furrer H et al. (1999) Clinical progression and virological failure on highly active antiretroviral therapy in HIV-1 patients: a prospective cohort study. *Lancet* 353: 863–868

7 Mocroft A, Phillips AN, Miller V, Gatell J, van Lunzen J, Parkin JM, Weber R, Roge B, Lazzarin A, Lundgren JD (2001) The use of and response to second-line protease inhibitor regimens: results from the EuroSIDA study. *AIDS* 15: 201–209

8 Yeni PG, Hammer SM, Carpenter CCJ, Cooper DA, Fischl MA, Gatell JM, Gazzard BG, Hirsch MS, Jacobsen DM, Katzenstein DA et al. (2002) Antiretroviral treatment for adult HIV infection in 2002: updated recommendations of the International AIDS Society – USA Panel. *JAMA* 288: 222–235

9 Bacheler LT, Anton ED, Kudish P, Baker D, Bunville J, Krakowski K, Bolling L, Aujay M, Wang XV, Ellis D et al. (2000) Human immunodeficiency virus type 1 mutations selected in patients failing efavirenz combination therapy. *Antimicrob Agents Chemother* 44: 2475–2484

10 Conway B, Wainberg MA, Hall D, Harris M, Reiss P, Cooper D, Vella S, Curry R, Robinson P, Lange JMA et al. (2001) Development of drug resistance in patients receiving combinations of zidovudine, didanosine and nevirapine. *AIDS* 15: 1269–1274

11 Briones C, Soriano V, Dona C, Barreiro P, Gonzalez-Lahoz J (2000) Can early failure with nevirapine be rescued with efavirenz? *J Acquir Immune Defic Syndr* 24: 76–78

12 Walmsley SL, Kelly DV, Tseng AL, Humar A, Harrigan PR (2001) Non-nucleoside reverse transcriptase inhibitor failure impairs HIV-RNA responses to efavirenz-containing salvage therapy. *AIDS* 15: 1581–1584

13 Deeks SG, Grant RM, Beatty GW, Horton C, Detmer J, Eastman S (1998) Activity of a ritonavir plus saquinavir-containing regimen in patients with virologic evidence of indinavir or ritonavir failure. *AIDS* 12: F97–F102

14 Deeks SG, Hellmann NS, Grant RM, Parkin NT, Petropoulos CJ, Becker M, Symonds W, Chesney M, Volberding PA (1999) Novel four-drug salvage treatment regimens after failure of a human

immunodeficiency virus type 1 protease inhibitor-containing regimen: antiviral activity and correlation of baseline phenotypic drug susceptibility and virologic outcome. *J Infect Dis* 179: 1375–1381

15 Reiser M, Salzberger B, Stiepel A, Hoetelmans R, Diehl V, Fatkenheuer G (1999) Virological efficacy and plasma drug concentrations of nelfinavir plus saquinavir as salvage therapy in HIV-infected patients refractory to standard triple therapy. *Eur J Med Res* 4: 54–58

16 Kemper CA, Witt MD, Keiser PH, Dube MP, Forthal DN, Leibowitz M, Smith DS, Rigby A, Hellmann NS, Lie YS et al. (2001) Sequencing of protease inhibitor therapy: insights from an analysis of HIV phenotypic resistance in patients failing protease inhibitors. *AIDS* 15: 609–615

17 Tebas P, Patick AK, Kane EM, Klebert MK, Simpson JH, Erice A, Powderly WG, Henry K (1999) Virologic responses to a ritonavir-saquinavir-containing regimen in patients who had previously failed nelfinavir. *AIDS* 13: F23–F28

18 Condra JH, Petropoulos CJ, Ziermann R, Schlief WA, Shivaprakash M, Emini EA (2000) Drug resistance and predicted virologic responses to human immunodeficiency virus type 1 protease inhibitor therapy. *J Infect Dis* 182: 758–765

19 Atkinson B, Isaacson J, Knowles M, Mazabel E, Patick AK (2000) Correlation between human immunodeficiency virus genotypic resistance and virologic response in patients receiving nelfinavir monotherapy or nelfinavir with lamivudine and zidovudine. *J Infect Dis* 182: 420–427

20 Deeks SG, Hecht FM, Swanson M, Elbeik T, Loftus R, Cohen PT, Grant RM (1999) HIV RNA and CD4 cell count response to protease inhibitor therapy in an urban AIDS clinic: response to both initial and salvage therapy. *AIDS* 13: F35–F43

21 Benson CA, Deeks SG, Brun SC, Gulick RM, Eron JJ, Kessler HA, Murphy RL, Hicks C, King M, Wheeler D et al. (2002) Safety and antiviral activity at 48 weeks of lopinavir/ritonavir plus nevirapine and 2 nucleoside reverse-transcriptase inhibitors in human immunodeficiency virus type 1-infected protease inhibitor-experienced patients. *J Infect Dis* 185: 599–607

22 Gulick RM, Hu XJ, Fiscus SA, Fletcher CV, Haubrich R, Cheng H, Acosta E, Lagakos SW, Swanstrom R, Freimuth W et al. (2000) Randomized study of saquinavir with ritonavir or nelfinavir together with delavirdine, adefovir, or both in human immunodeficiency virus-infected adults with virologic failure on indinavir: AIDS Clinical Trials Group Study 359. *J Infect Dis* 182: 1375–1384

23 Gulick RM, Smeaton LM, D'Aquila RT, Eron JJ, Currier JS, Gerber JG, Acosta E, Sommadossi JP, Tung R, Snyder S et al. (2001) Indinavir, nevirapine, stavudine, and lamivudine for human immunodeficiency virus-infected, amprenavir-experienced subjects: AIDS Clinical Trials Group Protocol 373. *J Infect Dis* 183: 715–721

24 Danner S, Brun S, Richards B, King M, Clumeck N, Lazzarin A, Girard PM, Rockstroh J, Becker S, Pantaleo G et al. (2002) Kaletra (lopinavir/ritonavir) and efavirenz: 72 week safety and efficacy evaluation in multiple PI-experienced patients. *41st Interscience Conference on Antimicrobial Agents and Chemotherapy*, Chicago, Illinois, 16–19 December 2001, Abstracts, 344

25 Para M, Weinstock M (1998) Retrospective analysis of protease inhibitor efficacy among patients failing a delavirdine regimen. In *Abstracts-on-Disk of the 12th World AIDS Conference*, Geneva, Switzerland, 28 June–3 July 1998, no 12236

26 Cohen C, Royal M, Melton D, Gathe J (2002) Efficacy and durability of nelfinavir after initial virological failure of an non-nucleoside reverse transcriptase inhibitor regimen. *Antiviral Ther*7: L72

27 Hall CS, Raines CP, Barnett SH, Moore RD, Gallant JE (1999) Efficacy of salvage therapy containing ritonavir and saquinavir after failure of single protease inhibitor-containing regimens. *AIDS* 13: 1207–1212

28 Havlir DV, Hellmann NS, Petropoulos CJ, Whitcomb JM, Collier AC, Hirsch MS, Tebas P, Sommadossi JP, Richman DD (2000) Drug susceptibility in HIV infection after viral rebound in patients receiving indinavir-containing regimens. *JAMA* 283: 229–234

29 Gallego O, de Mendoza C, Perez-Elias MJ, Guardiola JM, Pedreira J, Dalmau D, Gonzalez J, Moreno A, Arribas JR, Rubio A et al. (2001) Drug resistance in patients experiencing early virological failure under a triple combination including indinavir. *AIDS* 15: 1701–1706

30 Havlir DV, Bassett R, Levitan D, Gilbert P, Tebas P, Collier AC, Hirsch MS, Ignacio C, Condra J, Gunthard HF et al. (2001) Prevalence and predictive value of intermittent viremia with combination HIV therapy. *JAMA* 286: 171–179

31 Raboud JM, Rae S, Woods R, Harris M., Montaner JSG (2002) Consecutive rebounds in plasma viral load are associated with virological failure at 52 weeks among HIV infected patients. *AIDS* 16: 1627–1632

32 Grabar S, Le Moing V, Goujard C, Leport C, Kazatchkine MD, Costagliola D, Weiss L (2000) Clinical outcome of patients with HIV-1 infection according to immunologic and virologic response after 6 months of highly active antiretroviral therapy. *Ann Intern Med* 133: 401–410
33 Deeks SG, Barbour JD, Martin JN, Swanson MS, Grant RM (2000) Sustained CD4+ T cell response after virologic failure of protease inhibitor-based regimens in patients with human immunodeficiency virus infection. *J Infect Dis* 181: 946–953
34 Wood E, Hogg RS, Yip B, Tyndall MW, Sherlock CH, Harrigan RP, O'Shaughnessy MV, Montaner JS (2002) 'Discordant' increases in CD4 cell count relative to plasma viral load among a closely followed cohort of patients initiating antiretroviral therapy. *J Acquir Immune Defic Syndr* 30: 159–166
35 Deeks SG, Barbour JD, Grant RM, Martin JN (2002) Duration and predictors of CD4 T-cell gains in patients who continue combination therapy despite detectable plasma viremia. *AIDS* 16: 201–207
36 Deeks SG, Martin JN (2001) Reassessing the goal of antiretroviral therapy in the heavily pre-treated HIV-infected patient. *AIDS* 15: 117–119
37 Montaner JSG, Harrigan PR, Jahnke N, Raboud J, Castillo E, Hogg RS, Yip B, Harris M, Montessori V, O'Shaughnessy MV (2001) Multiple drug rescue therapy for HIV-infected individuals with prior virologic failure to multiple regimens. *AIDS* 15: 61–6
38 Montaner JSG, Harris M, Harrigan R, Hogg R, Wood E (2001) A compromise strategy for patients with multiple drug failure. *AIDS* 15: 2470
39 Durant J, Clevenbergh P, Halfon P, Delgiudice P, Porsin S, Simonet P, Montagne N, Boucher CA, Schapiro JM, Dellamonica P (1999) Drug resistance genotyping in HIV-1 therapy: the VIRADAPT randomized controlled trial. *Lancet* 353: 2195–2199
40 Baxter JD, Mayers DL, Wentworth DN, Nearon JD, Hoover ML, Winters MA, Mannheimer SB, Thompson MA, Abrams DI, Brizz BJ et al. (2000) A randomized study of antiretroviral management based on plasma genotypic antiretroviral resistance testing in patients failing therapy. CPCRA 046 Study Team for the Terry Beirn Community Programs for Clinical Research on AIDS. *AIDS* 14: F83–93
41 Larder B, De Vroey V, Dehertogh P, Kemp S, Bloor S, Hertogs K (1999) Predicting HIV-1 phenotypic resistance from genotype using a large phenotype-genotype relational database. *Antiviral Ther* 4 (Suppl 1): 41
42 Cohen C, Hunt S, Sension M, Farthing C, Conant M, Jacobson S, Nadler J, Verbiest W, Hertogs K, Ames M et al. (2000) Phenotypic resistance testing significantly improves response to therapy: a randomized trial (VIRA3001). In: *Program and Abstracts of the 7th Conference on Retroviruses and Opportunistic* Infections, San Francisco, CA, USA, 30 January–2 February 2000, Abstracts, 120
43 Melnick D, Rosenthal J, Cameron M, Snyder M, Griffith-Howard S, Hertogs K, Verbiest W, Graham N, Pham S (2000) Impact of phenotypic antiretroviral drug resistance testing on the response to salvage antiretroviral therapy in heavily experienced patients. In: *Program and Abstracts of the 7th Conference on Retroviruses and Opportunistic Infections*, San Francisco, CA, USA, 30 January– February 2000, Abstracts, 222
44 Meynard JL, Vray M, Morand-Joubert L, Race E, Descamps D, Peytavin G, Matheron S, Lamotte C, Guiramand S, Costagliola D et al. (2002) Phenotypic or genotypic resistance testing for choosing antiretroviral therapy after treatment failure: a randomized trial. *AIDS* 16: 727–736
45 Tesiorowski A, Harris M, Wood R, Singer J, O'Shaughnessy M, Harrigan R, Montaner J (2001) A comparative trial of virtual phenotypes (Virco) *versus* virtual phenotypes plus phenotype resistance testing in salvage therapy [abstract 580]. In: *Final Programme and Abstracts of the 1st International AIDS Society Conference on HIV Pathogenesis and Treatment*, Buenos Aires, Argentina, 8–11 July 2001, Abstracts, 252
46 Durant J, Clevenbergh P, Garraffo R, Halfon P, Icard S, Del Giudice P, Montagne N, Schapiro JM, Dellamonica P (2000) Importance of protease inhibitor plasma levels in HIV-infected patients treated with genotypic-guided therapy: pharmacological data from the Viradapt Study. *AIDS* 14: 1333–1339
47 Burger D, Hugen P, Droste J, Huitema A (2001) Therapeutic drug monitoring of nelfinavir and indinavir in treatment-naïve patients improves therapeutic outcome after 1 year: results from Athena. In: *Final Programme and Abstracts of the 1st International AIDS Society Conference on HIV Pathogenesis and Treatment*, Buenos Aires, Argentina, 8–11 July 2001, Abstracts, 88
48 Fletcher CV, Anderson PL, Kakuda TN, Schacker TW, Henry K, Gross CR, Brundage RC (2002) Concentration-controlled compared with conventional antiretroviral therapy for HIV infection.

AIDS 16: 551–560
49 Campo RE, Suarez G, Miller N, Moreno J, Kolber M, Holder DJ, Shivaprakash M, DeAngelis DM, Wright JL, Holmes K et al. (2000) Efficacy of indinavir/ritonavir-based regimens among patients with prior protease inhibitor failure. *Antiviral Ther.* 5 (suppl 2): 6
50 Grossman H, Luber A, Purdom D, Butcher D, Frechette G, Duong P, Markson L (2000) Salvage therapy with an indinavir 800 mg plus ritonavir 200 mg-based regimen in clinical practice. *Antiviral Ther* 5 (Suppl 2): 23
51 Wilson H, DeJesus E (2001) A multicenter, open-label, 24-week study to evaluate the efficacy and safety of indinavir sulfate 800 mg and ritonavir 200 mg b.i.d. plus 2 NRTIs b.i.d. in HIV-infected individuals who require early treatment intervention. In: *Final Programme and Abstracts of the 1st International AIDS Society Conference on HIV Pathogenesis and Treatment*, Buenos Aires, Argentina, 8–11 July 2001, Abstracts, 141
52 Low-Beer S, Yip B, O'Shaughnessy MV, Hogg RS, Montaner JSG (2000) Adherence to triple therapy and viral load response. *J Acquir Immune Defic Syndr* 23: 360–361
53 Kinter A, Fauci AS (1996) Interleukin-2 and human immunodeficiency virus infection: pathogenic mechanisms and potential for immunologic enhancement. *Immunol Res* 15: 1–15
54 Ghezzi S, Vicenzi E, Soldini L, Tambussi G, Murone M, Lazzarin A, Poli G (1997) Experiences in immune reconstitution. The rationale for interleukin-2 administration to HIV-infected individuals. *J Biol Regul Homeostc Agents* 11: 74–78
55 Nieto L, Angel J, Gazzard B, Cahn P, Ward D, Ramirez-Ronda C, Taglietti M, Greaves W (2001) Antiviral activity and tolerability of PEG-Intron in HIV patients failing HAART. In: *Final Programme and Abstracts of the 1st International AIDS Society Conference on HIV Pathogenesis and Treatment*, Buenos Aires, Argentina, 8–11 July 2001, Abstracts, 118
56 Miller V, Sabin C, Hertogs K, Bloor S, Martinez-Picado J, D'Aquila R, Larder B, Lutz T, Gute P, Weidmann E et al. (2000) Virological and immunological effects of treatment interruptions in HIV-1 infected patients with treatment failure. *AIDS* 14: 2857–2867
57 Delaugerre C, Valantin M-A, Mouroux M, Bonmarchand M, Carcelain G, Duvivier C, Tubiana R, Simon A, Bricaire F, Agut H et al. (2001) Re-occurrence of HIV-1 drug mutations after treatment re-initiation following interruption in patients with multiple treatment failure. *AIDS* 15: 2189–2191
58 Deeks SG, Wrin T, Liegler T, Hoh R, Hayden M, Barbour JD, Hellmann NS, Petropoulos CJ, McCune JM, Hellerstein MK et al. (2001) Virologic and immunologic consequences of discontinuing combination antiretroviral-drug therapy in HIV-infected patients with detectable viremia. *N Engl J Med* 344: 472–480
59 Katlama C (2002) Treatment interruption in salvage therapy. In: *Final Program of the XIV International AIDS Conference*, Barcelona, Spain, 7–12 July 2002, Abstract 208
60 Clotet B (2002) Treatment interruption in salvage therapy. In: *Final Program of the XIV International AIDS Conference*, Barcelona, Spain, 7–12 July 2002, Abstracts,208
61 Lori F, Malykh A, Cara A, Sun D, Weinstein JN, Lisziewicz J, Gallo RC (1994) Hydroxyurea as an inhibitor of human immunodeficiency virus type 1 replication. *Science* 266: 801–805
62 Margolis D, Heredia A, Gaywee J, Oldach D, Drusano G, Redfield R (1999) Abacavir and mycophenolic acid, an inhibitor of inosine monophosphate dehydrogenase, have profound and synergistic anti-HIV activity. *J Acquir Immune Defic Syndr* 21: 362–370
63 Lori F, Malykh AG, Foli A, Maserati R, De Antoni A, Minoli L, Padrini D, Degli Antoni A, Barchi E, Jessen H et al. (1997) Combination of a drug targeting the cell with a drug targeting the virus controls of human immunodeficiency virus type 1 resistance. *AIDS Res Hum Retroviruses* 13: 1403–1409
64 Montaner JSG, Zala C, Conway B, Raboud J, Patenaude P, Rae S, O'Shaughnessy MV, Schechter MT (1997) A pilot study of hydroxyurea among patients with advanced human immunodeficiency virus (HIV) disease receiving chronic didanosine therapy: Canadian HIV trials network protocol 080. *J Infect Dis* 175: 801–806
65 Havlir DV, Gilbert PB, Bennett K, Collier AC, Hirsch MS, Tebas P, Adams EM, Wheat LJ, Goodwin D, Schnittman S et al. (2001) Effects of treatment intensification with hydroxyurea in HIV-infected patients with virologic suppression. *AIDS* 15: 1379–1388
66 Moore RD, Keruly KC, Chaisson RE (2001) Incidence of pancreatitis in HIV-infected patients receiving nucleoside reverse transcriptase inhibitor drugs. *AIDS* 15: 617–620
67 Murphy RL, Katlama C, Belsey E, Berzins B, Montaner JSG, Pollard R, Hellinger J. (2001) Antiviral activity and CD4 count responses in patients treated with efavirenz, stavudine and didanosine plus hydroxyurea or placebo: 48 week final results from the 3D study. In: *Final*

Programme and Abstracts of the 1st International AIDS Society Conference on HIV Pathogenesis and Treatment, Buenos Aires, Argentina, 8–11 July 2001, Abstracts, 323

68 Squires K, Pierone G, Berger D, Steinhart C, Bellos N, Becker SL, Chen SS, Miller MD, Coakley DF, Cheng A. (2002) Tenofovir DF: A 48-week final analysis from a Phase III randomized, double blind placebo controlled study in antiretroviral experienced patients. In: *Program and Abstracts of the 9th Conference on Retroviruses and Opportunistic Infections*, Seattle, Washington, USA, 24–28 February 2002, Abstracts, 209

69 Clotet B, Lazzarin A, Cooper D, Reynes J, Arasteh K, Nelson M, Katlama C, Chung J, Fang L, Delehanty J et al. (2002) Enfuvirtide (T-20) in combination with an optimized background (OB) regimen *versus* OB alone in patients with prior experience or resistance to each of the three classes of approved antiretrovirals (ARVs) in Europe and Australia. In: *Program Supplement of the XIV International AIDS Conference*, Barcelona, Spain, 7–12 July 2002, Abstracts, 26

70 Henry K, Lalezari J, O'Hearn M, Trottier B, Montaner J, Piliero P, Walmsley S, Chung J, Fang L, Delehanty J et al. (2002) Enfuvirtide (T-20) in combination with an optimized background (OB) regimen *versus* OB alone in patients with prior experience or resistance to each of the three classes of approved antiretrovirals (ARVs) in North America and Brazil. In: *Program Supplement of the XIV International AIDS Conference*, Barcelona, Spain, 7–12 July 2002, Abstracts, 26

71 Montaner JSG, Mellors JW (2001) Antiretroviral therapy for previously treated patients. *N Engl J Med* 345: 452–455

Combination Therapy of AIDS
Edited by E.D.A. De Clercq and A.-M.I. Vandamme

Structured therapy interruptions (STIs): lessons from a therapeutic strategy

Felipe García, Joan Joseph and José M. Gatell

Clinic Institute of Infectious Diseases and Immunology. IDIBAPS (Institut d'Investigacions Biomèdiques August Pi I Sunyer), Hospital Clínic, Faculty of Medicine, University of Barcelona, Villarroel, 170, 08036 Barcelona, Spain

Introduction

Since the anecdotal case reports from the Franco Lori and Douglas Nixon groups [1, 2] the concept of interrupting therapy as a therapeutic strategy in HIV infection has gained interest both in research and in clinical practice [3]. Structured therapy interruption (STI) could be considered as the simplest method of active immuno-mediated therapy [3]. This strategy may be interpreted as an auto-vaccination with an autologous-attenuated virus, the attenuation being obtained with the reintroduction of HAART. An animal model has been described with a good virological outcome and strong helper and CTL anti-HIV specific responses [4].

STI has been studied in both primary infection (PHI) and chronically infected (CHI) patients. Bruce Walker's group have reported that in eight PHI patients who started HAART very soon (only a few days after primary infection) VL did not increase above 5,000 copies/ml after 6 months of interruption of HAART after performing from one to three cycles of STI. These patients had strong HIV-1-specific neutralizing antibody, T-cell helper and CTL responses [5, 6]. Conversely, different pilot studies in CHI patients provided discordant results [7–10]. In summary, it seems that helper and CTL HIV-1 specific responses are weaker in CHI than in PHI patients and only a minority of patients are able to re-set the viral load set-point after definitive interruption of therapy. Furthermore, it is no clear whether there is a relationship between the change in viral load set-point and the increase in helper and CTL HIV-1 specific responses. Lack of applicable results to daily clinical practice does not mean that these studies have not been important. In fact, structured therapy interruption has been useful to open new lines of investigation and to clarify some questions about HIV-1 itself and the immune response to HIV infection.

Although not covered in this article, STI has also been used in advanced failing patients in order to avoid toxicity and to re-sensitize the virus and in CHI responding patients in order to minimize exposure to antiretroviral agents without compromising long-term efficacy.

Lessons from a therapeutic strategy: the virus

A rebound in plasma viral load has been detected almost invariably in all studies of STI. The reported doubling time has been about 2 days [11]. This doubling time is almost identical to the half-life, described by Ho et al. [14], for the decay of viral load in more advanced patients who initiated ritonavir monotherapy. This rapid rebound (within 4 weeks) (Fig. 1) was similar to what has been observed after discontinuation of antiretroviral therapy after a short period of treatment (1 month) [15]. This further supports the hypothesis that many additional months of antiviral therapy and a marked suppression of HIV-1 RNA viremia does not seem to reduce the pool of latently infected cells as measured by integrated DNA [16, 17]. Based on the rapid viral load rebound, it is more likely that viral load increases are due to the presence of a low-level persistent virus replication, rather than a random reactivation from a pool of latently infected cells [18].

In some patients, it has also been observed that, after stopping therapy, the virus may rebound to levels above the viral load set-point prior to initiation of HAART (Fig. 1). The explanations for this finding are quite contradictory. Some authors have suggested that this may be explained by a higher viral replication due to an increase in CD4 lymphocytes associated with treatment (the named predator-prey hypothesis) [19]. In fact, this idea could be applicable also to primary infection situation. The decay of viral load after rebounding and reaching the peak is explained by the decrease of available CD4+

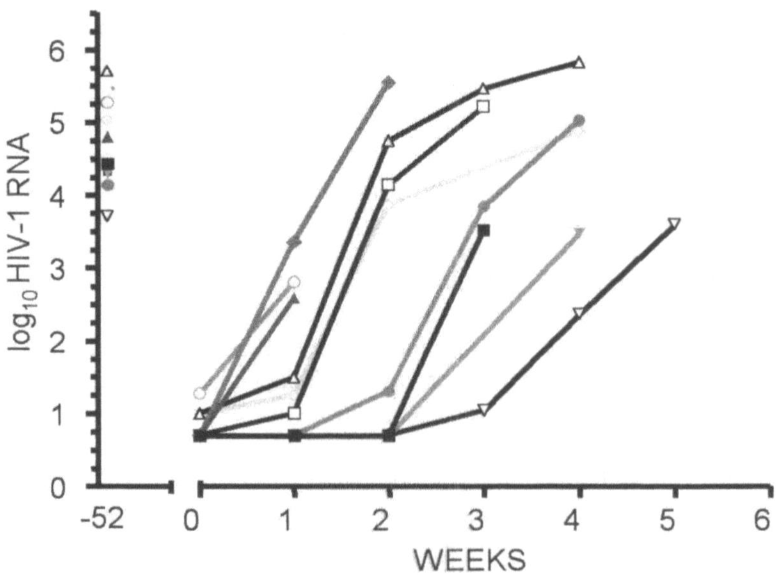

Figure 1. Rapid rebound of viral load after interruption of therapy. Some patients show viral load rebound to values higher than baseline before antiretroviral therapy.

T-cells due to the viral replication. This hypothesis can not explain why the overshoot of viremia levels it is not observed in all patients who discontinued HAART with a significant increase in the level of CD4 lymphocytes. Therefore, some other factors, such as the specific immune response, the viral fitness, the expression of entry co-receptors, altogether with the larger pool of proliferating CD4 cells, should play a role in the dynamics of viral rebound to explain the different dynamics observed in the patients.

On the other hand, Ruiz et al. and our group [8, 10, 20] have demonstrated a change in the dynamics of viral load rebound during the successive cycles of therapy interruptions (Fig. 2). A delay in the viral load rebound has been observed, with lower viral load peaks after the second or third interruption as compared with the first interruption. This could also be explained by the predator-prey hypothesis, if CD4+ T-cells levels were not similar before each interruption of therapy. The re-emergence of the virus from different reservoirs would be another alternative explanation, since different virus from different reservoirs would have different fitness. However, in this case, the dynamics of viral load rebound would be chaotic without a tendency to a delay in viral load rebound along the successive interruptions of HAART. These changes in the dynamics of viral rebound deserve further investigation, addressing the changes of the viral sequences of the *env* gene and addressing their relationship with viral fitness and specific HIV-1 immune-responses.

Finally, from the point of view of safety (Tab. 1), one of the criticisms to STI is the replenishment of viral reservoirs during the rebound of viremia. In fact, this phenomenon has been already reported [21]. However, the slow decay rate

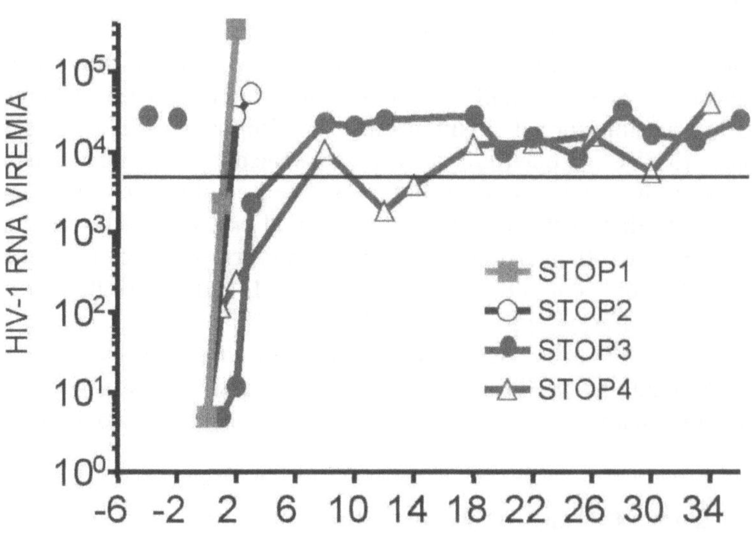

Figure 2. Changes in the dynamics of viral load rebound after interruption of therapy.

Table 1. Doubts and problems of STI

Doubts
• Number of cycles
• Duration of the cycles
• Time, viral load or CD4+ T-cells driven cycles
• End-points
• Kind of immunological response that is induced
Problems
• Selection of resistant mutations
• Drop in CD4+ T-cells
• Reservoir replenishment

of the latent reservoir would require a clinical challenge for 10 to 60 years of treatment in order to eradicate HIV-1 infection. Since the objective of eradication seems unrealistic, some other strategies should be tried. In any case the replenishment has to be demonstrated. In patients having had HAART + MMF, who discontinued therapy, a replenishment of reservoir could not been demonstrated. Recently Dybul et al. have studied 10 patients performing STI during a year with a 7 days on/7 days off therapy schedule and did not observe replenishment of the reservoirs [22]. Selection of resistance may be another potential weakness. Our group and others have not detected evidence for phenotypic or genotypic resistance after several cycles of STI [10, 22]. Moreover, in our study viral load dropped below 20 copies/ml in all patients 1 month after reintroduction of the same antiretroviral regimen and remained below 20 copies/ml after 6 months of treatment. Caution should be taken with those drugs with long intra-cellular or plasmatic half-life, in order to avoid exposure to a single drug if all drugs are interrupted simultaneously [8]. Reported cases of selection of resistance to 3TC or efavirenz may be explained by this mechanism [23]. Furthermore, the significant drop in CD4+ T-cell counts both during the first, second and further cycles of interruption should preclude such trials in more advanced patients. If intermittent therapy is performed in advanced patients, CD4+ T-cell counts could drop below 200×10^6/L and expose the patient to a higher risk of suffering opportunistic infections.

In summary, STI has raised a number of important issues about the dynamics of viral load, the evolution of the virus, the possibility of selection resistance mutants that would permit to learn new approaches to attack the virus.

Lessons from a therapeutic strategy: the immunity

If viral load rebounds stem from hidden reservoirs (in latently infected cells) or from ongoing replication is currently being actively investigated [16, 24,

25]. This rebound could be due, at least in part, to the lack of aspecific immune response against HIV-1 antigens in patients treated with HAART alone [26]. In addition, both CD4+ T-cell and CTL response decline after 6 months of effective HAART [27, 28].

There is growing evidence supporting the effectiveness of specific CD4+ T-cell and CTL response against HIV-1 antigens [28–34]. The more compelling data come from recent studies with SHIV- and SIV-infected rhesus macaques that support the contention that CD8 T-cells, by mechanism not yet elucidated, are required for control of HIV replication. With this model, the *in vivo* depletion of CD8 cells by monoclonal antibodies was transiently associated with significantly higher levels of viral replication [29, 31, 32]. With the recovery of CD8 cells, the SIV levels diminished [29, 32]. In addition, there are data that suggest that strong HIV-1 specific helper responses are necessary to maintain a certain level of CTL memory in the absence of HIV-1 antigens [5, 35]. Furthermore, stronger immune responses correlate with a lesser aggressive disease course. Although in most patients HIV eventually destroys the immune system, the rare long-term non-progressors (LTNP) tolerate HIV infection for many years without apparent ill effects. Their CD4 counts remain in the normal range and their plasma viral load is very low or undetectable. Most LTNPs show a strong and persisting cytotoxic lymphocyte (CTL) response towards HIV [28, 30, 36–39]. Such a response is transient in patients who are not LTNPs, allowing continued proliferation of HIV resulting in immunosuppression [40].

It has been hypothesized that cyclic interruptions of antiretroviral therapy may boost the HIV-1 specific immune responses and could be considered as an immune-based therapy for HIV-1 infection [1, 4, 41–43]. It seems that STI could reset viral load in approximately 20% of CHI patients, ranging from 10 to 40% [9, 10]. It is not yet clear if this resetting is associated with an increase in the HIV-1 specific cellular immune responses [7, 8, 10]. It seems that STI in chronically infected patients could increase the HIV-specific CD8 T-cells moderately (Figs 3 and 4). In addition, HIV-specific CD4 T-cell proliferative responses observed has been no vigorous and sustained [8, 10] (Figs 3 and 4). Some authors have suggested that the re-exposure to autologous virus induces expansion of only pre-existing dominant memory CD8 T-cells directed to HIV epitopes already recognized at baseline [8].

On the other hand, these immune-specific responses seem to be weaker in CHI patients than those obtained when STI is performed during very early stages of primary infection [5]. The pattern of viral replication after therapy interruption in these subjects seems to differ from primary HIV-1 infection (PHI) patients who stop drugs [1, 2]. Some PHI subjects lack VL rebound above 50 copies/ml after drug cessation. Assuming that HIV-1 specific immune responses are predominantly responsible for containment of viral replication, the differences between VL rebound in PHI or chronic HIV infection (CHI) subjects after STI may be explained by the degree of immunocompetence, viral sequence diversity, and the different patterns of HIV-1-specific immunity after the initiation of HAART. CHI subjects are considerably more immunosu-

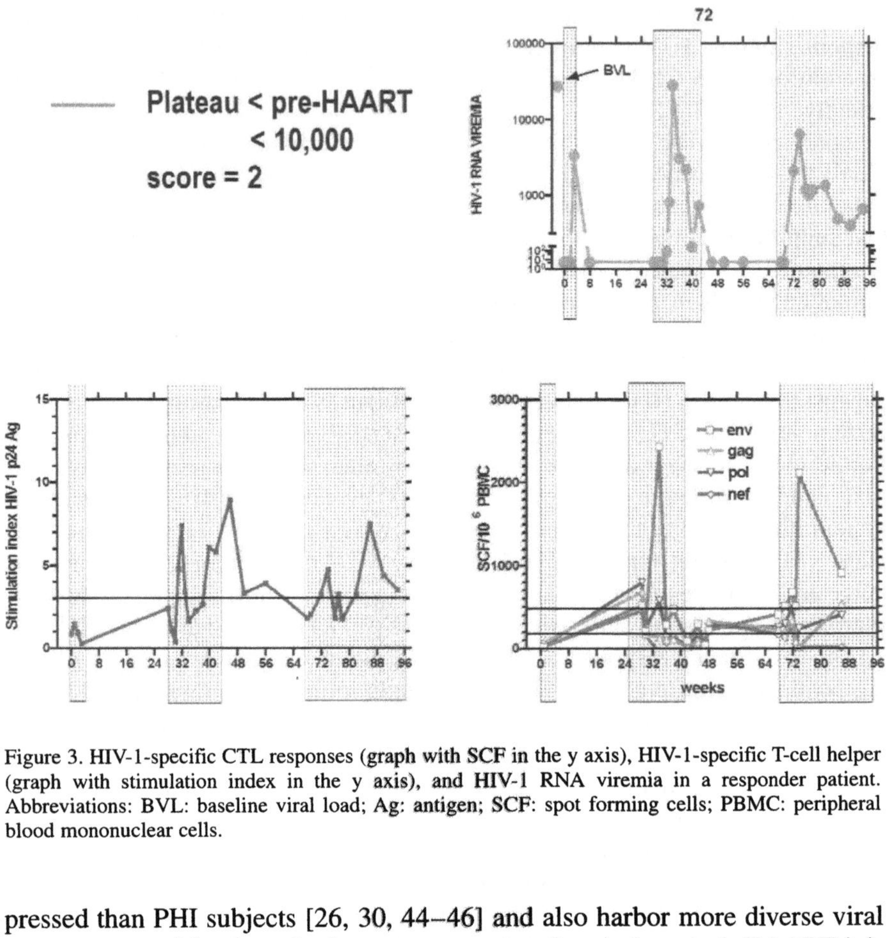

Figure 3. HIV-1-specific CTL responses (graph with SCF in the y axis), HIV-1-specific T-cell helper (graph with stimulation index in the y axis), and HIV-1 RNA viremia in a responder patient. Abbreviations: BVL: baseline viral load; Ag: antigen; SCF: spot forming cells; PBMC: peripheral blood mononuclear cells.

pressed than PHI subjects [26, 30, 44–46] and also harbor more diverse viral sequences requiring broader immune responses for viral control. HAART initiated during PHI may better preserve HIV-1-specific T-helper cell functions and CD8+ T-cell responses [30, 44, 47], which could maintain suppression of plasma viremia below 50 copies/ml after STI in PHI subjects [1, 2, 35].

Lessons from a therapeutic strategy: new drugs

Alternative immune-based interventions include antiretroviral drugs in combination with compounds that are able to modulate the function and reactivity of the immune system. The goal of these strategies is to achieve the long-term control of virus replication with a minimal or possibly no use of antiretroviral drugs. In this context, it is worth exploring the potential use of immune modulating drugs, such as mycophenolic acid

Hydroxyurea (HU) has been used for years in HIV-1 infection due to its antiviral and cytostatic effects [48, 49]. The antiviral effect is mediated by a

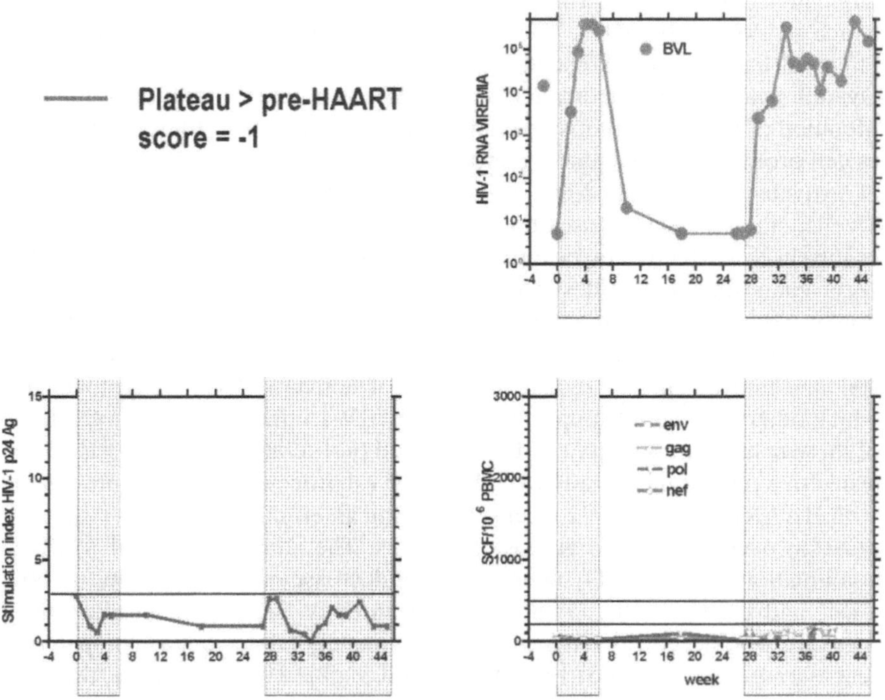

Figure 4. HIV-1-specific CTL responses (graph with SCF in the y axis), HIV-1-specific T-cell helper (graph with stimulation index in the y axis), and HIV-1 RNA viremia in a non-responder patient. Abbreviations: BVL: baseline viral load; Ag: antigen; SCF: spot forming cells; PBMC: peripheral blood mononuclear cells.

reduction of the levels of the deoxynucleotide triphosphates (dNTPs) that are needed for the HIV reverse transcription [50, 51]. Regarding immuno-modulating effect, the main cellular consequence of inhibiting the formation of dNTPs is a cytostatic effect [52]. The impact of this cytostatic action on the immune system cells remains unknown [53]. Since stopping HAART induces a sudden immune activation [12] and can boost HIV replication explaining the high peaks of viral replication [12], associating hydroxyurea to HAART and maintaining hydroxyurea after stopping HAART would help T-lymphocytes to remain quiescent, becoming refractory to productive HIV-1 infection and avoiding the high peaks of viral replication without blunting the HIV-1 specific immune responses. This possibility is being actively investigated.

Mycophenolate mofetil (MMF) is a well-known drug which is broadly used in renal transplantation for its ability to selectively inhibit lymphocyte division [54]. MMF is the morpholinoethyl ester of the active compound mycophenolic acid (MPA). Interest in MPA as an immunosuppressive agent in renal transplantation has been renewed by clinical studies showing that the proliferation of T- and B-lymphocytes is selectively blocked by MPA as their purine mononucleotide supply primarily relies on the *de novo* synthesis pathway which is blocked by MPA. The effect of MPA on cell activation and HIV-1 infection has been investigated in both *in vitro* and *in vivo* animal studies. Overall, available data indicate that: (a) MPA is a potent inhibitor of the proliferation of activated T-cells, even in the presence of IL-2; (b) MPA suppresses cell proliferation by inducing apoptosis and cell death of a large proportion of activated T-cells; (c) MPA is able to suppress HIV infection in CD4+ T-cells, since it likely causes the depletion of an important substrate for the occurrence of effective reverse transcription by impairing the *de novo* synthesis of guanosine nucleotides [55]. Therefore, MPA might inhibit HIV-1 infection by a dual mechanism, including an antiviral mechanism exerted by depleting the substrate (i.e., dGTP) for the reverse transcriptase and an immunological mechanism able to reduce the pool of activated $CD4^+$ T-lymphocytes which otherwise would support productive HIV-1 infection [56]. On the basis of these results, the activity of MPA on T-cell proliferation and HIV-1 has been tested in a pilot clinical study [57]. Eight HIV-1-infected subjects were treated with HAART for at least 72 weeks, achieving an efficient sustained virus suppression over a period of at least 36 weeks. The administration of MMF in combination with HAART induced a rapid decrease of the production of both $CD4^+$ and $CD8^+$ T-cells. Of note, the mean daily CD4 T-cell production in these eight subjects was reduced by approximately 60 percent as compared to that assessed in a cohort of 15 HIV-1 negative subjects. Furthermore, the virological effect induced by MMF treatment was evaluated in seven of the eight patients enrolled, assessing the ability to isolate virus from purified $CD4^+$ T-cell populations before and either 8 or 20 weeks after the addition of MMF to HAART. Of note, a remarkable reduction that reached 98 percent, of the titers of infectious units per million of $CD4^+$ T-cells (IUPM) was observed in five of the seven patients tested, while in one patient $CD4^+$ T-cell IUPM were below the limit of detection of the assay (<0.08 IUPM) both before and after the introduction of MMF. On the basis of these results, the question is whether MMF exerts an effect on the pool of resting latently infected $CD4^+$ T-cells. It is likely that MMF has no effect on resting cells, and thus does not affect directly the size of this pool of cells. However, once these cells are activated in the presence of MMF, MMF might in turn induce apoptosis and therefore cell death. Overall, these results represent the first piece of evidence that an immune modulation strategy may interfere with the HIV-1 lifecycle acting on the HIV-1 targeT-cell rather than interfering with the viral enzymes critical for virus replication [57]. On the other hand, the effects of associating MMF + HAART during STI and deferring MMF until after interruption of HAART on viral replication still need to be fully explored. This strategy might delay the viral load rebound both in plasma and tonsillar tissue after stopping HAART and improve the control of viral replication without HAART. This effect does not seem mediated either by a decrease in cell turn-over or an increase in apoptosis [58].

Lessons from a therapeutic strategy: HAART

New objectives and strategies can be tested by STI (Tab. 2). In a substantial minority of patients presently on HAART, especially those with low viral load when treatment was first initiated, STI can be safely managed without drugs for at least several months, potentially diminishing side-effects and costs of therapy. This has been the strategy proposed by Dybul et al. [22] as a way of avoiding medication. They initiated a protocol in which 10 CHI patients with CD4+ T-cell counts above 300 and plasma VL < 50 copies/ml received repeated cycles of 7 days off and 7 days on HAART. They did not find a significant increase in plasma VL, proviral DNA or lymph node viral load. CD4+ T-cells did not change significantly. There were no changes in helper responses, activation markers or CTL responses. Some of these patients developed resistance to efavirenz. Metabolic parameters such as cholesterol and triglyceride levels improved significantly. The authors concluded that STI could be an important strategy in the reduction of therapy costs and toxicity for HIV-infected patients.

Table 2. Future strategies for STI

Objectives
• Decrease of side-effects
• Saving medication
Improving the outcome
• Associate immunomodulators
• Associate cytokines
• Associate therapeutic vaccines

Acknowledgments
This study was supported in part by grants: FIPSE 3118/00*, SAF 98/0021, FIS 01/1595, SAF 01/2591.
* FIPSE is a non-profit Foundation including: Spanish Ministry of Health, Abbott Laboratories, Boehringer Ingelheim, Bristol Myers Squibb, GlaxoSmithKline, Merck Sharp and Dohme and Roche)

References

1 Lisziewicz J, Rosenberg E, Lieberman J, Jessen H, Lopalco L, Siliciano R, Walker B, Lori F (1999) Control of HIV despite the discontinuation of antiretroviral therapy [letter]. *N Engl J Med* 340: 1683–1684
2 Ortiz GM, Nixon DF, Trkola A, Binley J, Jin X, Bonhoeffer S, Kuebler PJ, Donahoe SM, Demoitie M-A, Kakimoto WM et al. (1999) HIV-1-specific immune responses in subjects who temporarily contain virus replication after discontinuation of highly active antiretroviral therapy. *J Clin Invest* 104: R13–R18

3 Montaner LJ (2001) Structured treatment interruptions to control HIV-1 and limit drug exposure. *Trends Immunol* 22: 92–96
4 Lori F, Lewis MG, Xu J, Varga G, Zinn DE Jr, Crabbs C, Wagner W, Greenhouse J, Silvera P, Yalley-Ogunro J et al. (2000) Control of SIV rebound through structured treatment interruptions during early infection. *Science* 290: 1591–1593
5 Rosenberg ES, Altfeld M, Poon SH, Phillips MN, Wilkes BM, Eldridge RL, Robbins GK, D'Aquila RT, Goulder PJ, Walker BD (2000) Immune control of HIV-1 after early treatment of acute infection. *Nature* 407: 523–526
6 Montefiori DC, Hill TS, Vo HT, Walker BD, Rosenberg ES (2001) Neutralizing antibodies associated with viremia control in a subset of individuals after treatment of acute human immunodeficiency virus type 1 infection. *J Virol* 75: 10200–10207
7 Carcelain G, Tubiana R, Samri A, Calvez V, Delaugerre C, Agut H, Katlama C, Autran B (2001) Transient mobilization of human immunodeficiency virus (HIV)-specific CD4 T-helper cells fails to control virus rebounds during intermittent antiretroviral therapy in chronic HIV type 1 infection. *J Virol* 75: 234–241
8 Ruiz L, Carcelain G, Martinez-Picado J, Frost S, Marfil S, Paredes R, Romeu J, Ferrer E, Morales-Lopetegi K, Autran B et al. (2001) HIV dynamics and T-cell immunity after three structured treatment interruptions in chronic HIV-1 infection. *AIDS* 15: F19–F27
9 Fagard C, Lebraz M, Gunthard H., Tortajada C, García F, Battegay, HJ Furrer, Vernazza P, Bernasconi E, Ruiz L et al. (2001) SSITT: A prospective trial of strategic treatment interruptions in 128 patients. 8th Conference on Retroviruses and Opportunistic Infections. Chicago, IL, USA, 4–8 February 2001, Abstract 357
10 Garcia F, Plana M, Ortiz GM, Bonhoeffer S, Soriano A, Vidal C, Cruceta A, Arnedo M, Gil C, Pantaleo G et al. (2001) The virological and immunological consequences of structured treatment interruptions in chronic HIV-1 infection. *AIDS* 15: F29–F40
11 Harrigan PR, Whaley M, Montaner JS (1999) Rate of HIV-1 RNA rebound upon stopping antiretroviral therapy. *AIDS* 13: F59–F62
12 Garcia F, Plana M, Vidal C, Cruceta A, O'Brien WA, Pantaleo G, Pumarola T, Gallart T, Miro JM, Gatell JM (1999) Dynamics of viral load rebound and immunological changes after stopping effective antiretroviral therapy. *AIDS* 13: F79–F86
13 Ruiz L, Martinez-Picado J, Romeu J, Paredes R, Zayat MK, Marfil S, Negredo E, Sirera G, Tural C, Clotet B (2000) Structured treatment interruption in chronically HIV-1 infected patients after long-term viral suppression. *AIDS* 14: 397–403
14 Ho DD, Neumann AU, Perelson AS, Chen W, Leonard JM, Markowitz M (1995) Rapid turnover of plasma virions and CD4 lymphocytes in HIV-1 infection. *Nature* 373: 123–126
15 Neumann AU, Tubiana R, Calvez V, Robert C, Li TS, Agut H, Autran B, Katlama C (1999) HIV-1 rebound during interruption of highly active antiretroviral therapy has no deleterious effect on reinitiated treatment. Comet Study Group. *AIDS* 13: 677–683
16 Chun TW, Carruth L, Finzi D, Shen X, DiGiuseppe JA, Taylor H, Hermankova M, Chadwick K, Marglick J, Quinn TC et al. (1997) Quantification of latent tissue reservoirs and total body viral load in HIV-1 infection [see comments]. *Nature* 387: 183–188
17 Finzi D, Hermankova M, Pierson T, Carruth LM, Buck C, Chaisson RE, Quinn TC, Chadwick K, Margolick J, Brookmeyer R et al. (1997) Identification of a reservoir for HIV-1 in patients on highly active antiretroviral therapy. *Science* 278: 1295–1300
18 Pantaleo G (1997) How immune-based interventions can change HIV therapy. *Nat Med* 3: 483–486
19 de Jong MD, de Boer RJ, de Wolf F, Foudraine NA, Boucher CA, Goudsmit J, Lange JM (1997) Overshoot of HIV-1 viraemia after early discontinuation of antiretroviral treatment. *AIDS* 11: F79–F84
20 Frost SD, Martinez-Picado J, Ruiz L, Clotet B, Brown AJ (2002) Viral dynamics during structured treatment interruptions of chronic human immunodeficiency virus type 1 infection. *J Virol* 76: 968–979
21 Ramratnam B, Mittler JE, Zhang L, Boden D, Hurley A, Fang F, Macken CA, Perelson AS, Markowitz M, Ho DD (2000) The decay of the latent reservoir of replication-competent HIV-1 is inversely correlated with the extent of residual viral replication during prolonged anti-retroviral therapy. *Nat Med* 6: 82–85
22 Dybul M, Chun TW, Yoder C, Hidalgo B, Belson M, Hertogs K, Larder B, Dewar RL, Fox CH, Hallahan CW et al. (2001) Short-cycle structured intermittent treatment of chronic HIV infection

with highly active antiretroviral therapy: effects on virologic, immunologic, and toxicity parameters. *Proc Natl Acad Sci USA* 98: 15161–15166

23 Martinez-Picado J, Morales-Lopetegi K, Wrin T, Prado JG, Frost SD, Petropoulos CJ, Clotet B, Ruiz L (2002) Selection of drug-resistant HIV-1 mutants in response to repeated structured treatment interruptions. *AIDS* 16: 895–899

24 Chun TW, Davey RT Jr, Engel D, Lane HC, Fauci AS (1999) Re-emergence of HIV after stopping therapy. *Nature* 401: 874–875

25 Chun TW, Davey R, Ostrowski MA, Engel D, Mullins JI, Lane C, Fauci, A (2000) Relationship between preexisting latent viral reservoirs and the reemergence of plasma viraemia following discontinuation of highly active antiretroviral therapy. 7th Conference on Retroviruses and Opportunistic Infections. San Francisco, CA, USA, 30 January – 2 February 2000, Abstracts

26 Plana M, Garcia F, Gallart T, Miro JM, Gatell JM (1998) Lack of T-cell proliferative response to HIV-1 antigens after 1 year of highly active antiretroviral treatment in early HIV-1 disease. Immunology Study Group of Spanish EARTH-1 Study. *Lancet* 352: 1194–1195

27 Pitcher CJ, Quittner C, Peterson DM, Connors M, Koup RA, Maino VC, Picker LJ (1999) HIV-1-specific CD4+ T cells are detectable in most individuals with active HIV-1 infection, but decline with prolonged viral suppression. *Nat Med* 5: 518–525

28 Ogg GS, Jin X, Bonhoeffer S, Dunbar PR, Nowak MA, Monard S, Segal JP, Cao Y, Rowland-Jones SL, Cerundolo V et al. (1998) Quantitation of HIV-1-specific cytotoxic T lymphocytes and plasma load of viral RNA. *Science* 279: 2103–2106

29 Schmitz JE, Kuroda MJ, Santra S, Sasseville VG, Simon MA, Lifton MA, Racz P, Tenner-Racz K, Dalesandro M, Scallon BJ et al. (1999) Control of viremia in simian immunodeficiency virus infection by CD8+ lymphocytes. *Science* 283: 857–860

30 Rosenberg ES, Billingsley JM, Caliendo AM, Boswell SL, Sax PE, Kalams SA, Walker BD (1997) Vigorous HIV-1-specific CD4+ T cell responses associated with control of viremia. *Science* 278: 1447–1450

31 Matano T, Shibata R, Siemon C, Connors M, Lane HC, Martin MA (1998) Administration of an anti-CD8 monoclonal antibody interferes with the clearance of chimeric simian/human immunodeficiency virus during primary infections of rhesus macaques. *J Virol* 72: 164–169

32 Jin X, Bauer DE, Tuttleton SE, Lewin S, Gettie A, Blanchard J, Irwin CE, Safrit JT, Mittler J, Weinberger L et al. (1999) Dramatic rise in plasma viremia after CD8(+) T cell depletion in simian immunodeficiency virus-infected macaques. *J Exp Med* 189: 991–998

33 Kalams SA, Walker BD (1998) The critical need for CD4 help in maintaining effective cytotoxic T lymphocyte responses. *J Exp Med* 188: 2199–2204

34 Brodie SJ, Lewinsohn DA, Patterson BK, Jiyamapa D, Krieger J, Corey L, Greenberg PD, Riddell SR (1999) *In vivo* migration and function of transferred HIV-1-specific cytotoxic T cells. *Nat Med* 5: 34–41

35 Wodarz D, May RM, Nowak MA (2000) The role of antigen-independent persistence of memory cytotoxic T lymphocytes. *Int Immunol* 12: 467–477

36 Dyer WB, Geczy AF, Kent SJ, McIntyre LB, Blasdall SA, Learmont JC, Sullivan JS (1997) Lymphoproliferative immune function in the Sydney Blood Bank Cohort, infected with natural nef/long terminal repeat mutants, and in other long-term survivors of transfusion-acquired HIV-1 infection. *AIDS* 11: 1565–1574

37 Greenough TC, Brettler DB, Somasundaran M, Panicali DL, Sullivan JL (1997) Human immunodeficiency virus type 1-specific cytotoxic T lymphocytes (CTL), virus load, and CD4 T cell loss: evidence supporting a protective role for CTL *in vivo*. *J Infect Dis* 176: 118–125

38 Cao Y, Qin L, Zhang L, Safrit J, Ho DD (1995) Virologic and immunologic characterization of long-term survivors of human immunodeficiency virus type 1 infection. *N Engl J Med* 332: 201–208

39 Pantaleo G, Menzo S, Vaccarezza M, Graziosi C, Cohen OJ, Demarest JF, Montefiori D, Orenstein JM, Fox C, Schrager LK et al. (1995) Studies in subjects with long-term nonprogressive human immunodeficiency virus infection. *N Engl J Med* 332: 209–216

40 Oldstone MB (1997) HIV *versus* cytotoxic T lymphocytes. The war being lost. *N Engl J Med* 337: 1306–1308

41 Haslett PA, Nixon DF, Shen Z, Larsson M, Cox WI, Manandhar R, Donahoe SM, Kaplan G (2000) Strong human immunodeficiency virus (HIV)-specific CD4+ T cell responses in a cohort of chronically infected patients are associated with interruptions in anti-HIV chemotherapy. *J Infect Dis* 181: 1264–1272

42 Binley JM, Schiller DS, Ortiz GM, Hurley A, Nixon DF, Markowitz MM, Moore JP (2000) The relationship between T cell proliferative responses and plasma viremia during treatment of human immunodeficiency virus type 1 infection with combination antiretroviral therapy. *J Infect Dis* 181: 1249–1263
43 Lori F, Maserati R, Foli A, Seminari E, Timpone J, Lisziewicz J (2000) Structured treatment interruptions to control HIV-1 infection. *Lancet* 355: 287–288
44 Oxenius A, Price DA, Easterbrook PJ, O'Callaghan CA, Kelleher AD, Whelan JA, Sontag G, Sewell AK, Phillips RE (2000) Early highly active antiretroviral therapy for acute HIV-1 infection preserves immune function of CD8+ and CD4+ T lymphocytes. *Proc Natl Acad Sci USA* 97: 3382–3387
45 Autran B, Carcelain G, Li TS, Blanc C, Mathez D, Tubiana R, Katlama C, Debre P, Leibowitch J (1997) Positive effects of combined antiretroviral therapy on CD4+ T cell homeostasis and function in advanced HIV disease. *Science* 277: 112–116
46 Plana M, García F, Gallart T, Tortajada C, Soriano A, Palou E, Maleno MJ, Barcelo JJ, Vidal C, Cruceta A et al. (2000) Immunological benefits of antiretroviral therapy in very early stages of asymtomatic chronic HIV-1 infection. *AIDS* 14: 1921–1933
47 Van Rompay KK, Miller MD, Marthas ML, Margot NA, Dailey PJ, Canfield DR, Tarara RP, Cherrington JM, Aguirre NL, Bischofberger N et al. (2000) Prophylactic and therapeutic benefits of short-term 9-[2-(R)-(phosphonomethoxy)propyl]adenine (PMPA) administration to newborn macaques following oral inoculation with simian immunodeficiency virus with reduced susceptibility to PMPA. *J Virol* 74: 1767–1774
48 Biron F, Lucht F, Peyramond D, Fresard A, Vallet T, Nugier F, Grange J, Malley S, Hamedi-Sangsari F, Vila J (1995) Anti-HIV activity of the combination of didanosine and hydroxyurea in HIV-1-infected individuals. *J Acquir Immune Defic Syndr Hum Retrovirol* 10: 36–40
49 Foli A, Maserati R, Minoli L, Wainberg MA, Gallo RC, Lisziewicz J, Lori F (1998) Therapeutic advantage of hydroxyurea and didanosine combination therapy in patients previously treated with zidovudine. *AIDS* 12: 1113–1114
50 Gao WY, Cara A, Gallo RC, Lori F (1993) Low levels of deoxynucleotides in peripheral blood lymphocytes: a strategy to inhibit human immunodeficiency virus type 1 replication. *Proc Natl Acad Sci USA* 90: 8925–8928
51 Lori F, Malykh A, Cara A, Sun D, Weinstein JN, Lisziewicz J, Gallo RC (1994) Hydroxyurea as an inhibitor of human immunodeficiency virus-type 1 replication. *Science* 266: 801–805
52 Maurer-Schultze B, Siebert M, Bassukas ID (1988) An *in vivo* study on the synchronizing effect of hydroxyurea. *Exp Cell Res* 174: 230–243
53 Lori F. Hydroxyurea and HIV: 5 years later–from antiviral to immuno-modulating effects. *AIDS* 13: 1433–1442
54 Brunet M, Oppenheimer F, Martorell J, Vilardell J, Carreno MC, Carrillo M, Corbella J (1999) Mycophenolic acid monitoring: evaluation of the EMIT MPA immunoassay in kidney and lung transplantation. *Transplant Proc* 31: 2275–2276
55 Brunet M, Martorell J, Oppenheimer F, Vilardell J, Millan O, Carrillo M, Rojo I, Corbella J (2000) Pharmacokinetics and pharmacodynamics of mycophenolic acid in stable renal transplant recipients treated with low doses of mycophenolate mofetil. *Transpl Int* 13 (Suppl 1): S301–S305
56 Margolis D, Heredia A, Gaywee J, Oldach D, Drusano G, Redfield R (1999) Abacavir and mycophenolic acid, an inhibitor of inosine monophosphate dehydrogenase, have profound and synergistic anti-HIV activity. *J Acquir Immune Defic Syndr* 21: 362–370
57 Chapuis AG, Paolo Rizzardi G, D'Agostino C, Attinger A, Knabenhans C, Fleury S, Acha-Orbea H, Pantaleo G (2000) Effects of mycophenolic acid on human immunodeficiency virus infection *in vitro* and *in vivo*. *Nat Med* 6: 762–768
58 Garcia F, Plana M, Brunet M, Arnedo M, Alcami J, Lopez-Galindez, C, Tenner-Racz K, Gil C, Vidal E, Mestre G et al. (2002) Effect of associating an immunosuppressive therapy (Mycophenolate Mofetil: MMF)+HAART during STI and holding the MMF drug after definitive interruption of HAART on viral replication. 9th Conference on Retroviruses and Opportunistic Infections, Seattle, Washington, USA, 24–28 February 2002, Poster session 72
59 British HIV Association (BHIVA) guidelines for the treatment of HIV-infected adults with antiretroviral therapy (July 2001) http://www aidsmap com/about/bhiva/guidelines pdf
60 Guidelines for the use of antiretroviral agents in HIV-infected adults and adolescents (NIH) (February 2002) http://hivatis org/trtgdlns html#Adult

Combination Therapy of AIDS
Edited by E.D.A. De Clercq and A.-M.I. Vandamme

Immune reconstitution in HIV infection

Brigitte Autran

Laboratoire d'Immunologie Cellulaire et Tissulaire, Unité INSERM 543, Hôpital Pitié-Salpétrière, 47-83, bld de l'hôpital, Bâtiment CERVI, 75013 Paris, France

Introduction

The course of the human immunodeficiency virus (HIV) infection has dramatically changed since the immune reconstitution induced by new antiretroviral therapeutic regimens, or highly active antiretroviral therapy (HAART) combining inhibitors of the HIV reverse transcriptase (RTI) and protease (PI). Indeed, by blocking virus production these agents allow the immune system to reconstitute the CD4 cell pools and to restore the host's protection against pathogens [1–5]. The success of antiviral therapies illustrates the key role played by HIV in the pleiomorphic immune alterations observed during the infection, i.e., the severe CD4 cell depletion and the profound immune defects leading to loss of immune control of opportunistic pathogens. The fast and efficient immune reconstitution obtained at any stage of the disease also demonstrates that these major immune deficiencies are not definitively altering the immune system allowing reconstitution due to its enormous plasticity. The mechanisms underlying these immune alterations as well as their correction under therapy are still heavily debated. Nevertheless, such immune reconstitution represents a real breakthrough for HIV-infected patients but also allows better understanding of the pathophysiology of the HIV disease as well as the homeostasis of the immune system. This chapter will briefly review the various mechanisms and aspects of the immune alterations induced by HIV and describe the kinetics and mechanisms, the current successes and limitations of the immune reconstitution.

HIV and immune defects

The major progressive CD4 T-cell depletion occurs via multiple mechanisms. Death of the infected CD4 cells is a consequence of the direct pathogenicity of the virus for CD4 cells with many HIV proteins being able to accelerate CD4 cell death processes. These losses mostly affect the main target cells of the virus that are the activated memory CD4 T-cells [6, 7]. Within this memory cell subset, the CD4 T-cells specific for HIV are more often infected than other memory T-cells [8], as a result of a massive exposure of these cells to the virus. Death of Infected CD4 cells that actively replicate the virus is also due to the cytotoxic T-lymphocyte- (CTL)-mediated cytolysis directed against HIV [9]. Finally the loss of "innocent" non-infected cells also results from an exaggerated activation-induced cell death or apoptosis linked to the massive inflammatory syndrome that affects the immune system [10–13]. Naïve CD4 T-cells, the reservoir for antigen-specific CD4 T-cells, become even more profoundly depleted than the memory subset during disease progression despite their low level of HIV infection [1, 2, 14–18], as part of the compensatory mechanisms that take place, at least transiently, to maintain CD4 cell numbers [19–22]. The thymic production of naïve CD4 T-cells is maintained, though decreased, throughout the course of the infection as assessed by direct measurement of the thymus size [15] and by molecular probes of thymus production (the TCR excision circles –TRECs-) present in non-proliferating naïve CD45RA+CD62L+ cells [16–18]. Homeostatic mechanisms play a major role by accelerating the rate of conversion of these naïve cells toward effector/memory T-cells in order to answer the enormous host's demand to control pathogens, thus maintaining for a while the host's memory against pathogens. Eventually, however, the thymus becomes unable to compensate for these accelerated losses in naïve T-cells. Increased CD4 cell proliferation certainly takes place also during the course of the disease, as assessed by an increased incorporation of deuterated glucose [22–23] and expression of cell-cycle associated markers, such as Ki67 [19–21]. Numbers of Ki67+ cells, however, do not provide an accurate measurement of the peripheral turn-over of CD4 cells, since most of these Ki67+ cells accumulate in the initial step of the cell cycle (G1) but are not in the S or G2-M proliferative stages [24, 25]. In addition, molecular measurements of telomere lengths in CD4 T-cells do not show a very high replacement rate [26]. In fact, proliferation of memory CD4 T-cells is rapidly limited during disease progression [27] and cannot compensate for the cell losses. Many alterations in CD4 cell subsets, dendritic cells and the overactivation of the immune system contribute to this "anergy" [28–31]. Finally, a preferential compartmentalization of the CD4 T-cells within the lymph nodes leads to underevaluation of the actual CD4 T-cell compartment when performed on the peripheral blood [32, 33]. Indeed, an active infection in the lymphoid tissues induces alterations of the micro-environment and a sequestration of the lymphocytes at the site of infection, which leads to substantial losses in the circulating cells since the peripheral lymphocytes represent only 2% of the whole lymphocyte compartment.

All defenses against all pathogens become altered during the course of the HIV infection [34, 35]. Indeed, the CD4 T helper-1 cells produce cytokines such as interleukin-2 and interferon-gamma, which are central for the initiation and maintenance of the CD8 cell-mediated cytolysis of cells infected by viruses such as HIV, CMV, EBV or other herpesviruses, as well as for the macrophage-mediated elimination of intra-cellular pathogens such as mycobacteria. The CD4 T helper-2 cells are also instrumental by controlling the production of pathogen-specific antibodies. The defenses directed against HIV itself are also affected despite the major HIV-specific T- and B-cell responses induced during the primary infection. The CD4 Th1 responses against HIV disappear very rapidly after the infection is fully established and before CD4 defenses against other pathogens [36, 37]. The CD8 T-cell responses to HIV persist at very high levels for years and display a very diverse repertoire against all HIV antigens, thus containing HIV for years [38, 40]. Finally, the permanent replication of the virus and the emergence of new variants eventually induce CTL exhaustion with the onset of AIDS.

Mono- or bi-therapy with nucleoside reverse transcriptase inhibitors (NrtIs) were inefficient in reversing the course of these numerous and major immune defects. Only potent antiretroviral regimens (i.e., HAART) introduced in 1996 were successful in inducing a sustained CD4 cell increases and in restoring the host's immune defenses and protection against HIV and opportunistic pathogens [41, 43]

The kinetics of CD4 cell reconstitution

The consequences of HAART on the immune system were first demonstrated in patients naïve for any antiretroviral drugs who received at an advanced stage of the disease (CD4 cell counts below 250/mm^3) a combination of one PI and two NRTIs [1]. In this setting a very early increase in CD4 cell counts was observed during the first 2 months with strong slopes of CD4 cell increase at about 1–4 CD4 cell/mm^3/day, concurrently with the rapid reduction in virus load (Fig. 1). The CD8 cells usually increase in parallel suggesting that some T-cells sequestered in the lymphoid organs prior to treatment were allowed to recirculate when virus was controlled. The rapid arrest of local virus production [44–46] in lymphoid tissues is indeed supposed to halt the alterations of the microenvironment and vascularization, thereby allowing cell release in the periphery [1, 2]. Since peripheral blood lymphocytes represent only 2% of the total lymphocyte compartment, a mobilization of the sequestered CD4 cells might allow substantial increases in peripheral blood cell numbers in patients with low CD4 counts. Such rapid changes in the CD4 cell counts last for about 3 months but might not represent a net gain in total CD4 cell numbers during this first phase of immune reconstitution. The virus production plays only a minimal role in these rapid changes, which are mostly influenced by the CD4 cell counts at baseline and by the amounts of cells sequestered in the tissue [47]. Thus a modest reduction in virus load can induce an important redistribution of CD4 cells. When patients are treated early with subnormal CD4 cell counts, the amount of redistributed CD4 cells represents only a minimal change in their blood CD4 counts and a moderate increase is observed [48].

The kinetics of CD4 cell expansion is reduced after 2 or 3 months of treatment with slopes down to 0.1–0.5 CD4 cell/mm^3/day. The magnitude of plasma viral load reduction and its stability over time are the main factors influ-

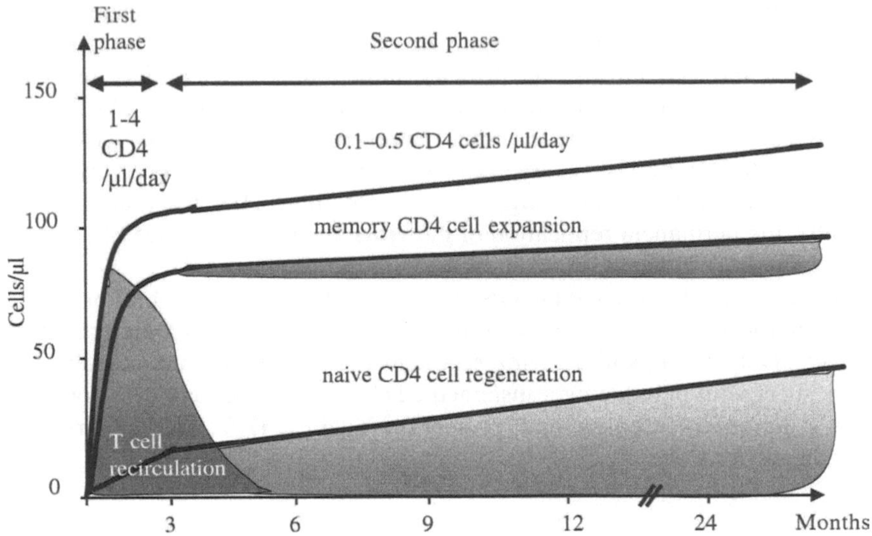

Figure 1. Kinetics and mechanisms of CD4 T-cell reconstitution HAART initiation during late disease. A first increase, usually rapid and large, represents re-distribution of memory CD4 T-cells, and is followed by a second, slow but steady, increase that reflects regeneration of naïve CD4 T-cells. The slopes are calculated from data published [3, 77]. The slopes of CD4 increase in the long term depend on the magnitude of the viral load reduction [calculated from ref. 47].

encing this second phase of immune reconstitution [47–49]. Indeed, a 4 log reduction in the plasma viral load can be associated to a 40% increment slope during the 2 years following HAART initiation, while a 1 log viral load reduction is usually associated with a 5% slope only. In contrast, the baseline CD4 cell depletion plays only a minor role in this CD4 cell recovery. Such a slow kinetics of recovery is observed for years provided the viral load is kept under tight control. An identical slope of CD4 recovery is observed for a given reduction in viral load, whatever the stage of the CD4 depletion, demonstrating that the severe immune alterations that had taken place before introduction of HAART are not definitive (Fig. 2).

A full recovery and normalization of the CD4 compartment is feasible but can take years. The normalization of the CD4 counts requires stable virus control and is a linear function of CD4 counts at initiation of therapy. A plateau might be observed, however, probably reflecting an incomplete viral suppression rather than a limit in T-cell reconstitution. Delays rather than failures of CD4 reconstitution can occur in rare occasions, usually in late AIDS patients representing less than 5% of all cases [47]. Addition of interleukin-2 can significantly help at accelerating the CD4 counts reconstitution above the zone of $200/mm^3$ with high risks for opportunistic pathogens [50, 51]. Large phase III trials are required, however, to test whether this strategy will bring significant help in preventing opportunistic events.

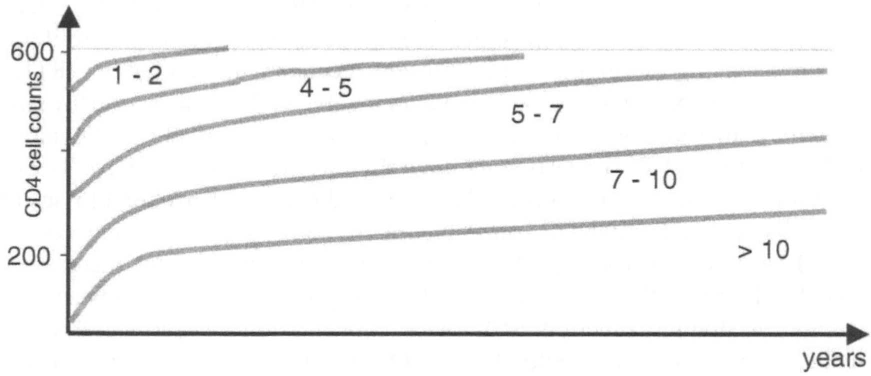

Figure 2. Kinetics of CD4 count normalization during HAART as a function of the CD4 depletion at time of HAART initiation. The slopes of reconstitution are roughly paralleled in early and late disease stages. Data are calculated from 45 patients receiving HAART, none of them being drug-experienced, with viral loads successfully controlled below 200 copies/ml) ([3, 47, 48, 77] and unpublished data).

The mechanisms of CD4 reconstitution: naïve and memory CD4 T-cells and immune de-activation

Memory CD4 T-cells bearing the CD45RO isoform [1–3] are responsible for the first rapid wave of CD4 T-cells. Markers of cell cycle and early activation such as Ki67 rapidly decrease during this early period, suggesting that the rapid CD4 changes are not due to an active CD4 proliferation of memory T-cells but only reflects memory cell redistribution in the peripheral blood [1–3, 19–21, 24].

Regeneration of naïve CD4 T-cells is the major actor of the second phase of CD4 reconstitution. Indeed naïve CD4 T-cells significantly increase both in proportions and absolute numbers after the third month of treatment when introduced at late stages of CD4 cell depletion [1–3, 20] (Fig. 1), or earlier if treatment is introduced at earlier stages of the disease, or even at primary infection [48]. Such increase in naïve T-cells was the major surprise of immune reconstitution considering the dogma of thymus involution in adults [52], and raised the hypothesis that the thymus might maintain its ability to regenerate naïve T-cells during HIV-infection even in adults. Computerized tomography confirmed the preservation of the thymic tissue in HIV-infected untreated adults and its progressive increase in parallel to naïve CD4 T-cell amplification during treatment [15]. The detection of markers of thymic origin in these naïve T-cells brought direct evidence for the thymus participation in the CD4 T-cell regeneration during antiretroviral therapy [16–18]. Indeed, the DNA episomal circles produced within the thymus during rearrangements of the T-cell receptor (TCR) segments do not replicate with the cell genome. Naïve cells that emigrate from the thymus recirculate in the periphery until they encounter specific antigens, then proliferate and convert into memory

T-cells. Thus, detection of TRCs in naïve CD45RA+CD62L+ CD4 T-cells would assess those cells that have not undergone multiple cycles of proliferation since leaving the thymus and represent recent thymic emigrants (RTE), though intervals between two cell cycles in a naïve cell might reach 10 years. An increase in TRCs numbers in CD4 cells is observed in parallel to the naïve T-cell subset [17–19], suggesting *de novo* thymus production of naïve T-cells after virus reduction. Alternatively, however, reduction in peripheral losses of CD4 cells with virus control can decrease the excessive conversion rate of naïve T-cells toward memory T-cells and allow for an apparent increase in naïve T-cells and TRCs numbers [53]. Certainly these two mechanisms, i.e., increase in thymus function and lower consumption of peripheral naïve T-cells, act in concert and help at reconstituting the major reservoir of naïve T-cells. Overall, the consequence for the patient is the reconstitution of a vast repertoire of naïve CD4 T-cells as shown by direct analysis of the T-cell repertoire diversity [54], thus restoring the host's defenses against all pathogens. Finally, a true, though limited, proliferation of the memory CD4 T-cells also takes place, resulting from the antigenic stimulation from environmental and endogenous pathogens [20].

The abnormal activation of CD4 T-cells characteristic of the infection also rapidly resolves with all markers of T-cell activation decreasing in parallel to the plasma virus load reduction. During the first 2 weeks, however, the peripheral blood memory CD4 cell subset displays activation markers such as CD25 or HLA-DR [1, 2] but this activated subset rapidly decreases in lymphoid tissues [44–46], thus reflecting the recirculation of memory cells that had been previously activated and sequestered in the tissues rather than a concurrent activation. Several weeks after initiation of HAART, when the virus becomes undetectable most of the T-cell activation markers are gone. The strong correlation observed between the decay of virus production and immune activation indicate the strong link between these two phenomena. The cell surface expression of death molecules such as Fas-Fas-ligand decreases with the activation markers resulting in a decreased sensitivity to apoptosis of the memory CD4 T-cells. A similar decrease is observed for the pro-inflammatory cytokines TNF-α and IL-6. These phenomena occur both in the peripheral blood and in the lymphoid tissues and help at reducing the abnormal rate of CD4 T-cell death observed in HIV infection and at restoring CD4 T-cell numbers [44–46]. They also have important impacts on other immune cells such as CD8 T-cells, B cells or polymorphonuclear cells.

Reducing the abnormal immune activation and losses in the memory T-cell compartment prolongs their survival, reduces the consumption of naïve T-cells and helps to restore their numbers while facilitating recirculation of the CD4 T-cell subset. Such anti-inflammatory effects might be even as important as the thymus for the immune recovery. Thus antiretroviral drugs, by blocking virus production, reduce the virus-driven immune activation and play a potent anti-inflammatory influence on the immune system.

Reconstitution of host defenses against pathogens and HIV

Whether restoration of CD4 numbers, phenotypes and repertoire could restore the host's immune defenses lost with AIDS is a key question. Indeed, a restoration of CD4 T-cell responses against CMV and mycobacterium tuberculosis is usually observed after 3 months of HAART in AIDS patients who had previously lost these responses [1–3, 56, 57] (Fig. 3). Such reconstitution appears concurrent to the early memory CD4 cell increase and to the normalization of the inflammatory syndrome, before regeneration of naïve CD4 T-cells, but requires a stable control over time of HIV replication [3]. The clinical benefits of such restoration of host defenses against opportunistic pathogens were rapidly confirmed by the dramatic reduction in the mortality and the morbidity related to AIDS [4, 5]. The incidence of opportunistic infections was shown to be similar in treated patients recovering CD4 cell numbers above 200/mm^3 and in untreated patients below 200/mm^3 [58]. These findings suggest the CD4 cell counts have the same value as indicators of the immune deficiency whether patients were on the declining CD4 slope during natural disease or on the increasing CD4 slope during treatment. Immune reconstitution also allows to stop prophylaxis of opportunistic infections above 200 CD4 cell counts/mm^3 [59–62].

In some cases, however, recurrences of CMV retinitis were reported within 3 months after introduction of HAART [63, 64]. This observation cannot be interpreted as a failure of immune reconstitution but indicate that treatment of opportunistic pathogens has to be maintained during the first months of anti-

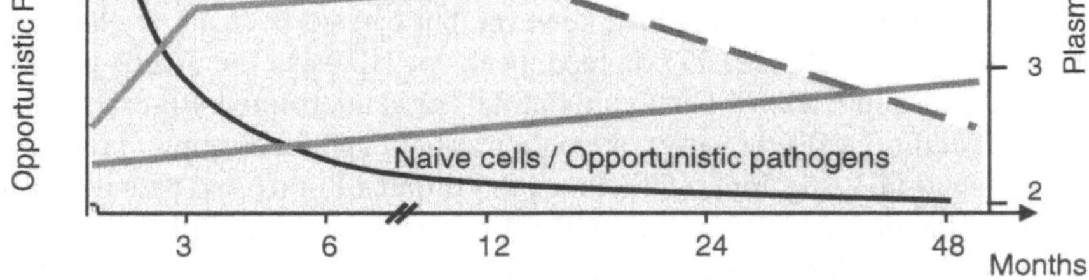

Figure 3. Memory CD4 T-cell re-expansion with HAART requires re-exposure to antigens. The rapid viral load reduction allows an early reappearance of CD4 cell responses to opportunistic antigens that contrasts with the lack of restoration of CD4 T-cell responses to HIV antigens. This phenomenon can be explained by the distinct exposure of these various T-cells to their respective pathogens HAART directly reduces production of HIV antigens but does not affect directly production of opportunistic pathogens that are present in quantities sufficient to allow re-expansion of memory T-cells.

retroviral therapy before reconstitution of the host's immune defenses becomes truly efficient. These observations also mean that initiating treatment at very low CD4 cell counts leaves advanced patients with a prolonged exposure to the risk of opportunistic infections. Paradoxical exacerbations of infectious events such as tuberculosis or other mycobacterial infections, or of autoimmune diseases, can also occur during the first 3 to 6 months of treatment [65, 66]. These phenomena do not assess failure of immune reconstitution but, on the contrary, assess the strength of the restoration of the T helper-1 cell function that allows rapid reconstitution of Th1-dependent granulomas or other immuno-pathological lesions. These events, baptized the "*Immune Reconstitution Syndrome*" (IRS), impose to treat tuberculosis or other mycobacterial infections for 3 months to diminish the mycobacterial burden before introducing the antiretroviral medications.

The restoration of host defenses against HIV, however, is really limited. Poor restoration of a CD4 cell reactivity against HIV itself occurs despite the potent immune reconstitution described above [1–3, 55–57], except when HAART is initiated at time of primary infection [36]. The HIV-specific CD8 T-cells and antibody titers decline during efficient antiretroviral therapy of established infection and are even delayed if treatment is introduced before seroconversion takes place [67–70]. Such a dissociation between a strong and protective immunity rapidly restored against opportunistic pathogens and a waning immunity against HIV is explained by the differences in the exposure to antigenic stimuli. Antigen exposure is required to allow appropriate stimulation, proliferation and re-expansion of antigen-specific T-cells. Lymphocytes specific for opportunistic pathogens which are over-represented when HAART is introduced can easily expand when HIV production is stopped. In contrast, the lack of HIV production during treatment cannot restimulate HIV-specific T-cells. Thus, a perfectly controlled patient is left with weak immunity against HIV despite a strong restoration of his immune defenses against other pathogens. One of the best hallmarks of this unrestored immunity against HIV is the rebound of virus replication observed when HAART is stopped [71, 72].

Memory against HIV is not lost, however, but persists even at very low levels and is re-inducible [73–77]. Indeed, peaks of HIV-specific T-cells parallel peaks of virus replication when treatment fails or is interrupted. Re-expansions of HIV-specific T-cells, however, are only transient and involve only pre-existing dominant HIV-specific T-cell clones in chronically-infected patients [69]. Again, restoration of HIV-specific CD4 T-cell responses might be more sustained over time during treatment interruptions at time of primary infection [76].

These treatment interruptions, even if unsuccessful at restoring a durable and protective immunity to HIV, help at demonstrating that restoration of an HIV-specific immunity is feasible but might require strategies other than the simple "switch-on and switch-off " of antiretroviral therapy. Another and safer strategy is to re-immunize treated patients against HIV with safe HIV antigens, i.e., vaccines, in order to rise an immune barrier to the virus before treatment is interrupted and before virus relapses [77] (Fig. 4). Several clinical trials are currently in progress to evaluate whether such immune-based strategies will generate protective CD4 and CD8 T-cell responses against HIV and allow to interrupt therapy for prolonged periods of time.

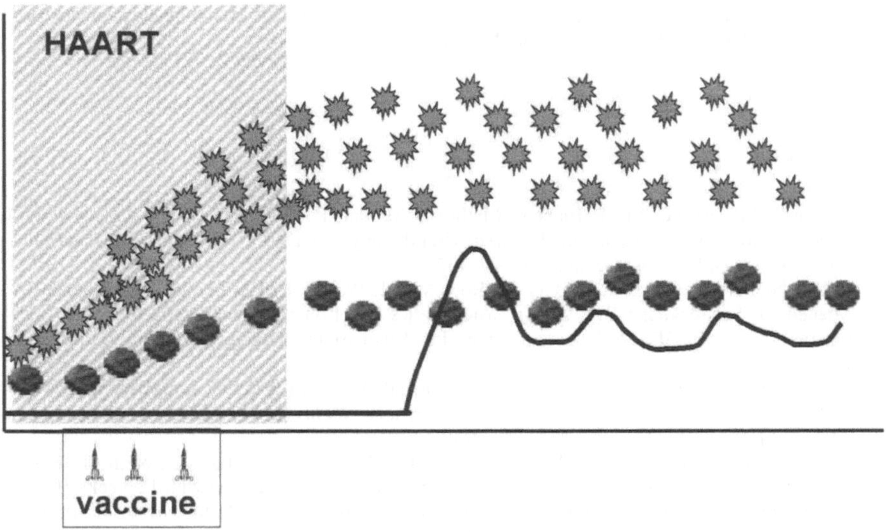

Figure 4. The rationale for therapeutic immunization. The working hypothesis of these new therapies is to add anti-HIV vaccines to the antiretroviral treatment in order to provide a source of non pathogenic HIV antigens capable to restimulate HIV-specific CD4 (black circles) and CD8 T-cells (jagged circles) before stopping HAART. After HAART interruption the HIV-specific immunity should be capable to control levels of HIV replication under a certain threshold that would permit to delay reintroduction of HAART as much as possible.

Conclusion

HAART allows immune restoration and protection of the host against opportunistic pathogens. The condition required for immune reconstitution is an efficient and durable inhibition of virus replication. These positive effects can be obtained at late stages of the disease even when the patients have been heavily pre-treated and demonstrate that HIV does not definitively alter the lymphoid tissues or the immune defenses, even after years of infection and severe immune suppression. Initiating HAART in late or in early disease might differ however by 1) the time required to reach normalized immune functions and cell counts and 2) the lack of amplification of HIV-specific T-cells in late disease. The delay to recovery or the lack of reconstitution of a solid immunity against HIV itself is currently increasing the need for additive therapeutic strategies based upon immune interventions. At a time where virus eradication does not appear to be at hand and where chronic antiviral therapy becomes too toxic, the ultimate goal of therapeutic strategies is to restore an immunity to HIV to help at controlling the virus with the host's recovering immune system.

Acknowledgments

The studies reported in this review could not have been performed without the valuable collaboration with Guislaine Carcelain, Li Taisheng, Lucile Mollet, Guy Gorochov, Catherine Blanc, Marc Renaud, in the Laboratoire d'Immunologie Cellulaire, CNRS UMR 7627, directed by Pr. Patrice Debré; Roland Tubiana, Pr. Christine Katlama and colleagues in the Service des Maladies Infectieuses, Hôpital Pitié-Salpétrière, SIDACTION and the ANRS

References

1 Autran B, Carcelain G, Li TS, Blanc C, Mathez D, Tubiana R, Katlama C, Debre P, Leibowitch J (1997) Positive effects of combined antiretroviral therapy on CD4+ T cell homeostasis and function in advanced HIV disease. *Science* 277: 112–116

2 Pakker NG, Notermans DW, de Boer RJ, Roos MT, de Wolf F, Hill A, Leonard JM, Danner SA, Miedema F, Schellekens PT (1998) Biphasic kinetics of peripheral blood T cells after triple combination therapy in HIV-1 infection: a composite of redistribution and proliferation. *Nat Med* 42: 208–214

3 Li TS, Tubiana R, Katlama C, Calvez V, Ait Mohand H, Autran B (1998) Long-lasting recovery in CD4 T-cell function and viral-load reduction after highly active antiretroviral therapy in advanced HIV disease. *Lancet* 351: 1682–1686

4 Lederman MM, Connick E, Landay A, Kuritzkes DR, Spritzler J, St Clair M, Kotzin BL, Fox L, Chiozzi MH, Leonard JM et al. (1998) Immunologic responses associated with 12 weeks of combination antiretroviral therapy consisting of zidovudine, lamivudine, and ritonavir: results of AIDS clinical trials group protocol 315. *J Infect Dis* 178: 70–79

5 Palella FJ Jr, Delaney KM, Moorman AC, Loveless MO, Fuhrer J, Satten GA, Aschman DJ, Holmberg SD (1998) Declining morbidity and mortality among patients with advanced human immunodeficiency virus infection. HIV Outpatient Study Investigators. *N Engl J Med* 338: 853–860

6 Schnittman SM, Lane HC, Greenhouse J, Justement JS, Baseler M, Fauci AS (1990) Preferential infection of CD4+ memory T cells by human immunodeficiency virus type 1: evidence for a role in the selective T-cell functional defects observed in infected individuals. *Proc Natl Acad Sci USA* 87: 6058–6062

7 Spine CA, Prince HE, Richman DD (1997) Preferential replication of HIV-1 in the CD45RO memory cell subset of primary CD4 lymphocytes *in vitro*. *J Clin Invest* 99: 1774–1785

8 Douek DC, Brenchley JM, Betts MR, Ambrozak DR, Hill BJ, Okamoto Y, Casazza JP, Kuruppu J, Kunstman K, Wolinsky S et al. (2002) HIV preferentially infects HIV-specific CD4+ T cells. *Nature* 417: 95–98

9 Cheynier R, Henrichwark S, Hadida F, Pelletier E, Oksenhendler E, Autran B, Wain-Hobson S (1994) HIV and T cell expansion in splenic white pulps is accompanied by infiltration of HIV-specific cytotoxic T lymphocytes. *Cell* 78: 373–387

10 Groux H, Torpier G, Monte D, Mouton Y, Capron A, Ameisen JC (1992) Activation-induced death by apoptosis in CD4+ T cells from human immunodeficiency virus-infected asymptomatic individuals. *J Exp Med* 175: 331–340

11 Meyaard L, Otto SA, Jonker RR, Mijnster MJ, Keet RP, Miedema F (1992) Programmed death of T cells in HIV-1 infection. *Science* 257: 217–219

12 Badley AD, Dockrell DH, Algeciras A, Ziesmer S, Landay A, Lederman MM, Connick E, Kessler H, Kuritzkes D, Lynch DH et al. (1998) *In vivo* analysis of Fas/FasL interactions in HIV-infected patients. *J Clin Invest* 102: 79–87

13 Ho DD, Neumann AU, Perelson AS, Chen W, Leonard JM, Markowitz M (1995) Rapid turnover of plasma virions and CD4 lymphocytes in HIV-1 infection. *Nature* 373: 123–126

14 Hellerstein MK, McCune JM (1997) T cell turnover in HIV-1 disease. *Immunity* 7: 583–589

15 McCune JM, Loftus R, Schmidt DK, Carrell P, Webster D, Swor-Yim LB, Francis IR, Gross BH, Grant RM (1998) High prevalence of thymic tissue in adults with human immunodeficiency virus-1 infection. *J Clin Invest* 101: 2301–2308
16 Douek DC., McFarlard RD, Keiser PH, Gage EA, Massey JM, Haynes BF, Polis MA, Haase AT, Feinberg MB, Sullivan JL et al. (1998) Changes in thymic function with age and during the treatment of HIV infection. *Nature* 396: 690–695
17 Poulin JF, Viswanathan MN, Harris JM, Komanduri KV, Wieder E, Ringuette N, Jenkins M, Mc Cune JM, Sekaly RP (1999) Direct evidence for thymic function in adult humans. *J Exp Med* 190: 479–486
18 Zhang L, Lewin SR, Markowitz M, Lin HH, Skulsky E, Karanicolas R, He Y, Jin X, Tuttleton S, Vesanen M et al. (1999) Measuring recent thymic emigrants in blood of normal and HIV-1-infected individuals before and after effective therapy. *J Exp Med* 190: 725–732
19 Sachsenberg N, Perelson AS, Yerly S, Schockmel GA, Leduc D, Hirschel B, Perrin L (1998) Turnover of CD4+ and CD8+ T lymphocytes in HIV-1 infection as measured by Ki-67 antigen. *J Exp Med* 187: 1295–1303
20 Fleury S, de Boer RB, Rizzardi GP, Wolthers KC, Otto SA, Welbon CC, Grasiozi C, Knabenhans C, Soudeyns H, Bart PA et al. (1998) Limited CD4+ T-cell renewal in early HIV-1 infection: effect of highly active antiretroviral therapy. *Nat Med* 4: 794–801
21 Fleury S, Rizzardi GP, Chapuis A, Tambussi G, Knabenhans C, Simeoni E, Meuwly J, Corpataux J, Lazzarin A, Miedema F et al. (2000) Long-term kinetics of T cell production in HIV-infected subjects treated with highly active antiretroviral therapy. *Proc Natl Acad Sci USA* 97: 5393–5398
22 Hellerstein M, Hanley MB, Cesar D, Siler S, Papageorgopoulos C, Wieder E, Schmidt D, Hoh R, Neese R, Macallan D et al. (1999) Directly measured kinetics of circulating T lymphocytes in normal and HIV-1 infected humans. *Nat Med* 5: 83–89
23 Mohri H, Perelson AS, Tung K, Ribeiro RM, Ramratnam B, Markowitz M, Kost R, Hurley A, Weinberger L, Cesar D, Hellerstein MK et al. (2001) Increased turnover of T lymphocytes in HIV-1 infection and its reduction by antiretroviral therapy. *J Exp Med* 194: 1277–1287
24 Combadere B, Blanc C, Li T, Carcelain G, Delaugerre C, Calvez V, Tubiana R, Debre P, Katlama C, Autran B (2000) CD4+Ki67+ lymphocytes in HIV-infected patients are effector T cells accumulated in the G1 phase of the cell cycle. *Eur J Immunol* 30: 3598–3603
25 Sieg SF, Harding CV, Lederman MM (2001) HIV-1 infection impairs cell cycle progression of CD4(+) T cells without affecting early activation responses. *J Clin Invest* 108: 757–764
26 Wolthers KC, Bea G, Wisman A, Otto SA, de Roda Husman AM, Scaft N, de Wolf F, Goudsmit J, Coutinho RA, van der Zee AG et al. (1996) T cell telomere length in HIV-1 infection: no evidence for increased CD4+ T cell turnover. *Science* 274: 1543–1547
27 Tissot O, Viard JP, Rabian C, Ngo N, Burgard M, Rouzioux C, Penit C (1998) No evidence for proliferation in the blood CD4+ T cell pool during HIV-1 infection and triple combination therapy. *AIDS* 12: 879–884
28 Autran B, Legac E, Blanc C, Debré P (1995) A TH0/TH2 function of CD4+CD7- T helper cells from normal donors and HIV-infected patients. *J Immunol* 154 3: 1408–1417
29 Grassi F, Hosmalin A, McIlroy D, Calvez V, Debre P, Autran B (1999) Depletion in blood CD11c-positive dendritic cells from HIV-infected patients. *AIDS* 13: 759–766
30 Borthwick NJ, Bofill M, Gombert WM, Akbar AN, Medina E, Sagawa K, Lipman MC, Johnson MA, Janossy G (1994) Lymphocyte activation in HIV-1 infection. II. Functional defects of CD28- T cells. *AIDS* 8: 431–441
31 Sousa AE, Carneiro J, Meier-Scellersheim M, Grossman Z, Victorino RM (2002) CD4 T cell depletion is linked directly to immune activation in the pathogenesis of HIV-1 and HIV-2 but only indirectly to the viral load. *J Immunol* 169: 3400–3406
32 Rosok BI, Bostad L, Voltersvik P, Olofsson J, Asjö B, Brinchmann JE (1996) Reduced CD4 cell counts in blood do not reflect CD4 cell depletion in tonsillar tissue in asymptomatic HIV-1 infection. *AIDS* 10: 35–38
33 Tenner-Racz K, Stellbrink HJ, van Lunzen J, Schneider C, Jacobs JP, Raschdorff B., Grosschupff G, Steinman RM, Racz P (1998) The unenlarged lymph nodes of HIV-1-infected, asymptomatic patients with high CD4 T cell counts are sites for virus replication and CD4 T cell proliferation. The impact of highly active antiretroviral therapy. *J Exp Med* 187: 949–959
34 Lane HC, Masur H, Gelmann EP, Longo DL, Steis RG, Chused T, Whalen G, Edgar LC, Fauci AS (1985) Correlation between immunologic function and clinical subpopulations of patients with the acquired immune deficiency syndrome. *Am J Med* 78: 417–422

35 Clerici M, Stocks NI, Zajac RA, Boswell RN, Lucey DR, Via CS, Shearer GM (1989) Detection of three distinct patterns of T helper cell dysfunction in asymptomatic, HIV-seropositive patients: Independence of CD4+ cell numbers and clinical staging. *J Clin Invest* 84: 1892–1899
36 Rosenberg ES, Billingsley JM, Caliendo AM, Boswell SL, Sax PE, Kalams SA, Walker BD (1997) Vigorous HIV-1 specific CD4+ T cell responses associated with control of viremia. *Science* 278: 1447–1450
37 Pitcher CJ, Quittner C, Peterson DM, Connors M, Koup RA, Maino VC, Picker LJ (1999) HIV-1-specific CD4+ T cells are detectable in most individuals with active HIV-1 infection, but decline with prolonged viral suppression. *Nat Med* 5: 518–525
38 Hoffenbach A, Langlade-Demoyen P, Dadaglio G, Vilmer E, Michel F, Mayaud C, Autran B, Plata F (1989) Unusually high frequencies of HIV-specific cytotoxic T lymphocytes in humans. *J Immunol* 142: 452–462
39 Altman JD, Moss PA, Goulder PJ, Barouch DH, McHeyzer-Williams MG, Bell JI, McMichael AJ, Davis MM (1996) Phenotypic analysis of antigen-specific T lymphocytes. *Science* 274: 94–96
40 Altfeld M, Rosenberg ES, Shankarappa R, Mukherjee JS, Hecht FM, Eldridge RL, Addo MM, Poon SH, Phillips MN, Robbins GK et al. (2001) Cellular immune responses and viral diversity in individuals treated during acute and early HIV-1 infection. *J Exp Med* 193: 169–180
41 Collier AC, Coombs RW, Schoenfeld DA, Bassett RL, Timpone J, Baruch A, Jones M, Facey K, Whitacre C, McAuliffe VJ et al. (1996) Treatment of human immunodeficiency virus infection with saquinavir, zidovudine and zalcitabine. *N Engl J Med* 334: 1011–1017
42 Katzenstein DA, Hammer SM, Hughes MD, Gundacker H, Jackson JB, Fiscus S, Rasheed S, Elbeik T, Reichman R, Japour A (1996) The relation of virologic and immunologic markers to clinical outcomes after nucleoside therapy in HIV-infected adults with 200 to 500 CD4 cells per cubic millimeter. AIDS Clinical Trials Group Study 175 Virology Study Team. *N Engl J Med* 335: 1091–1098
43 Hammer SM, Squires KE, Hughes MD, Grimes JM, Demeter LM, Currier JS, Eron JJ Jr, Feinberg JE, Balfour HH Jr, Deyton LR et al. (1997) A controlled trial of two nucleoside analogues plus indinavir in persons with human immunodeficiency virus infection and CD4 cell counts of 200 per cubic millimeter or less. AIDS Clinical Trials Group 320 Study Team. *N Engl J Med* 337: 725–739
44 Bucy RP, Hockett RD, Derdeyn CA, Saag MS, Squires K, Sillers M, Mitsuyasu RT, Kilby JP (1999) Initial increase in blood CD4(+) lymphocytes after HIV retroviral therapy reflects redistribution from lymphoid tissues. *J Clin Invest* 103: 1391–1398
45 Andersson J, Fehniger TE, Patterson BK, Pottage J, Agnoli M, Jones P, Behbahani H, Landay A (1998) Early reduction of immune activation in lymphoid tissue following highly active HIV therapy. *AIDS* 12: F123–F129
46 Cavert W, Notermans DW, Staskus K, Wietgrefe SW, Zupancic M, Gebhard K, Henry K, Zhang ZQ, Mills R, McDade H (1997) Kinetics of response in lymphoid tissues to antiretroviral therapy of HIV-1 infection. *Science* 176: 960–964
47 Renaud M, Katlama C, Mallet A, Calvez V, Carcelain G, Tubiana R, Jouan M, Caumes E, Agut H, Bricaire F (1999) Determinants of paradoxical CD4 cell reconstitution after protease inhibitor-containing antiretroviral regimen. *AIDS* 13: 669–676
48 Carcelain G, Blanc C, Leibowitch J, Mariot P, Mathez D, Schneider V, Saimot AG, Damond F, Simon F, Debre P (1999) T-cell changes after combined nucleoside analogue therapy in HIV primary infection. *AIDS* 13: 1077–1081
49 Wong JK, Gunthard HF, Havlir DV, Zhang ZQ, Haase AT, Ignacio CC, Kwok S, Emini E, Richman DD (1997) Reduction in HIV-1 in blood and lymph nodes following potent antiretroviral therapy and the virologic correlates of treatment failure. *Proc Natl Acad Sci USA* 94: 12574–12579
50 Natarajan V, Lempicki RA, Sereti I, Badralmaa Y, Adelsberger JW, Metcalf JA, Prieto DA, Stevens R, Baseler MW, Kovacs JA (2002) Increased peripheral expansion of naïve CD4+ T cells *in vivo* after IL-2 treatment of patients with HIV infection. *Proc Natl Acad Sci USA* 99: 10712–10717
51 Katlama C, Duvivier C, Chouquet C, Autran B, Carcelain G, de Sa L, Zagury M, Costagliola D, Tubiana R (2002) A randomized comparative open-label study of interleukin (IL2) in patients with CD4 < 200/mm^3 despite effective HAART. 7th Conference on Retroviruses and Opportunistic Infections, San Francisco, CA, USA, 30 January–2 February 2000. Poster no. 543
52 Mackall CL, Granger L, Sheard MA, Cepeda R, Gress RE (1993) T-cell regeneration after bone marrow transplantation: differential CD45 isoform expression on thymic-derived *versus* thymic-independent progeny. *Blood* 82: 2585–2894

53 Hazenberg MD, Otto SA, Cohen Stuart JW, Verschuren MC, Borleffs JC, Boucher CA, Coutinho RA, Lange JM, Rinke de Wit TF, Tsegaye A et al. (2000) Increased cell division but not thymic dysfunction rapidly affects the T-cell receptor excision circle content of the naïve T cell population in HIV-1 infection. *Nat Med* 6: 1036–1042
54 Gorochov G, Neumann AU, Kereveur A, Parizot C, Li TS, Katlama C, Karmochkine M, Raguin G, Autran B, Debre P (1998) Perturbation of CD4+ and CD8+ T-cell repertoires during progression to AIDS and regulation of the CD4+ repertoire during antiviral therapy. *Nature Med* 4: 215–221
55 Rinaldo Cr Jr, Liebmann JM, Huang XL, Fan Z, Al-Shboul Q, McMahon DK, Day RD, Riddler SA, Mellors JW (1999) Prolonged suppression of human immunodeficiency virus type 1 (HIV-1) viremia in persons with advanced disease results in enhancement of CD4 T cell reactivity to microbial antigens but not to HIV-1 antigens. *J Infect Dis* 179: 329–336
56 Pontesilli O, Kerkhof-Garde S, Notermans DW, Foudraine NA, Roos MT, Klein MR, Danner SA, Lange JM, Miedemaa F (1999) Functional T cell reconstitution and human immunodeficiency virus-1-specific cell-mediated immunity during highly active antiretroviral therapy. *J Infect Dis* 180: 76–86
57 Plana M, Garcia F, Miro JM, Gatell JM (1998) Lack of T-cell proliferative response to HIV-1 antigens after 1 year of highly active antiretroviral treatment in early HIV-1 disease. Immunology Study Group of Spanish EARTH-1 Study. *Lancet* 352: 1194–1195
58 Miller V, Mocroft A, Reiss P, Katlama C, Papadopoulos AI, Katzenstein T, van Lunzen J, Antunes F, Phillips AN, Lundgren JD (1999) Relations among CD4 lymphocyte count nadir, antiretroviral therapy, and HIV-1 disease progression: results from the EuroSIDA study. *Ann Int Med* 130: 570–577
59 Schneider MM, Borleffs JC, Stolk RP, Jaspers CA, Hoepelman AI (1999) Discontinuation of prophylaxis for *Pneumocystis carinii* pneumonia in HIV-1-infected patients treated with highly active antiretroviral therapy. *Lancet* 353: 201–203
60 Weverling GJ, Mocroft A, Ledergerber B, Kirk O, Gonzalez-Lahoz J, d'Arminio Monforte A, Proenca R, Phillips AN, Lundgren JD, Reiss P (1999) Discontinuation of *Pneumocystis carinii* pneumonia prophylaxis after start of highly active antiretroviral therapy in HIV-1 infection. EuroSIDA Study Group. *Lancet* 353: 1293–1298
61 Furrer H, Egger M, Opravil M, Bernasconi E, Hirschel B, Battegay M, Telenti A, Vernazza PL, Rickenbach M, Flepp M (1999) Discontinuation of primary prophylaxis against *Pneumocystis carinii* pneumonia in HIV-1-infected adults treated with combination antiretroviral therapy. Swiss HIV Cohort Study. *N Engl J Med* 340: 1301–1306
62 Jouan M, Saves M, Tubania R, Carcelain G, Cassoux N, Aubron-Olivier C, Fillet AM, Nciri M, Senechal B, Chene G (2001) Discontinuation of maintenance therapy for cytomegalovirus retinitis in HIV-infected patients receiving highly active antiretroviral therapy. *AIDS* 15: 23–31
63 Jacobson MA, Zegans M, Pavan PR, O'Donnell JJ, Sattler F, Rao N, Owens S, Pollard R (1997) Cytomegalovirus retinitis after initiation of highly active antiretroviral therapy. *Lancet* 349: 1443–1445
64 Gilquin J, Piketty C, Thomas V, Gonzales-Canali G, Belec L, Kazatchkine MD (1997) Acute cytomegalovirus infection in AIDS patients with CD4 counts above 100×10^6 cells/l following combination antiretroviral therapy including protease inhibitors. *AIDS* 11: 1659–1660
65 Price P, Mathiot N, Krueger R, Stone S, Keane NM, French MA (2001) Immune dysfunction and immune restoration disease in HIV patients given highly active antiretroviral therapy. *J Clin Virol* 22: 279–287
66 Monsuez JJ, Vittecoq D, Musset L, Alemanni M, Dussaix E, Autran B (1998) Arthralgias and cryoglobulinemia during protease inhibitor therapy in a patient infected with human immunodeficiency virus and hepatitis C virus. *Arthritis Rheum* 41: 740–743
67 Ogg GS, Jin X, Bonhoeffer S, Moss P, Nowak MA, Monard S, Segal JP, Cao Y, Rowland-Jones SL, Hurley A (1999) Decay kinetics of human immunodeficiency virus-specific effector cytotoxic T lymphocytes after combination antiretroviral therapy. *J Virol* 73: 797–800
68 Dalod M, Harzic M, Pellegrin I, Dumon B, Hoen B, Sereni D, Deschemin JC, Levy JP, Venet A, Gomard E (1998) Evolution of cytotoxic T lymphocytes responses to HIV-1 in patients with symptomatic primary infection receiving antiretroviral triple therapy. *J Infect Dis* 178: 61–69
69 Mollet L, Li TS, Samri A, Tournay C, Tubiana R, Calvez V, Debre P, Katlama C, Autran B (2000) Dynamics of HIV-specific CD8+ T lymphocytes with changes in viral load. The RESTIM and COMET Study Groups. *J Immunol*, 165: 1692–1704

70 Morris L, Binley JM, Clas BA, Bonhoeffer S, Astill TP, Kost R, Hurley A, Cao, Markowitz M, Ho DD (1998) HIV-1 antigen-specific and –nonspecific B cell responses are sensitive to combination antiretroviral therapy. *J Exp Med* 188: 233–245
71 Neumann AU, Tubiana R, Calvez V, Robert C, Li TS, Agut H, Autran B, Katlama C (1999) HIV-1 rebound during interruption of HAART has no deleterious effect on reinitiated treatment. Comet Study Group. *AIDS* 13: 677–683
72 Lisziewicz J, Rosenberg E, Lieberman J, Jessen H, Lopalco L, Siliciano R, Walker B, Lori F (1999) Control of HIV despite the discontinuation of antiretroviral therapy. *N Engl J Med* 340: 1683–1684
73 Ortiz GM, Nixon DF, Trkola A, Binley J, Jin X, Bonhoeffer S, Kuebler PJ, Donahoe SM, Demoitie MA, Kakimoto WM et al. (1999) HIV-1-specific immune responses in subjects who temporarily contain virus replication after discontinuation of highly active antiretroviral therapy. *J Clin Invest* 104: R13–18
74 Carcelain G, Tubiana R, Samri A, Calvez V, Delaugerre C, Agut H, Katlama C, Autran B (2001) Transient mobilization of human immunodeficiency virus (HIV)-specific CD4 T-helper cells fails to control virus rebounds during intermittent antiretroviral therapy in chronic HIV type 1 infection. *J Virol.* 75: 234–241
75 Ruiz L, Carcelain G, Martinez-Picado J, Frost S, Marfil S, Paredes, R, Romeu J, Ferrer E, Morales-Lopetegi K, Autran B et al. (2001) HIV dynamics and T-cell immunity after three structured treatment interruptions in chronic HIV-1 infection. *AIDS* 15: 19–27
76 Rosenberg ES, Altfeld M, Poon SH, Phillips MN, Wilkes BM, Eldridge RL, Robbins GK, D'Aquila RT, Goulder PJ, Walker BD (2000) Immune control of HIV-1 after early treatment of acute infection. *Nature* 407: 523–526
77 Autran B, Carcelain G (2000) AIDS. Boosting immunity to HIV–can the virus help? *Science* 290: 946–949

Combination Therapy of AIDS
Edited by E.D.A. De Clercq and A.-M.I. Vandamme

Highly active antiretroviral treatment (HAART) of pediatric HIV-1 infection

Nicole H. Tobin[1] and Lisa M. Frenkel[1, 2]

University of Washington, Departments of [1]Pediatrics and [2]Laboratory Medicine, 4800 Sand Point Way, NE, Seattle, WA 98105, USA

Introduction

Implementing successful HAART of HIV-1-infected children poses several unique challenges. Children appear to have less virologic success from HAART than adults. However, careful consideration of multiple factors may assist in maximizing the virologic efficacy of HAART, including its duration [1]. Obstacles to successful HAART of children include: difficulties in administering drugs; difficulties adhering to complicated and even simple regimens; variable, erratic or unknown pharmacokinetics for certain drugs, especially in young infants; adverse reactions to drugs and drug-related toxicities; and the selection of drug-resistant virus. Drug-resistant mutants are selected when drug combinations are insufficient to inhibit HIV-1 replication yet remain at levels that exert selective pressure. The goal of health care providers should be to prescribe sufficiently potent HAART to avoid selection of drug-resistant virus on the basis of inadequate drug pressure, while minimizing adverse sequelae from the drugs and unnecessary expense by utilization of an excessive number of drugs. Achieving these therapeutic goals with the currently available antiretrovirals is often difficult due to the extremely high degree of adherence to therapy (>95%) that appears to be required to preclude the emergence of drug-resistant virus [2], the persistence of drug-resistant virus in the child [3], HIV-1 cross-resistance to drugs within each class and adverse reactions associated with antiretroviral therapy.

In this chapter, the initiation of HAART is discussed and studies of three- and four-drug HAART regimens and salvage therapy in children are reviewed. Additionally, data pertaining to the pharmacokinetics of antiretrovirals in children are reviewed including absorption, distribution, and rationale for monitoring of drug levels, challenges relating to antiretroviral resistance and recommendations for treatment of HIV-1-infected children.

Initiation of HAART

The optimal time to initiate HAART in children has not been rigorously investigated. Untreated HIV-1 infection progresses to AIDS in a mean of 4.7 years [4–7]. However, this figure obscures what is effectively two patterns of HIV-1 disease progression in children: rapid progression to AIDS in the first year or two of life occurs in approximately 10 to 25% of infected infants, and slower disease progression, over 6–9 years, in the remaining children. The rapidity at which disease can progress led many experts to promote the initiation of treatment early in the course of infection (*Guidelines for the Use of Antiretroviral Agents in Pediatric HIV Infection* on the AIDSinfo Website; http://www.aidsinfo.nih.gov. More recently, treatment of primary infection has been advocated to assist in the maintenance of HIV-1-specific immune responses [8]. Some children, however, enjoy good health and development for many years without treatment. Thus, objective parameters that accurately indicate the risk and rate of disease progression would assist in guiding the initiation of HAART in children. Plasma viral load and CD4 levels independently predict the rapidity of HIV-1 disease progression in children [9–11]; however, the predictive value of plasma HIV-1 RNA is less precise than observed in adults [12, 13]. HIV-1 RNA levels are higher in children compared to adults during primary infection and these high levels decay slowly during the first 2 years of life [10]. Enhancing the power of viral and immune parameters in predicting the rate of disease progression could prove difficult as multiple viral [14, 15], host [16, 17] (CCR2 [18, 19], CD38 [20], a maturing immune system [20, 21]) and environmental factors (nutrition [22, 23], and Vitamin A [24]) appear to interact in a complex fashion to affect the rate of disease progression. Ideally, predictors of disease progression and the risk of adverse reactions and long-term toxicities associated with antiretrovirals could be weighed by clinicians and families when developing a plan to administer HAART to a child. At the present time available data does not allow such precision in the development of a treatment plan. Rather, the clinician is advised to engage in an ongoing assessment of the risk of disease progression while educating the family as to the risks and benefits of HAART and together elect a therapeutic plan.

Once initiated, HAART has been associated with CD4 reconstitution, often to levels within the norms for age [1]. In the mid-1990s as little as 3 years of HAART was predicted to be sufficient to eliminate HIV-1 infection [25]; thus, once initiated continuous HAART was advocated. Multiple studies have since indicated that HAART may not completely stop viral replication and low-level viral replication could replenish viral reservoirs [26–37]. In addition, latently infected cells appear to survive for a longer time than initially predicted in adults [38] and in children [39], markedly prolonging the estimated "time to cure," and casting doubt that utilization of currently available antiretrovirals could eliminate infection [40]. Data suggesting that HAART will not cure HIV-1 infection has led to studies of alternative treatment regimens. Episodic treatments, aimed at lessening drug toxicities and expenses attributable to antiretrovirals, have been investigated in adults [41], and intermittent therapy with treatment duration determined by each subject's CD4 cell counts is under evaluation by the NIH-sponsored Adult AIDS Clinical Trials Group (Protocol A5102). These studies will help in ascertaining the relative efficacy of continuous and intermittent therapy in adults. Similar studies in young children seem warranted because the immune regenerative capacity of children is potentially greater and the cumulative long-term adverse reactions may be more pronounced that those found in adults. Clearly, the optimal time to initiate HAART and whether to administer therapy continuously or intermittently remains to be determined.

Three-drug HAART

Early comparative trials of HAART composed of one or two nucleoside analogs (nucleoside reverse transcriptase inhibitors (NRTIs)) and a protease inhibitor (PI) demonstrated virologic benefits in children compared to two nucleoside analogs [42]. The rate of viral suppression (plasma HIV-1 RNA < 400 copies/ml) by dual nucleoside therapy (zidovudine plus lamivudine) was inferior compared to zidovudine plus lamivudine plus ritonavir and to stavudine plus ritonavir in this randomized comparative trial, with maximal virologic benefit seen in association with the three-drug regimen. In this and other studies of three-drug HAART, plasma HIV-1 RNA was suppressed to <400 copies/ml in 25–42% of nucleoside-experienced children [42–45]. The PENTA five trial compared three regimens composed of two nucleosides (zidovudine plus lamivudine, zidovudine plus abacavir, and abacavir plus lamivudine) in previously untreated children [46]. Symptomatic children, which included 80% of the study population, were also given nelfinavir. The abacavir containing regimens resulted in a significantly greater reduction in HIV-1 RNA levels compared to zidovudine plus lamivudine at 24 weeks. The lamivudine plus abacavir combination resulted in the most durable reduction in HIV-1 RNA load with 56% of patients <50 copies/ml and 71% of patients <400 copies/ml at 48 weeks (Tab. 1). A short-lived additive effect of nelfinavir was apparent at study week 24, but not at study week 48. The nelfinavir dose during the first part of the study was low and nelfinavir powder rather than tablets was administered; both factors may have contributed to sub-optimal drug levels and to the transient effect on suppression of viral replication. Also of note was that abacavir hypersensitivity reactions did not appear to be increased in children.

Four-drug HAART regimens

To date comparative trials of HAART in children have been few. However, in reviewing both the comparative and observational studies, four-drug HAART regimens in children have demonstrated higher rates of suppression of viral

Table 1. Randomized comparative efficacy of HAART in children

ACTG 338 [42]**	Percentage of children with viral load < 400 copies/ml			
	Week 12	*Week 24*		
Zidovudine/Lamivudine ($n = 100$)	12			
Zidovudine/Lamivudine/Ritonavir ($n = 100$)	52	42		
Stavudine/Ritonavir ($n = 97$)	54	27		
	$p < 0.001^x$	$p = 0.04$		
ACTG 377 [47, 48]**	**Percentage of children with viral load < 400 copies/ml**			
	Week 4	*Week 8*	*Week 12*	*Week 24*
Stavudine/Nevirapine/Ritonavir ($n = 41$)	44	49	44*	41
Stavudine/Nevirapine/Nelfinavir ($n = 44$)	39	52	50	41
Stavudine/Lamivudine/Nevirapine/Nelfinavir ($n = 41$)	39	59	68*^	61
Stavudine/Lamivudine/Nelfinavir ($n = 50$)	34	42	44^	46
Stavudine/Lamivudine/Nelfinavir BID ($n = 11$)	55	55	55	64
Saquinavir Study [83]***	**Percentage of children with viral load < 50 copies/ml**			
	Week 48	*Week 72*		
Two NRTIs plus Saquinavir+ ($n = 14$)		36		
One or Two NRTIs Saquinavir/Ritonavir ($n = 13$)	62			

(Continued on next page)

Table 1. (Continued)

PENTA 5 [46]	Percentage of children with viral load < X copies/ml			
	Week 24	*Week 24*	*Week 48*	*Week 48*
	<50	<400	<50	<400
Zidovudine/Lamivudine ± Nelfinavir ($n = 36$)	39	44	31	44
Zidovudine/Abacavir ± Nelfinavir ($n = 43$)	49	74	44	60
Lamivudine/Abacavir ± Nelfinavir ($n = 45$)	49	76	56	71
Adjusted p-value comparing efficacy of all three regimens^^	$p = 0.65$	$p = 0.10$	$p = 0.006$	$p = 0.005$
Concentration-Controlled Therapy [68]	**Percentage of children with viral load < 50 copies/ml**			
	Week 52			
Zidovudine/Lamivudine/Indinavir – SOC# ($n = 16$)	59			
Zidovudine/Lamivudine/Indinavir – CC# ($n = 17$)	94			
	$p = 0.017$			

* $p = 0.045$; ^ $p = 0.03$
x comparison of ritonavir-containing regimens to zidovudine plus lamivudine
** Patients were nucleoside-experienced
*** Patients were protease-inhibitor naïve but were significantly nucleoside-experienced with eight children in each group with known nucleoside-resistance
+ Pharmocokinetics studies found saquinavir levels were suboptimal
^^ adjusted for nelfinavir, age, baseline CD4 percentage and plasma HIV-1 RNA; abacavir at week 24 and lamivudine/abacavir at week 48 appeared most potent
# SOC = standard of care;
CC = concentration-controlled dosing of antiretrovirals
ACTG = AIDS Clinical Trials Group
PENTA = Pediatric European Network for Treatment of AIDS
SOC = standard of care group
CC = concentration-controlled group
BID = twice daily dosing
NRTIs = nucleoside reverse transcriptase inhibitors

replication compared to three-drug regimens. Most studies of HAART have included children who have had previous treatment with nucleoside analogs. The only randomized study that compared three- to four-drug regimens in children is Pediatric AIDS Clinical Trials Group (PACTG) 377 [47, 48]. In this study children were randomized to the four-drug combination of two NRTIs (stavudine and lamivudine), one non-nucleoside reverse transcriptase inhibitor (NNRTI) (nevirapine) and one PI (nelfinavir), or to one of three three-drug combinations: stavudine, nevirapine and ritonavir; stavudine, nevirapine and nelfinavir; or stavudine, lamivudine and nelfinavir. Plasma HIV-1 RNA levels were <400 copies/ml in 68%, 61% and 52% of children on the four-drug regimen after 12, 24 and 48 weeks of therapy, respectively (Tab. 1). The rate of viral suppression for the four-drug regimen was statistically greater than two of the three-drug arms, but not different from the group receiving the combination of stavudine, lamivudine and nelfinavir. The nelfinavir dose was administered either twice or thrice daily to these children with more of the twice daily nelfinavir having plasma HIV-1 RNA levels <400 copies/ml after 24 weeks compared to those prescribed nelfinavir three times a day (64% of $n = 11$ *versus* 46% of $n = 52$). This difference was not statistically significant, due to the small size of the study groups. It is likely that the convenience of twice daily administration of the stavudine, lamivudine and nelfinavir regimen increased adherence and therefore the rate of suppression of viral replication.

In another trial (PACTG 382) four-drug HAART was administered with all enrollees receiving efavirenz, nelfinavir and one ($n = 36$) or two nucleosides ($n = 61$) in children that may have had previous experience with nucleoside analogs [49]. This trial focused on the pharmacokinetics of efavirenz, and the antiviral and immune benefits of this combination of antiretrovirals as part of a Phase I/II study. Children in this study had a slightly greater rate of suppression of HIV-1 replication compared to those in ACTG 377, with 76% having plasma RNA levels <400 copies/ml and 63% <50 copies/ml after 48 weeks taking the regimen, with response rates not evaluated by the number of drugs administered.

In an observational study of lopinavir/ritonavir combined with two NRTIs 84% of previously untreated children and 88% of NRTI-pretreated children had viral loads <400 copies/ml at 48 weeks in an intent-to-treat analysis [50]. A comparison of three- to four-drug HAART is ongoing in sequential cohorts of antiretroviral-naïve infants (PACTG Protocol 356), and preliminary reports similarly suggest that four-drug therapy produces a greater antiviral effect [51].

Many experts argue against the use of four drugs in the initial HAART regimen based on several reasonable suppositions, namely, that adherence to therapy could decrease as the pill burden increases; that adverse reactions and toxicities could be increased; and that the initial use of four-drug HAART leaves the child with less of a chance for successful "salvage" therapy, particularly if all three (NRTI, NNRTI, PI) classes of antiretrovirals are included in the initial regimen. While the data examining these issues is meager, it does not clearly support these suppositions. Toxicities were not increased among children taking four-drug compared to those taking three-drug HAART in PACTG

377 study [47]. The rate of treatment cessation was on average more than two-fold greater with three- compared to four-drug HAART in this study, primarily due to lack of suppression of viral replication. Adherence to the regimens was not evaluated; however, the superior antiviral effect of the four-drug HAART suggests that adherence to the former was not significantly compromised, in spite of a greater pill burden.

Second HAART and "salvage" therapy

When the first HAART regimen no longer suppresses viral replication many clinicians then prescribe a second regimen often consisting of four- or more drugs, often followed by a third regimen. To simplify discussion we will refer to all regimens following the initial HAART as "salvage" therapy. Viral suppression during salvage therapy has been inferior to the initial therapy of untreated children. At the present time, treatment options following "failure" of HAART are meager, due to the limited number of available antiretrovirals and a high degree of cross-resistance between agents within each drug class. The relative success of an initial four-drug HAART regimen *versus* three-drug HAART followed by four-drug salvage has not been evaluated in children. Preliminary analysis of "salvage" therapy for children demonstrated little success at viral suppression in children previously treated with three-drug HAART that included protease inhibitors (PACTG 366); however, CD4 cell counts increased in many children [52]. On entering the study 100% of the children had "primary" mutations associated with resistance to protease inhibitors (Frenkel, unpublished data). Among these children approximately 10% had their plasma HIV-1 RNA replication suppressed to <400 copies/ml after 12 weeks of a four-drug regimen including nucleoside analogs not previously used, nevirapine and a protease inhibitor.

The outcome of a second HAART regimen or "salvage" therapy can potentially be improved by limiting the time "failing" therapy is continued after viral "rebound." When the viral genotype was determined shortly after virologic failure (i.e., shortly after viral replication was detected) during HAART, the virus had few drug-resistance mutations [53]. HIV-1 drug resistance was essentially limited to the mutation conferring resistance to lamivudine, the one agent in the regimen that requires a single mutation for high-level resistance. Importantly, mutations associated with resistance to protease inhibitors were not detected. HIV-1 resistance to protease inhibitors generally requires multiple mutations and at least several months of ongoing replication to select for highly drug-resistant virus. The loss of suppression of viral replication, or "virologic treatment failure," in children has been associated with the selection of drug-resistant viruses and the accumulation of additional mutations if treatment was continued [76]. Thus, frequent monitoring of plasma HIV-1 RNA can detect resumption or "rebound" of viral replication early on when HIV-1 resistance is likely limited to a single agent in the regimen and there is the potential for achieving suppression of viral replication by changing or adding agents. In addition, so-called “mega-HAART”, which includes treatment with five or more antiretrovirals, can, in our experience [1], suppress viral replication in antiretroviral-experienced children, if they do not have high-level resistance to protease inhibitors, assuming a high rate of adherence (>95%) to the regimen.

“Salvage” therapies have not yet been systematically studied in children. It is possible that the sequencing of the different classes of drugs might allow the currently available agents to be configured into several effective HAART regimens, and thus, improve the duration of virologic response to therapy.

HIV-1 drug resistance

Viral drug-resistance is an increasingly common challenge to the effective treatment of HIV-1 infected children. Drug-resistant mutants are rapidly selected when antiretroviral drug levels are inadequate to inhibit viral replication, but exert selective pressure. In children achieving drug levels that suppress viral replication is problematic due to both the pharmacokinetic properties of the drugs and the difficulties in administering them. Additionally, infection with drug-resistant virus has been reported in infants [54–57], the frequency of which appears to be increasing [58]. Consequently, it is anticipated that HIV-1 drug resistance will become a major obstacle to the effective treatment of children.

As most studies of HAART have focused on the proportion of individuals that demonstrate suppression of HIV-1 replication to below the limits of detection, the immunologic benefits conferred by partial suppression of replication may have been overlooked. HAART resulting in incomplete but sustained suppression of viral replication to RNA levels below their “viral set point” has been associated with significant increases in CD4 cell counts [59]. Importantly, the duration of these benefits has not been characterized. The determinants of immunologic recovery have not been defined in children but are likely to be age-dependent. In young children, we have observed sustained CD4 cell increases (>3 years) despite high-level viral resistance and return of plasma HIV-1 RNA levels to the individual’s “set-point” (Frenkel, unpublished data). These observations suggest two phenomena that require definition. First, the spectrum of immunologic benefits provided by lowering the level of viral replication, and second, the parameters affecting immune reconstitution of children, and once achieved, its natural history.

Pharmacokinetics potentially affecting pediatric HAART

Newborn infants have immature hepatic and renal function, which undergo rapid maturational changes in the first few months of life. Glomerular filtration generally doubles during the first 4 weeks of life and secretion capacity reaches adult levels around 30 months of age. These changes in renal function, coupled with variable drug absorption at all ages, result in different dosing requirements between neonates, young children and older children and argue for therapeutic monitoring of certain drugs. Interactions between drugs also suggest changes in dosing. Unfortunately, pharmacokinetic data upon which to determine dosing in neonates and young infants is sparse or absent for many drugs.

The plasma concentrations of efavirenz and nelfinavir were lower than expected in young children participating in PACTG 382 [49] and PENTA 7 [60], replaying a common theme with antiretrovirals in infants and young children, including ritonavir [61, 62], nelfinavir [46, 63] and nevirapine [64–66]. Didanosine pharmacokinetics are very variable in children [67]. Drug-drug interactions in adults significantly lower certain drug-concentrations. Until studies are completed in children the dosing adjustments recommended for drug interactions in adults should be applied to children (http://www.aidsinfo.nih.gov). Pharmacokinetic variability, drug-drug interactions and an absence of dosing data for young infants suggest that therapeutic monitoring of drug levels might be valuable for optimal management of HAART in children. The first study of concentration-controlled therapy in children strongly supports this supposition [68]. Among 40 antiretroviral-naïve children administered zidovudine, lamivudine and indinavir, 15/16 children with concentration-controlled therapy *versus* 9/17 patients receiving "standard-of-care" achieved HIV-1 RNA levels < 50 copies/ml at study week 52. Concentration-controlled therapy significantly increased indinavir concentrations without an increase in adverse events in a study of adults [69], and indinavir trough and didanosine area under the curve concentrations have correlated with changes in HIV-1 RNA levels in children between baseline and study week 24 [70]. While these data argue for therapeutic drug monitoring, studies in adults do not invariably demonstrate an effect on viral outcome [71], indicating that additional studies are needed to define its role with selected drugs and ages of children. Drugs that inhibit the cytochrome P450 system, including the protease inhibitor ritonavir, can substantially increase levels of other protease inhibitors. Use of low-dose ritonavir as a pharmacologic "booster" would seem reasonable when constructing HAART regimens for children, until studies defining optimal drug-doses and plasma or intracellular drug levels are completed.

Pediatric and adolescent dosing

Data concerning pediatric dosing is rapidly accumulating. When prescribing antiretrovirals, one can assume that newborns generally excrete drugs more slowly than older infants, and that infants and children generally have lower drug concentrations when dosed by weight than would be anticipated from studies in adults, due to poor absorption and/or more rapid excretion. Data for dosing of infants is sparse, however, recommendations for dosing are continuously updated and can be found in the *Guidelines for the Use of Antiretroviral Agents in Pediatric HIV Infection* on the AIDSinfo Website (http://www.aidsinfo.nih.gov).

Dosing of adolescents is generally based on Tanner staging of puberty, and not on age. During early puberty (Tanner Stage I and II) pediatric dosing is recommended, while during late puberty (Tanner Stage V) adult dosing is recommended. During the adolescent growth spurt (Tanner Stages III, female, and IV, male) close monitoring is recommended, both for efficacy in suppressing viral replication and for manifestations of toxicity. We recommend continuing with pediatric dosing during the growth spurt, unless toxicities occur, as maintenance of suppression of viral replication during this time is tenuous due to social and psychological factors.

Drug administration

Administering protease inhibitors can present challenges. Amprenavir suspension contains propylene glycol and should be avoided in children <3 years old. Indinavir is manufactured only as capsules, although formulation into a suspensions has been explored [72]. Lopinavir/ritonavir and ritonavir suspensions are unpalatable to most children. Nelfinavir powder is difficult to dissolve and the amount is larger than most young children will tolerate, the tablets dissolved in water have been better accepted [46]. Administration of medications via a gastrostomy-tube has been lifesaving for many infants and children unable to swallow pills. Unpalatable drug suspensions and certain pills dissolved in water can be successfully administered using gastrostomy tubes [1, 46, 73].

A practical approach to HAART in children

As basic questions pertaining to the initiation and management of HAART in children remain unanswered, we recommend that whenever possible children join studies designed to define optimal HAART regimens and strategies. When studies are not available we recommend HAART of HIV-1-infected infants, even though comparative trials have not been done to evaluate the efficacy and toxicities of this approach compared to delayed treatment. We would continue therapy beyond infancy if toxicities are not evident and if continuation is acceptable to the family. In children with adequate CD4 cell levels we would agree with suspension of therapy if adherence is sub-optimal or if the family requests that HAART be suspended. Due to the rapidity with which drug-resistant viruses are selected when adherence to HAART is sub-optimal, we do not recommend therapy of children beyond infancy until the family is highly motivated to treat the child's HIV-1 infection. In most instances this will be the case when the child has a moderate or severe degree of immunodeficiency or HIV-1-related symptoms.

Data are not available to determine the best initial HAART regimen. An algorithm for initial therapy in the *Guidelines for the Use of Antiretroviral Agents in Pediatric HIV Infection* on the AIDSinfo Website, http://www.aidsinfo.nih.gov suggests either one highly active protease inhibitor (nelfinavir or ritonavir) or lopinavir/ritonavir plus two nucleoside analogue reverse transcriptase inhibitors. If children can swallow pills, efavirenz plus two NRTIsS, or efavirenz plus nelfinavir and one NRTI is suggested; however, these regimens have not been studied in children and their absolute and relative efficacy are not known. As only 40–65% of children and adults have continued suppression of HIV-1 replication to below the limits of detection after 6 months or more of treatment [47, 49, 74, 75], we believe that more effective HAART regimens are required. The loss of suppression of viral replication, or "virologic treatment failure," in children has been associated with the selection of drug-resistant viruses and the accumulation of additional mutations if treatment was continued [76]. In addition, the pharmacokinetic properties are frequently unreliable in young children [77]. Taken together, these data lead us to advocate treating antiretroviral-naïve infants and children with four-drug HAART, including two NRTIs, one NNRTI and one PI, as the majority of children received in PACTG 382 [49], or with a combination that we and others [50] have found quite effective and well-tolerated (namely, lopinavir/ritonavir, lamivudine and stavudine). Adequate levels of protease inhibitors clearly impact on the success of HAART. While ritonavir-boosting generally leads to higher concentrations of protease inhibitors, some experts advocate therapeutic drug monitoring due to the large variability in the effects achieved [68].

Frequent monitoring of plasma HIV-1 RNA (i.e., monthly) early after treatment is initiated provides the opportunity to modify, intensify or stop therapy if plasma RNA levels do not progressively decrease or if RNA "rebounds." Using this strategy the antiretroviral regimen can usually be intensified with additional drugs, prior to the selection of multi-drug resistant mutants, to a regimen that generally will suppress HIV-1 RNA levels to <50 copies/ml [1].

We recommend HIV-1 drug-resistance testing of infants infected by antiretroviral-treated mothers; infants, children or adolescents failing their HAART regimen; and in adolescents infected in communities known to have primary infections with drug-resistant HIV-1. Testing for drug-resistant mutants in infants infected by antiretroviral-treated mothers and in adolescents may indicate drugs to which the virus is resistant that are best avoided when designing the HAART regimen. Genotypic tests appear more sensitive than phenotypic assays in detecting drug-resistant virus transmission, in preliminary studies [78, 79]. Point mutation assays are more sensitive than consensus sequencing [80, 81] and may be preferable when mutants are likely to be outcompeted by wild-type viruses, providing that relevant mutants are detected in the assay [82].

Salvage regimens can achieve marked suppression of viral replication for certain children [1] and immunologic benefits with minimal suppression of viral replication for others [52]. We recommend the utilization of regimens dosed twice daily that include five or more drugs, if possible including antiretrovirals that were previously not utilized together and to which testing and treatment history do not indicate a high-level of virus-drug resistance.

Conclusion

Multiple factors appear to influence whether HAART provides long-term suppression of viral replication in children, including dosing frequency and adherence, variable pharmacokinetics, presence of drug-resistant virus and superior suppression of viral replication by potent HAART regimens. Recognition of these factors by families and health care providers can foster the cooperation critical for successful HAART. Furthermore, acknowledgment of large knowledge gaps in our optimally implementing HAART should provide the rationale to treat children within protocols addressing these critical questions.

References

1 Melvin AJ, Lewis PF, Mohan KM, Naugler WS, Frenkel LM (2002) Efficacy and toxicity of antiretroviral therapy using 4 or more agents: application of a strategy for antiretroviral management in human immunodeficiency virus-infected children. *Arch Pediatr Adolesc Med* 156: 568–573

2 Paterson DL, Swindells S, Mohr J, Brester M, Vergis EN, Squier C, Wagener MM, Singh N (2000) Adherence to protease inhibitor therapy and outcomes in patients with HIV infection. *Ann Intern Med* 133: 21–30

3 Hermankova M, Ray SC, Ruff C, Powell-Davis M, Ingersoll R, D'Aquila RT, Quinn TC, Siliciano JD, Siliciano RF, Persaud D (2001) HIV-1 drug resistance profiles in children and adults with viral load of <50 copies/ml receiving combination therapy. *JAMA* 286: 196–207

4 Auger I, Thomas P, De Gruttola V, Morse D, Moore D, Williams R, Truman B, Lawrence CE (1988) Incubation periods for paediatric AIDS patients. *Nature* 136: 575–577

5 Duliege AM, Messiah A, Blanche S, Tardieu M, Griscelli C, Spira A (1992) Natural history of human immunodeficiency virus type 1 infection in children: prognostic value of laboratory tests on the bimodal progression of the disease. *Pediatr Infect Dis J* 11: 630–635

6 Mayaux MJ, Burgard M, Teglas JP, Cottalorda J, Krivine A, Simon F, Puel J, Tamalet C, Dormont D, Masquelier B et al. (1996) Neonatal characteristics in rapidly progressive perinatally acquired HIV-1 disease. The French Pediatric HIV Infection Study Group. *JAMA* 275: 606–610

7 Aldrovandi GM (1999) Natural history of pediatric HIV disease. In: Zeichner SL, Read JS, ed. *Handbook of Pediatric HIV Care*. Philadelphia: Lippincott Williams and Wilkins

8 Rosenberg ES, Billingsley JM, Caliendo AM, Boswell SL, Sax PE, Kalams SA, Walker BD (1997) Vigorous HIV-1-specific CD4+ T cell responses associated with control of viremia. *Science* 278: 1447–1450

9 Mofenson LM, Korelitz J, Meyer WA 3rd, Bethel J, Rich K, Pahwa S, Moye J Jr, Nugent R, Read J (1997) The relationship between serum human immunodeficiency virus type 1 (HIV-1) RNA level, CD4 lymphocyte percent, and long-term mortality risk in HIV-1-infected children. National Institute of Child Health and Human Development Intravenous Immunoglobulin Clinical Trial Study Group. *J Infect Dis* 175: 1029–1038

10 Shearer WT, Quinn TC, LaRussa P, Lew JF, Mofenson L, Almy S, Rich K, Handelsman E, Diaz C, Pagano M et al. (1997) Viral load and disease progression in infants infected with human immunodeficiency virus type 1. Women and Infants Transmission Study Group. *N Engl J Med*

336: 1337–1342
11 Palumbo PE, Raskino C, Fiscus S, Pahwa S, Fowler MG, Spector SA, Englund JA, Baker CJ (1998) Predictive value of quantitative plasma HIV RNA and CD4+ lymphocyte count in HIV-infected infants and children. *JAMA* 279: 756–761
12 Mellors JW, Rinaldo CR Jr, Gupta P, White RM, Todd JA, Kingsley LA (1996) Prognosis in HIV-1 infection predicted by the quantity of virus in plasma. *Science* 272: 1167–1170
13 Mellors JW, Muñoz A, Giorgi JV, Margolick JB, Tassoni CJ, Gupta P, Kingsley LA, Todd JA, Saah AJ, Detels R et al. (1997) Plasma viral load and CD4+ lymphocytes as prognostic markers of HIV-1 infection. *Ann Intern Med* 126: 946–954
14 Blanche S, Mayaux MJ, Rouzioux C, Teglas JP, Firtion G, Monpoux F, Ciraru-Vigneron N, Meier F, Tricoire J, Courpotin C et al. (1994) Relation of the course of HIV infection in children to the severity of the disease in their mothers at delivery. *N Engl J Med* 330: 308–312
15 Dickover RE, Dillon M, Leung KM, Krogstad P, Plaeger S, Kwok S, Christopherson C, Deveikis A, Keller M, Stiehm ER et al. (1998) Early prognostic indicators in primary perinatal human immunodeficiency virus type 1 infection: importance of viral RNA and the timing of transmission on long-term outcome. *J Infect Dis* 178: 375–387
16 Wang J, Crawford K, Yuan M, Wang H, Gorry PR, Gabuzda D (2002) Regulation of CC chemokine receptor 5 and CD4 expression and human immunodeficiency virus type 1 replication in human macrophages and microglia by T helper type 2 cytokines. *J Infect Dis* 185: 885–897
17 McCarthy M, He J, Auger D, Geffin R, Woodson C, Hutto C, Wood C, Scott G (2002) Cellular tropisms and co-receptor usage of HIV-1 isolates from vertically infected children with neurological abnormalities and rapid disease progression. *J Med Virol* 67: 1–8
18 Lathey JL, Tierney C, Chang SY, D'Aquila RT, Bettendorf DM, Alexander HC, Santini CD, Hughes AM, Barroga CF, Spector SA et al. (2001) Associations of CCR5, CCR2, and stromal cell-derived factor 1 genotypes with human immunodeficiency virus disease progression in patients receiving nucleoside therapy. *J Infect Dis* 184: 1402–1411
19 Tang J, Shelton B, Makhatadze NJ, Zhang Y, Schaen M, Louie LG, Goedert JJ, Seaberg EC, Margolick JB, Mellors J et al. (2002) Distribution of chemokine receptor CCR2 and CCR5 genotypes and their relative contribution to human immunodeficiency virus type 1 (HIV-1) seroconversion, early HIV-1 RNA concentration in plasma, and later disease progression. *J Virol* 76: 662–672
20 McCloskey TW, Kohn N, Lesser M, Bakshi S, Pahwa S (2001) Immunophenotypic analysis of HIV-infected children: alterations within the first year of life, changes with disease progression, and longitudinal analyses of lymphocyte subsets. *Cytometry* 46: 157–165
21 Rich KC, Brambilla D, Pitt J, Moye J, Cooper E, Hillyer G, Mendez H, Fowler MG, Landay A (1997) Lymphocyte phenotyping in infants: maturation of lymphocyte subpopulations and the effects of HIV infection. *Clin Immunol Immunopathol* 85: 273–281
22 Raiten DJ, Talbot JM (1996) Nutrition in pediatric HIV infection: setting the research agenda. *J Nutr* 126: 2599S–2696S
23 Salomon J, De TP, Melchior JC (2002) Nutrition and HIV infection. *Br J Nutr* 87 (Suppl 1): S111–S119
24 Duggan C, Fawzi W (2001) Micronutrients and child health: studies in international nutrition and HIV infection. *Nutr Rev* 59: 358–369
25 Perelson AS, Essunger P, Cao Y, Vesanen M, Hurley A, Saksela K, Markowitz M, Ho DD (1997) Decay characteristics of HIV-1-infected compartments during combination therapy. *Nature* 387: 188–191
26 Chun TW, Stuyver L, Mizell SB, Ehler LA, Michan JA, Baseler M, Lloyd AL, Nowak MA, Fauci AS (1997) Presence of an inducible HIV-1 latent reservoir during highly active antiretroviral therapy. *Proc Natl Acad Sci USA* 94: 13193–13197
27 Finzi D, Hermankova M, Pierson T, Carruth LM, Buck C, Chaisson RE, Quinn TC, Chadwick K, Margolick J, Brookmeyer R et al. (1997) Identification of a reservoir for HIV-1 in patients on highly active antiretroviral therapy. *Science* 278: 1295–1300
28 Zhang L, Ramratnam B, Tenner-Racz K, He Y, Vesanen M, Lewin S, Talal A, Racz P, Perelson AS, Korber BT et al. (1999) Quantifying residual HIV-1 replication in patients receiving combination antiretroviral therapy. *N Engl J Med* 340: 1605–1613
29 Natarajan V, Bosche M, Metcalf JA, Ward DJ, Lane HC, Kovacs JA (1999) HIV-1 replication in patients with undetectable plasma virus receiving HAART. Highly active antiretroviral therapy. *Lancet* 353: 119–120

30 Dornadula G, Zhang H, VanUitert B, Stern J, Livornese L Jr, Ingerman MJ, Witek J, Kedanis RJ, Natkin J, DeSimone J et al. (1999) Residual HIV-1 RNA in blood plasma of patients taking suppressive highly active antiretroviral therapy. *JAMA* 282: 1627–1632
31 Furtado MR, Callaway DS, Phair JP, Kunstman KJ, Stanton JL, Macken CA, Perelson AS, Wolinsky SM (1999) Persistence of HIV-1 transcription in peripheral-blood mononuclear cells in patients receiving potent antiretroviral therapy. *N Engl J Med* 340: 1614–1622
32 Gunthard HF, Frost SD, Leigh-Brown AJ, Ignacio CC, Kee K, Perelson AS, Spina CA, Havlir DV, Hezareh M, Looney DJ et al. (1999) Evolution of envelope sequences of human immunodeficiency virus type 1 in cellular reservoirs in the setting of potent antiviral therapy. *J Virol* 73: 9404–9412
33 Hockett RD, Kilby JM, Derdeyn CA, Saag MS, Sillers M, Squires K, Chiz S, Nowak MA, Shaw GM, Bucy RP (1999) Constant mean viral copy number per infected cell in tissues regardless of high, low, or undetectable plasma HIV RNA. *J Exp Med* 189: 1545–1554
34 Sharkey ME, Teo I, Greenough T, Sharova N, Luzuriaga K, Sullivan JL, Bucy RP, Kostrikis LG, Haase A, Veryard C et al. (2000) Persistence of episomal HIV-1 infection intermediates in patients on highly active anti-retroviral therapy. *Nat Med* 6: 76–81
35 Yerly S, Perneger TV, Vora S, Hirschel B, Perrin L (2000) Decay of cell-associated HIV-1 DNA correlates with residual replication in patients treated during acute HIV-1 infection. *AIDS* 14: 2805–2812
36 Ramratnam B, Mittler JE, Zhang L, Boden D, Hurley A, Fang F, Macken CA, Perelson As, Markowitz M, Ho DD (2000) The decay of the latent reservoir of replication-competent HIV-1 is inversely correlated with the extent of residual viral replication during prolonged anti-retroviral therapy. *Nat Med* 6: 82–85
37 Martinez-Picado J, DePasquale MP, Kartsonis N, Hanna GJ, Wong J, Finzi D, Rosenberg E, Gunthard HF, Sutton L, Savara A et al. (2000) Antiretroviral resistance during successful therapy of HIV type 1 infection. *Proc Natl Acad Sci USA* 97: 10948–10953
38 Finzi D, Blankson J, Siliciano JD, Margolick JB, Chadwick K, Pierson T, Smith K, Lisziewicz J, Lori F, Flexner C et al. (1999) Latent infection of CD4+ T cells provides a mechanism for lifelong persistence of HIV-1, even in patients on effective combination therapy. *Nat Med* 5: 512–517
39 Persaud D, Pierson T, Ruff C, Finzi D, Chadwick KR, Margolick JB, Ruff A, Hutton N, Ray S, Siliciano A (2000) A stable latent reservoir for HIV-1 in resting $CD4^+$ T lymphocytes in infected children. *J Clin Invest* 105: 995–1003
40 Siliciano R (2000) Can antiretroviral therapy ever be stopped? *AIDS Read* 10: 224–229
41 Dybul M, Chun TW, Yoder C, Hidalgo B, Belson M, Hertogs K, Larder B, Dewar RL, Fox CH, Hallahan CW et al. (2001) Short-cycle structured intermittent treatment of chronic HIV infection with highly active antiretroviral therapy: effects on virologic, immunologic, and toxicity parameters. *Proc Natl Acad Sci USA* 98: 15161–15166
42 Nachman SA, Stanley K, Yogev R, Pelton S, Wiznia A, Lee S, Mofenson L, Fiscus S, Rathore M, Jimenez E et al. (2000) Nucleoside analogs plus ritonavir in stable antiretroviral therapy- experienced HIV-infected children: a randomized controlled trial. Pediatric AIDS Clinical Trials Group 338 Study Team. *JAMA* 283: 492–498
43 Kline MW, Fletcher CV, Harris AT, Evans KD, Brundage RC, Remmel RP, Calles NR, Kirkpatrick SB, Simon C (1998) A pilot study of combination therapy with indinavir, stavudine (d4T), and didanosine (ddI) in children infected with the human immunodeficiency virus. *J Pediatr* 132: 543–546
44 Rutstein RM, Feingold A, Meislich D, Word B, Rudy B (1997) Protease inhibitor therapy in children with perinatally acquired HIV infection. *AIDS* 11: F107–F111
45 Krogstad P, Wiznia A, Luzuriaga K, Dankner W, Nielsen K, Gersten M, Kerr B, Hendricks A, Boczany B, Rosenberg M et al. (1999) Treatment of human immunodeficiency virus 1-infected infants and children with the protease inhibitor nelfinavir mesylate. *Clin Infect Dis* 28: 1109–1118
46 Paediatric European Network for Treatment of AIDS (PENTA) (2002) Comparison of dual nucleoside-analog reverse-transcriptase inhibitor regiments with and without nelfinavir in children with children with HIV-1 who have not previously been treated: the PENTA 5 randomised trial. *Lancet* 359: 733–740
47 Wiznia A, Stanley K, Krogstad P, Johnson G, Lee S, McNamara J, Moye J, Jackson JB, Mendez H, Aguayo R et al. (2000) Combination nucleoside analog reverse transcriptase inhibitor(s) plus nevirapine, nelfinavir, or ritonavir in stable antiretroviral therapy- experienced HIV-infected children: week 24 results of a randomized controlled trial–PACTG 377. Pediatric AIDS Clinical Trials

Group 377 Study Team. *AIDS Res Hum Retroviruses* 16: 1113–1121
48 Krogstad P, Lee S, Johnson G, Stanley K, McNamara J, Moye J, Jackson JB, Aguayo R, Dieudonne A, Khoury M et al. (2002) Nucleoside-analogue reverse-transcriptase inhibitors plus nevirapine, nelfinavir, or ritonavir for pretreated children infected with human immunodeficiency virus type 1. *Clin Infect Dis* 34: 991–1001
49 Starr SE, Fletcher CV, Spector SA, Yong FH, Fenton T, Brundage RC, Manion D, Ruiz N, Gersten M, Becker M et al. (1999) Combination therapy with efavirenz, nelfinavir, and nucleoside reverse-transcriptase inhibitors in children infected with human immunodeficiency virus type 1. Pediatric AIDS Clinical Trials Group 382 Team. *N Engl J Med* 341: 1874–1881
50 Saez-Llorens X, Renz C, Deetz C, Jiang P, Cahn P, Violar A, Gomez P, Handelsman E, Pelton S, Ramilo O et al. (2001) Kaletra (lopinavir/ritonavir) in HIV-infected children at 48 weeks. 8th Conference on Retroviruses and Opportunistic Infections, Chicago, IL, USA, 4–8 February 2001. Abstracts, no 680
51 Luzuriaga K (2001) Early antiretroviral therapy and the immunopathogenesis of vertical HIV-1 infection. 8th Conference on Retroviruses and Opportunistic Infections, Chicago, IL, USA, 4–8 February 2001. Abstracts, no S14
52 Kovacs A, Burchett S, Khoury M, Carey V, Pahwa S, McIntosh K, Oyomopito R, Smith E, Mofenson L, the PACTG 366 Team (2001) Virologic and immunologic responses in children with advanced HIV disease on a new HAART regimen. 8th Conference on Retroviruses and Opportunistic Infections, Chicago, IL, USA, 4–8 February 2001. Abstracts, no 684
53 Havlir DV, Hellmann NS, Petropoulos CJ, Whitcomb JM, Collier AC, Hirsch MS, Tebas P, Sommadossi JP, Richman DD (2000) Drug susceptibility in HIV infection after viral rebound in patients receiving indinavir-containing regimens. *JAMA* 283: 229–234
54 Frenkel LM, Wagner LE 2nd, Demeter LM, Dewhurst S, Coombs RW, Murante BL, Reichman RC (1995) Effects of zidovudine use during pregnancy on resistance and vertical transmission of human immunodeficiency virus type 1. *Clin Infect Dis* 20: 1321–1326
55 Colgrove RC, Pitt J, Chung PH, Welles SL, Japour AJ (1998) Selective vertical transmission of HIV-1 antiretroviral resistance mutations. *AIDS* 12: 2281–2288
56 Yerly S, Kaiser L, Race E, Bru JP, Clavel F, Perrin L (1999) Transmission of antiretroviral-drug-resistant HIV-1 variants. *Lancet* 354: 729–733
57 Johnson VA, Petropoulos CJ, Woods CR, Hazelwood JD, Parkin NT, Hamilton CD, Fiscus SA (2001) Vertical transmission of multidrug-resistant human immunodeficiency virus type 1 (HIV-1) and continued evolution of drug resistance in an HIV-1-infected infant. *J Infect Dis* 183: 1688–1693
58 Kamkamidze G, Sullivan T, Charbonneau T (2001) Occurrence of HIV-1 reverse transcriptase gene mutation at codon 215 in HIV-infected infants. *J Clin Virol* 22: 143–148
59 Deeks SG, Wrin T, Liegler T, Hoh R, Hayden M, Barbour JD, Hellmann NS, Petropoulos CJ, McCune JM, Hellerstein MK et al. (2001) Virologic and immunologic consequences of discontinuing combination antiretroviral-drug therapy in HIV-infected patients with detectable viremia. *N Engl J Med* 344: 472–480
60 Litalien C, Giaquinto C, Faye A, Mechinaud F, Grosch I (2000) Nelfinavir doses should be increased in infants (The PENTA 7 Study). XIII International AIDS Conference, Durban, South Africa, 9–14 July 2000. Abstracts, no MoPeB2213
61 Mueller BU, Nelson RP Jr, Sleasman J, Zuckerman J, Heath-Chiozzi M, Steinberg SM, Balis FM, Brouwers P, Hsu A, Saulis R et al. (1998) A phase I/II study of the protease inhibitor ritonavir in children with human immunodeficiency virus infection. *Pediatrics* 101: 335–343
62 Rodman J, Chadwick W, Palumbo P, Abrams E, Hsu A, Yogev R (1999) Ritonavir (RTV) pharmacokinetics and dose requirements in HIV infected children less than two years of age. 6th Conference on Retroviruses and Opportunistic Infections, Chicago, IL, USA, 31 January- 4 February 1999. Abstracts, no 421
63 Capparelli EV, Sullivan JL, Mofenson L, Smith E, Graham B, Britto P, Becker MI, Holland D, Connor JD, Luzuriaga K (2001) Pharmacokinetics of nelfinavir in human immunodeficiency virus-infected infants. *Pediatr Infect Dis* J 20: 746–751
64 Mirochnick M, Fenton T, Gagnier P, Pav J, Gwynne M, Siminski S, Sperling RS, Beckerman K, Jimenez E, Yogev R et al. (1998) Pharmacokinetics of nevirapine in human immunodeficiency virus type 1-infected pregnant women and their neonates. Pediatric AIDS Clinical Trials Group Protocol 250 Team. *J Infect Dis* 178: 368–374
65 Mirochnick M, Clarke DF, Dorenbaum A (2000) Nevirapine: pharmacokinetic considerations in

children and pregnant women. *Clin Pharmacokinet* 39: 281–293
66 Mirochnick M, Siminski S, Fenton T, Lugo M, Sullivan JL (2001) Nevirapine pharmacokinetics in pregnant women and in their infants after *in utero* exposure. *Pediatr Infect Dis J* 20: 803–805
67 Stevens RC, Rodman JH, Yong FH, Carey V, Knupp CA, Frenkel LM (2000) Effect of food and pharmacokinetic variability on didanosine systemic exposure in HIV-infected children. Pediatric AIDS Clinical Trials Group Protocol 144 Study Team. *AIDS Res Hum Retroviruses* 16: 415–421
68 Fletcher CV, Anderson PL, Kakuda TN, Schacker TW, Henry K, Gross CR, Brundage RC (2002) Concentration-controlled compared with conventional antiretroviral therapy for HIV infection. *AIDS* 16: 551–560
69 Kakuda TN, Page LM, Anderson PL, Henry K, Schacker TW, Rhame FS, Acosta EP, Brundage RC, Fletcher CV (2001) Pharmacological basis for concentration-controlled therapy with zidovudine, lamivudine, and indinavir. *Antimicrob Agents Chemother* 45: 236–242
70 Fletcher CV, Brundage RC, Remmel RP, Page LM, Weller D, Calles NR, Simon C, Kline MW (2000) Pharmacologic characteristics of indinavir, didanosine, and stavudine in human immunodeficiency virus-infected children receiving combination therapy. *Antimicrob Agents Chemother* 44: 1029–1034
71 Bossi P, Peytavin G, Delaugerre C, David DJ, Bonmarchand M, Cacace R, Kortza N, Ait Mohand H, Simon A, Laroche C et al. (2002) GENOPHAR: A randomized study of plasmatic drug measurements associated with genotypic resistance testing in patients failing antiretroviral therapy. 9th Conference on Retroviruses and Opportunistic Infections, Seattle, WA, USA, 24–28 February 2002. Abstracts, no 585-T
72 Mueller BU, Sleasman J, Nelson RP Jr, Smith S, Deutsch PJ, Ju W, Steinberg SM, Balis FM, Jarosinski PF, Brouwers P et al. (1998) A phase I/II study of the protease inhibitor indinavir in children with HIV infection. *Pediatrics* 102: 101–109
73 Shingadia D, Viani RM, Yogev R, Binns H, Dankner WM, Spector SA, Chadwick EG (2000) Gastrostomy tube insertion for improvement of adherence to highly active antiretroviral therapy in pediatric patients with human immunodeficiency virus. *Pediatrics* 105: E80
74 Hammer SM, Squires KE, Hughes MD, Grimes JM, Demeter LM, Currier JS, Eron JJ Jr, Feinberg JE, Balfour HH Jr, Deyton LR et al. (1997) A controlled trial of two nucleoside analogues plus indinavir in persons with human immunodeficiency virus infection and CD4 cell counts of 200 per cubic millimeter or less. *N Engl J Med* 337: 725–733
75 Staszewski S, Morales-Ramirez J, Tashima KT, Rachlis A, Skiest D, Stanford, J, Stryker R, Johnson P, Labriola DF, Farina D et al. (1999) Efavirenz plus zidovudine and lamivudine, efavirenz plus indinavir, and indinavir plus zidovudine and lamivudine in the treatment of HIV-1 infection in adults. Study 006 Team. *N Engl J Med* 341: 1865–1873
76 Eshleman SH, Krogstad P, Jackson JB, Wang YG, Lee S, Wei LJ, Cunningham S, Wantman M, Wiznia A, Johnson G et al. (2001) Analysis of human immunodeficiency virus type 1 drug resistance in children receiving nucleoside analogue reverse-transcriptase inhibitors plus nevirapine, nelfinavir, or ritonavir (Pediatric AIDS Clinical Trials Group 377). *J Infect Dis* 183: 1732–1738
77 Fletcher CV, Brundage RC, Fenton T et al. (1999) Efavirenz (EFV) and nelfinavir (NFV) pharmacokinetics (PK) in HIV-infected children participating in and area under the curve (UC) controlled trial. 6th Conference on Retroviruses and Opportunistic Infections, Chicago, IL, USA, 31 January–4 February 1999. Abstracts, no 366
78 Little SJ, Daar ES, Holte S, Frost SDW, Routy JP, Markowitz M, Collier AC, Margolick J, Koup R, Conway B et al. (2002) Persistance of transmitted drug resistance among subjects with primary HIV infection not receiving antiretroviral therapy. 9th Conference on Retroviruses and Opportunistic Infections, Seattle, WA, USA, 24–28 February 2002. Abstracts, no 95
79 Daar ES, Frost SDW, Wong JK, Hellmann NS, Wrin T, Petropoulos CJ, Richman DD, Pitt JA, Little SJ, Leigh Brown AJ (2002) Mixed infection with multidrug resistant (MDR) and wild type HIV strains in primary HIV infection (PHI): early viral rebound suggests loss of immune control. 9th Conference on Retroviruses and Opportunistic Infections, Seattle, WA, USA, 24–28 February 2002. Abstracts, no 96
80 Schmit JC, Ruiz L, Stuyver L, Van Laethem K, Vanderlinden I, Puig T, Rossau R, Desmyter J, De Clercq E, Clotet B et al. (1998) Comparison of the LiPA HIV-1 RT test, selective PCR and direct solid phase sequencing for the detection of HIV-1 drug resistance mutations. *J Virol Methods* 73: 77–82
81 Beck IA, Mahalanabis M, Pepper G, Wright A, Hamilton s, Langston E, Frenkel LM (2002) Rapid and sensitive oligonucleotide ligation assay for detection of mutations in human immunodeficien-

cy virus type 1 associated with high-level resistance to protease inhibitors. *J Clin Microbiol* 40: 1413–1419
82 Mahalanabis M, Beck IA, Pepper G et al. (2003) Improved sensitivity of oligoligase assay in detection of HIV-1 drug-resistance mutations compared to consensus dideoxynucleotide terminator sequencing. *In Review*
83 Kline MW, Brundage RC, Fletcher CV, Schwarzwald H, Calles NR, Buss NE, Snell P, DeLora P, Eason M, Jorga K et al. (2001) Combination therapy with saquinavir soft gelatin capsules in children with human immunodeficiency virus infection. *Pediatr Infect Dis J* 20: 666–671

Combination Therapy of AIDS
Edited by E.D.A. De Clercq and A.-M.I. Vandamme

Causes of HIV treatment failure

Roy M. Gulick

Weill Medical College of Cornell University, 525 East 68th Street, New York, NY 10021, USA

Introduction

What is HIV treatment failure?

Currently, there is no consensus on the definition of HIV treatment failure. Current guidelines [1–3] suggest considering virologic, immunologic and clinical responses in assessing treatment failure (Tab. 1). Virologic failure may be defined as a lack of an initial HIV RNA response to therapy or an increase or rebound in HIV RNA levels following an initial response. Immunologic failure may be defined as a decrease or a lack of increase in CD4 cell count in response to therapy. Clinical failure may be defined as the occurrence or recurrence of HIV-related signs or symptoms or AIDS-defining illnesses. There are multiple possible causes of antiretroviral treatment failure (Tab. 2).

However, discordant virologic, immunologic, and clinical responses occur. For example, CD4 cell counts may stabilize or continue to increase in the set-

Table 1. Criteria for changing therapy: U.S. Department of Health and Human Services Guidelines [1]

Virologic
• Less than a 0.5–0.75 log10 reduction in plasma HIV RNA by 4 weeks following initiation of therapy
• Less than a 1 log10 reduction in plasma HIV RNA by 8 weeks
• Failure to suppress plasma HIV RNA to undetectable levels within 4–6 months of initiating therapy
• Repeated detection of virus in plasma after initial suppression to undetectable levels, suggesting the development of resistance
• Any reproducible significant increase, defined as three-fold or greater, from the nadir plasma HIV RNA not attributable to intercurrent infection, vaccination, or test methodology
• Undetectable viremia in the patient receiving double nucleoside therapy
Immunologic
• Persistently declining CD4+ T-cell numbers, as measured on at least two separate occasions
Clinical
• Clinical deterioration

Table 2. Causes of treatment failure

- Demographic factors (e.g., age)
- Baseline parameters (e.g., HIV RNA level, CD4 cell count)
- Baseline antiretroviral resistance
- Prior antiretroviral treatment experience
- Toxicity/tolerability
- Regimen choice
- Regimen potency
- Physician experience
- Virologic response (e.g., HIV RNA nadir level)
- Immunologic response
- Clinical response
- Adherence
- Pharmacokinetics (e.g., drug concentrations, tissue reservoir penetration, drug-drug interactions)
- Other unknown reasons

ting of virologic rebound [4–6]. Clinical events may occur, sometimes with unusual manifestations [7], despite appropriate virologic and immunologic responses. Thus, virologic, immunologic, and clinical responses must all be considered in assessing treatment failure.

In addition, discontinuation of treatment alone may not signify treatment failure in many cases. Discontinuation of antiretroviral therapy occurs commonly in both clinical trials and clinical cohorts. In fact, treatment discontinuation is designated as treatment failure in clinical trials using "missing data equals failure" or "noncompleter equals failure" analyses. However, these analytic approaches may underestimate the response to therapy since some study subjects who discontinue initial treatment will go on to respond to other, subsequent regimens. Discontinuation for toxicity, non-adherence or other reasons must be distinguished from virologic, immunologic and clinical treatment failure.

How common is treatment failure?

Initial antiretroviral treatment regimens using one or two nucleoside analogue reverse transcriptase inhibitors (NRTIs) conferred transient virologic, immunologic and clinical responses [8, 9], but treatment failure occurred eventually for most patients. With the availability of potent three-drug combination antiretroviral regimens, more durable virologic, immunologic and clinical responses were demonstrated [10–15]. Despite this, treatment failure continues to occur commonly.

Bartlett and colleagues performed a review of 23 clinical trials representing 19 different three-drug antiretroviral regimens and 3257 patients [16]. The overall estimated percentage of patients on clinical studies with HIV RNA

<400 copies per milliliter (cpm) at 24 weeks was 64% (95% confidence interval, CI, 60–67%) (Fig. 1). The percentage of patients with HIV RNA <50 cpm at 48 weeks was 45–51% and the CD4 cell count increases were +123/mm^3 at 24 weeks and +160/mm^3 at 48 weeks and did not differ among the regimen classes. In this meta analysis of clinical studies, treatment failure occurred commonly.

It was also appreciated early on that treatment failure rates from clinical cohorts were unexpectedly high. One of the first reports of treatment failure from a clinical cohort was a retrospective study from Fatkenheuer and colleagues [17]. They studied 198 patients treated with protease inhibitor (PI)-containing regimens in 1996. Overall, the virological failure rate was 44%. Casado et al. [18] described a prospective study of 400 patients who started a PI from March 1996 to March 1997. In this largely treatment-experienced group, at 1 year only 45% had HIV RNA levels <200 cpm and 59% had an increase of CD4 cell count of >100 cells/mm^3.

D'Arminio Monforte and colleagues [19] evaluated 250 heavily treatment-experienced patients who started a PI-containing regimen. Over a median follow-up of 8 months, 75 (30%) had treatment failure due to an AIDS-defining event or death ($n = 24$, 32%), increasing HIV RNA level or decreasing CD4 cell count ($n = 24$, 32%), or other causes. Using data from the Swiss HIV Cohort Study, Ledergerber et al. [20] analyzed prospective data from 2674 outpatients who started potent antiretroviral treatment from 1995–1998 and found that 91% of previously untreated patients had HIV RNA <400 cpm after 12

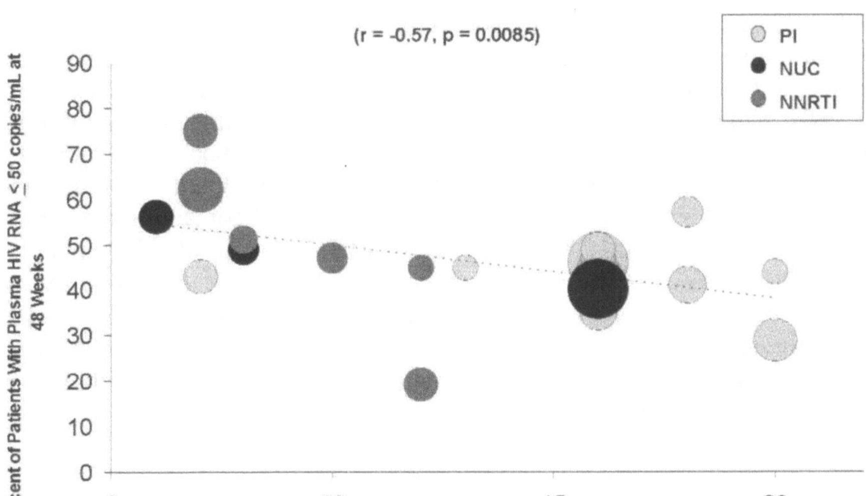

Figure 1. Virologic failure rates in selected clinical studies and relationship to drug class and pill burden. Size of symbol is directly proportional to weight of the data point in the analysis (Bartlett et al. [16]).

months of therapy, compared with 70–79% of previously treated patients. At 30 months, AIDS or death occurred in 7% of patients with sustained HIV RNA levels <400 cpm, compared with 9% who experienced virologic rebound and 20% who never had an HIV RNA level <400 cpm.

Lucas and colleagues [21] reported retrospective results from 273 PI-naïve patients followed in the Johns Hopkins HIV Clinic who started a PI and at least one other new antiretroviral agent. Overall, only 37% of the cohort had HIV RNA <500 cpm at 7–14 months. Staszewski et al. [22] reported their findings in 901 patients from the Frankfurt HIV Cohort who started a PI-containing regimen. Overall there was a 79% probability that HIV RNA was <500 cpm by 24 weeks of therapy. In the Amsterdam cohort study, of 271 treatment-naïve patients who started HAART, virologic treatment failure (HIV RNA > 1000 cpm) occurred in 40%, 6% had at least one new AIDS-defining event and 3% died over 48 weeks of follow-up [23].

In a report from the University-affiliated HIV clinic in Cleveland, Valdez and colleagues [24] identified 310 patients treated with PI-containing regimens from June 1995-December 1997 who had virological follow-up data. With a median follow-up of 335 days, only 47% had HIV RNA levels <400 cpm. In the San Francisco cohort study, Deeks et al. [25] analyzed 337 patients who received at least 16 continuous weeks of the first PI-containing regimen. A total of 170 (50%) had HIV RNA <500 cpm after 48 weeks of treatment. However, the CD4 cell counts remained above baseline levels through week 48 in most patients, regardless of the virologic response. Of the 99 patients who experienced prior virologic failure and changed to a salvage regimen, only 22 (22%) had HIV RNA <500 cpm after 24 weeks of therapy.

Paredes and colleagues [26] reported results from 1469 patients followed as part of the EuroSIDA cohort who first started a potent antiretroviral regimen. In total, 80% of patients had HIV RNA <500 cpm sometime during follow-up and 60% had HIV RNA <500 cpm at 6 months after starting therapy. Of those with HIV RNA <500 cpm, 24% experienced virologic rebound at 6 months.

In a more recent cohort study from the Aproco study group, Le Moing and colleagues [27] reported 1283 patients who started a PI-containing regimen in 1997–1999. In this study 830 (74%) had an HIV RNA level <500 cpm at 4 months and 32% of these patients demonstrated virological rebound over a median of 20 months of follow-up. Taken together, these cohort studies demonstrate significant rates of virological failure in patients starting potent antiretroviral therapy regimens and even higher rates of virological failure in treatment-experienced patients.

Toxicity, treatment discontinuation and treatment failure

Adverse events occur commonly in patients taking antiretroviral treatments. For example, Fellay and colleagues [28] reported data from the Swiss HIV Cohort Study that assessed the prevalence of adverse events in 1160 patients who were receiving antiretroviral treatment and found that 47% (545 of 1160) of patients presented with clinical adverse events and 27% (194 of 712) with laboratory adverse events that were probably or definitely attributed to antiretroviral treatment. Toxicity from antiretroviral regimens leads commonly to treatment discontinuation (Tab. 3).

D'Arminio Monforte and colleagues [29] studied a cohort of 862 treatment-naïve patients from the ICONA cohort who started antiretroviral therapy and then discontinued. Over a median 45 weeks of follow-up, 312 (36%) discontinued treatment, including 182 (21%) due to toxicity and 44 (5%) due to treatment failure. In a multivariate analysis, gender (relative hazard, RH 0.51 for men compared with women), choice of PI (RH 1.9 for indinavir and 3.8 for ritonavir each compared with saquinavir hgc) and treatment duration (RH 0.89 for each additional month) were independently associated with discontinuation for toxicity. Similarly, current HIV RNA level (RH 3.2 for each $\log_{10}$ cpm higher) and choice of PI (RH 0.21 for indinavir and 0.23 for ritonavir compared with saquinavir) were independently associated with discontinuation due to treatment failure. In this study, toxicity represented the major cause of treatment discontinuation over the first year of therapy. In a follow-up report

Table 3. Causes of treatment discontinuation in selected studies

Study (reference)	Total N	Follow-up time	Cohort Patients	Follow-up
ICONA [29]	862	~45 weeks	Starting potent antiretroviral therapy	312 (36%) discontinued regimen 21% due to toxicity 5% due to failure 4.2% virologic, 0.6% immunologic, 0.3% clinical 7% due to non-adherence 3% due to other reasons
Royal Free Hospital, London [31]	556	~57 weeks	Starting potent antiretroviral therapy	247 (44%) discontinued regimen 29% due to toxicity, patient choice, poor adherence 13% due to virologic or immunologic failure 2% due to other reasons
WIHS [33]	1058	~3 years	Women starting potent antiretroviral therapy	512 (48%) discontinued regimen 40% due to side-effects (subset analysis)
Sacco Hospital, Milan [34]	263	~69 weeks	Experiencing failure or intolerant of first potent antiretroviral therapy regimen; starting 2nd regimen	154 (59%) discontinued second regimen 33% due to intolerance/toxicity 18% due to virologic, immunologic or clinical failure 8% due to drop out/patient request

from the ICONA cohort study of 2002 treatment-naïve patients who started potent antiretroviral regimens, a total of 857 (43%) discontinued after a median duration of follow-up of about 10 months and 366 (43% of those who discontinued) did so for toxicity [30].

Mocroft and colleagues [31] analyzed 556 patients followed at the Royal Free Hospital in London who started combination antiretroviral therapy. During a median of 14.2 months of follow-up, 247 (44%) of patients modified their regimen: 159 (64%) due to toxicity, patient choice, or poor adherence and 72 (29%) due to virologic or immunologic failure. A total of 148 (27%) discontinued their antiretrovirals while over half (i.e., 83,56%) of them later restarted antiretroviral regimens. Less likely to modify their regimens were older patients (RH 0.73 per 10 years of age, $p = 0.0008$), previously treatment-naïve patients (RH 0.64, $p = 0.027$), patients who took nelfinavir (RH 0.57, $p = 0.035$) and clinical trial participants (RH 0.64, $p = 0.027$). More likely to modify their regimens were patients who took four or more drugs (RH 2.2, $p < 0.0001$), patients who took ritonavir in their initial regimen (RH 1.4, $p = 0.035$) and patients with higher HIV RNA levels during follow-up (RH 1.5 per $\log_{10}$ cpm increase, $p < 0.0001$).

Crystal and colleagues [32] assessed 2459 New Jersey adult Medicaid beneficiaries with AIDS for their use of antiretroviral therapy. Of 1739 patients who survived the full observation period (1996–1998), 1261 (67%) had at least one paid claim for an NNRTI or PI, and 35% had discontinued the drug. African Americans and Latinos used NNRTI/PI for less of the observation time than whites (64% *versus* 72%, $p < 0.05$). Factors positively associated with sustained NNRTI/PI use in a multivariate analysis were age >40, case-management participation, and residence near New York City or Philadelphia.

Grant et al. [33] studied 1058 women who initiated potent antiretroviral therapy between 1996 and 1999 in the Women's Interagency HIV Study (WIHS). At 6 months, 7% overall had discontinued their regimen and the proportion discontinuing increased from 3% in 1996 to 9% in 1999. A higher HIV RNA level was associated with discontinuation ($p < 0.05$). The most common reason for discontinuation was side-effects (40%).

Bini et al. [34] studied 263 patients who had experienced virologic failure (56%) or intolerance/toxicity (44%) on their first HAART regimen who went on to take a second HAART regimen. A total of 154 (59%) discontinued their second HAART regimen, including 86 (33%) due to intolerance/toxicity. Among the factors associated with a decreased risk of virologic failure on the second regimen was intolerance/toxicity on a first HAART regimen (RH 0.58 *versus* virologic failure on the first regimen, 95% CI 0.35–0.98). Taken together, these studies show that intolerance or toxicity occurs commonly in patients starting antiretroviral regimens and leads to treatment discontinuation in a significant proportion. However, because these patients often have other treatment options, discontinuation for intolerance or toxicity does not always signify treatment failure.

Baseline factors and treatment failure

In one of the first studies evaluating treatment failure, Fatkenheuer and colleagues [17] described a number of independent risk factors for virologic failure in 198 patients taking PI in 1996, including the baseline CD4 cell count, prior antiretroviral therapy, and the choice of PI. Casado et al. [18] performed a logistic regression analysis of 400 patients and found that treatment failure was associated with a higher baseline HIV RNA (relative risk, RR, 2.0, $p < 0.01$), prior antiretroviral therapy (RR 2.1, $p < 0.01$), and the use of saquinavir (RR 1.6, $p = 0.03$) [18]. In their cohort study, d'Arminio Monforte and colleagues [19] found that treatment failure was predicted by baseline severe immunosuppression (as indicated by the CD4 cell counts), AIDS diagnosis and the use of saquinavir (RR 2.5 *versus* indinavir) [19].

In the report from the Swiss HIV Cohort Study, Ledergerber and colleagues [20] reported that a decreased probability of HIV RNA <400 cpm was associated with clinical stage C HIV disease, a higher baseline HIV RNA level, prior treatment experience and the use of saquinavir. For example, 91% of treatment naïve patients had HIV RNA <400 cpm, compared with 79% of treatment-experienced patients starting three new drugs, 74% starting two new drugs and 70% starting one new drug. In this study, a higher baseline CD4 cell count and more recent introduction of therapy (1997–98 *versus* 1995–96) were associated with an increased probability of HIV RNA <400 cpm.

In the Hopkins cohort study, factors associated with lack of virologic suppression included non-white ethnicity, age 40 years or younger, injection drug use, lower baseline CD4 cell counts, higher baseline HIV RNA level, and higher rates of missed clinic appointments (all $p < 0.05$) [21]. In a multivariate model, only higher rates of missed clinic appointments were independently associated with lack of suppression of HIV RNA to <500 cpm at 1 year. In the Frankfurt HIV Cohort study, in a multiple regression model, the probability of achieving an HIV RNA level <500 cpm was associated with a lower baseline HIV RNA level (RH 0.72, $p < 0.0001$), a higher CD4 cell count (RH 1.07, $p = 0.002$), and with being treatment-naïve, using indinavir or nelfinavir in the regimen and starting other new drugs at the same time as the PI (RH 1.5 for two new drugs *versus* none, $p = 0.003$) [22].

In the Amsterdam cohort study, risk factors for treatment failure were baseline HIV RNA level (odds ratio, OR, 1.35 per 1.7 $\log_{10}$ cpm increase), baseline CD4 cell counts (OR, 1.35 per 100 CD4 cell/mm^3 increase) and the use of saquinavir (OR 3.21 *versus* the other PI) [23]. In the Cleveland cohort study, the likelihood of virologic suppression was decreased in women ($p < 0.02$), in patients who missed two or more clinic visits in the prior year ($p < 0.001$), in those who started the treatment regimen earlier in the evaluation period ($p < 0.4$), and in patients with a lower nadir CD4 cell count ($p < 0.04$) and a higher peak HIV RNA level ($p < 0.001$) [24].

In the San Francisco cohort study, independent predictors of virological failure in PI-naïve patients starting a PI-containing regimen were a higher baseline HIV RNA level, a lower baseline CD4 cell, and failure to start at least one new nucleoside analogue simultaneously with the PI [25]. Independent predictors of virologic failure in PI-experienced patients starting a salvage regimen were an HIV RNA level >10,000 cpm and failure to use an NNRTI in the regimen. In the EuroSIDA cohort study, patients were less likely to achieve virologic suppression with higher baseline HIV RNA levels (RH 0.76 per log higher, $p < 0.001$) or taking saquinavir hard gelatin capsule (RH 0.62, $p < 0.001$) [26, 35]. A higher probability of virologic suppression was associated with a higher baseline CD4 cell count (RH 1.09 per 50% higher, $p = 0.008$) and the initiation of three or more new antiretroviral drugs (RH 1.3, $p = 0.2$). After suppression, virologic rebound was less likely in antiretroviral naïve patients (RH 0.5, $p = 0.1$), older patients (RH 0.86 per year older, $p = 0.04$) and those using a PI other than saquinavir (RH 0.66, $p = 0.04$).

In the Aproco cohort, virologic rebound to >500 cpm was associated with prior antiretroviral treatment (adjusted RH 2.4, $p < 0.0001$), younger age (RH 1.4 per each 10 years younger, $p < 0.0001$), baseline CD4 cell count <500/mm^3 (RH 2.3, $p < 0.001$), higher baseline HIV RNA level (RH 1.4 per $\log_{10}$ cpm higher, $p < 0.001$) and reported <80% adherence or discontinuation of PI therapy at 4 months [27].

In their review of 23 three-drug clinical trials, Bartlett and colleagues [16] found that baseline HIV RNA level, baseline CD4 cell counts and treatment regimen were not associated with the probability of HIV RNA <50 cpm at week 48, although a higher pill count of the regimen was negatively associated ($p = 0.0085$) [16]. Conversely, in a meta analysis of 30 studies representing 36 three-drug treatment arms, Skowron et al. [36] found in a multivariate regression analysis that baseline CD4 cell counts, but not baseline HIV RNA levels, correlated with virologic suppresion at 6 ($p = 0.008$) and 12 months ($p = 0.01$) [36]. Table 4 summarizes baseline factors associated with treatment failure.

Table 4. Baseline factors associated with treatment failure

- Baseline drug resistance
- Prior antiretroviral treatment
- Higher baseline (or maximum) HIV RNA level
- Lower baseline (or nadir) CD4 cell count
- AIDS diagnosis
- Treatment regimen
  - Antiretroviral drug and regimen choice
  - Pill count
  - Use and number of new drugs/new classes in the regimen
- Earlier year of starting therapy
- Missed clinic visits
- Self-reported poor adherence
- Demographics (e.g., younger age)

Baseline resistance

Acquisition of HIV with drug-resistance mutations would be expected to lead to treatment failure. Hecht and colleagues [37] first reported a case of sexual HIV transmission to a treatment-naïve patient with a virus with multiple mutations that conferred resistance to both reverse transcriptase inhibitors and PI (i.e., multidrug-resistant HIV). In 1999, two groups reported the prevalence of reduced drug susceptibility in treatment-naïve patients. Boden and colleagues [38] studied 80 patients with recent HIV infection and found that 13 (16%) had genotypic markers of resistance to at least one antiretroviral agent and 4% had multi-drug resistant virus. Little and colleagues analyzed [39] 141 patients with recent HIV infection and found 36 (26%) with 2.5–10 fold and three (2%) with >10-fold reduced susceptibility to one or more drugs using phenotypic testing.

The United Kingdom Group on Monitoring the Transmission of HIV Drug Resistance studied 69 patients who acquired HIV infection from 1994–2000 and found genotypic resistance to at least one agent in 14% and multidrug resistance in 3% [40]. They further noted an increase in the prevalence of drug resistant variants over that same time period, increasing to 27% in 2000. Little et al. [41] updated their findings in 2001 showing that of 408 treatment-naïve subjects with recent infection, 8% had >10-fold reduced susceptibility to at least one drug. They further noted that time to complete treatment-induced virological suppression to <400 cpm was significantly increased in patients with baseline reduced antiretroviral susceptibility (>5-fold).

Regimen choice

Treatment-naïve patients

A number of studies demonstrated that three-drug regimens were superior to monotherapy or dual drug regimens, both in terms of virologic suppression rates [11–15] and clinical endpoints [10] and this led to the current treatment recommendations [1–3]. In addition, triple-drug therapy started simultaneously led to superior virologic responses compared to introducing the agents sequentially, over a period of months [42]. Starting a three- or four-drug combination regimen and then reducing to a one- or two-drug maintenance regimen after 3–6 months was associated with increased virologic rebound compared to continuing the original regimen [43–45]. From a follow-up report on a small number of patients, it appeared that extending the duration of the induction regimen to 26–50 weeks of therapy did not improve the virologic responses associated with maintenance regimens [46].

The choice of drugs comprising combination antiretroviral regimens is also associated with virologic outcome. Early cohort studies revealed that the choice of saquinavir (hard gelatin capsule formulation) in a PI-containing regimen was associated with an increased risk of virologic failure compared to other PIs [17–20, 22, 23, 26].

Relatively few prospective, comparative clinical trials have been conducted that had adequate power to detect differences in virologic responses among current first-line treatment regimens: Staszewski et al. [47] reported that an efavirenz-based regimen was superior, in terms of virologic response rate, to an indinavir-based regimen, although this open-label study had increased discontinuations in the indinavir arm. In a double-blind study, Walmsley and colleagues [48] found that a lopinavir/ritonavir-based regimen had a significantly better virologic response rate than a nelfinavir-based regimen.

In more recent cohort studies, the choice of regimens has been linked to discontinuation rates and treatment failure. Investigators of the ICONA cohort reported that of 2002 patients who started potent combination antiretroviral regimens containing either a PI or an NNRTI, 857 (43%) discontinued their regimen, and of those, 244 (28% of those who discontinued) did so for treatment failure [29]. The adjusted relative hazard of discontinuing for treatment failure was 0.61 for those taking an NNRTI-containing regimen, compared to those taking a PI-based regimen, but this was not statistically different (95% CI, 0.31–1.2).

Matthews et al. [49] reported their experience for a treatment-naïve patient cohort from two large clinics in London. A total of 888 patients began therapy with either a PI- (n = 484,55%) or NNRTI- (n = 404,45%) containing regimen and were followed for virologic responses. Overall, >78% had HIV RNA levels <500 cpm at 6 months. After stratification by center, year, and baseline variables, only baseline HIV RNA and treatment group were associated significantly with HIV RNA <500 cpm at 6 months. Patients taking an efavirenz-containing regimen had a greater virologic suppression rate at 6 months than those taking PI- (RH 0.74, $p = 0.01$) or nevirapine- (RH 0.77, $p = 0.02$) based regimens.

Barreiro et al. [50] recently reported their findings assessing CD4 cell response in 100 patients who started a PI- or NNRTI-based regimen with CD4 cell counts <500/mm^3, good treatment adherence and suppression of HIV RNA on therapy to <50 cpm. In this cohort, a mean increase from baseline CD4 counts of 42 cells/mm^3/month occurred over the first 12 weeks, regardless of treatment. However, between 12 and 48 weeks, the PI group had a mean increase of 15 cells/mm^3/month *versus* the NNRTI group who increased 10 cells/mm^3/month. This difference reached statistical significance in the subset of patients with CD4 cell counts <300/mm^3 (+17 *versus* +6 cells/month for PI and NNRTI regimens, respectively, $p < 0.05$).

Treatment-experienced patients

In data from the Swiss HIV Cohort Study, Ledergerber and colleagues [20] showed that in treatment-experienced patients, the number of new drugs in the regimen was associated with the probability of virologic suppression to <400 cpm. In this study, 79% of treatment-experienced patients starting three new drugs, compared with 74% starting two new drugs and 70% starting one new drug had HIV RNA levels <400 cpm over 1 year.

Bini and colleagues [34] studied a group of 263 patients who either experienced virologic failure (148, 56%) or intolerance (115, 44%) on their first potent antiretroviral regimens. Over a median follow-up of 483 days, 154 (59%) discontinued their second regimen: 135 (51%) because of virologic failure and 86 (33%) because of intolerance or toxicity. Among the independent factors associated with virologic failure were >6 months of prior HAART.

Prospective clinical trials of therapy for HAART-experienced patients also have demonstrated that using a new class of drugs (e.g., NNRTI in NNRTI-naïve, PI-experienced patients) is associated with an improved virologic response rate on a second regimen [51, 52] as is using a dual PI combination in PI-experienced patients [53]. Other studies have suggested that starting therapy with certain protease inhibitors (e.g., saquinavir, nelfinavir, amprenavir) may be associated with improved virologic response rates on subsequent regimens containing PI or dual PI combinations [34, 54, 55]. Additional comparative antiretroviral studies are in progress.

Regimen potency and treatment failure

Several groups have suggested that the potency of current combination antiretroviral therapy regimens is suboptimal by demonstrating that viral replication occurs even when viremia is suppressed below the limits of detection of current HIV RNA assays [56–58]. Martinez-Picado and colleagues [59] have shown that new HIV drug resistance mutations may be selected during treatment-induced suppression of HIV RNA levels to <50 cpm, even in the absence of virologic rebound. On the basis of these findings, more aggressive therapy regimens with a greater number of drugs have been proposed and tested in clinical studies.

Weverling and colleagues [60] conducted an open-label study of a three-class, five-drug regimen *versus* a standard three-drug regimen (containing a PI and two nucleoside analogues) in 24 treatment-naïve patients with baseline HIV RNA levels >10,000 cpm. They found that the median time to HIV RNA <50 cpm was 4 weeks in the five-drug group and 12 weeks in the three-drug group ($p = 0.0005$), although the initial HIV RNA elimination rate constants were not different between the two groups.

Other studies comparing regimens with four or more drugs to standard three-drug regimens have shown mixed results. For example, Eron and colleagues [61] randomized antiretroviral-naïve subjects to receive a four-drug regimen of zidovudine, lamivudine, abacavir and amprenavir or a standard three-drug regimen of zidovudine, lamivudine and nelfinavir. The primary study analysis found that the time to treatment failure using a combined virologic and toxicity endpoint actually was earlier in the four-drug arm than the three-drug arm ($p = 0.017$), due to a greater number of adverse events and study withdrawals in the four-drug arm.

Fischl and colleagues [62] recently reported the results of ACTG 388 which was a comparison of two four-drug regimens with a standard three-drug regimen in treatment-naïve patients with advanced HIV disease. A total of 517 patients received zidovudine/lamivudine combined with efavirenz and indinavir, nelfinavir and indinavir, or indinavir alone. With a median follow-up of 2.1 years, compared to the standard three-drug regimen, a lower rate of virologic failure occurred with the efavirenz/indinavir regimen ($p = 0.04$), but a higher rate occurred with the nelfinavir/indinavir regimen ($p = 0.006$). While the use of four or more antiretroviral drugs may enhance virologic suppression rates over that of three-drug regimens, such regimens may also increase rates of adverse events and adversely impact adherence. Further studies are in progress.

Physician experience

Prior to the availability of potent antiretroviral therapy, Kitahata and colleagues [63] assessed 403 patients who developed their first AIDS-defining illness from 1984-mid-1994 and were followed by 125 primary care physicians with varying degrees of HIV experience. In their analysis, the median survival of patients under the care of physicians with the least experience was 14 months, compared to 26 months for those under the care of physicians with the most experience ($p < 0.001$). However, the use of antiretroviral therapy was not associated with survival in this patient cohort during this time period.

Brosgart and colleagues [64] mailed questionnaires to 999 healthcare providers in more than 20 cities in the United States in May 1996 to assess their preferred management of 30 HIV-related medical conditions. The investigators found a statistically significant association between the number of HIV-infected patients cared for and the report of prescribing potent antiretroviral therapy: 46% of those taking care of 1–5 patients reported prescribing three-drug combination therapy for a patient with a CD4 cell count $<50/mm^3$ compared with 86% of those taking care of >20 patients ($p = 0.001$). Providers taking care of fewer patients also were more likely to prescribe antiretroviral monotherapy or dual therapy.

Stone and colleagues [65] surveyed 2478 internal medicine and infectious disease physicians to assess the relationship of the choice of antiretroviral therapy to physician specialty and HIV/AIDS experience. The questionnaire was mailed to physicians in four states in May 1998 and a total of 1233 physicians (52%) responded. Choices of antiretroviral regimens were categorized as recommended or not recommended according to 1997 treatment guidelines. Of the responders, 73% chose antiretroviral regimens that conformed to the guidelines. A total of 88% of infectious disease physicians, compared to 57%

of internal medicine physicians ($p = 0.001$), chose recommended regimens. When the analysis was restricted to the physicians who would assume primary care themselves, the differences in specialty were no longer significant.

In a recent letter to the editor, Edelstein and Wilson [66] reported a retrospective review of 71 randomly-selected hospital admissions of HIV-infected patients at one teaching hospital from 1998–2000. Of the 71 patients identified, 34 patients were prescribed antiretroviral medications during a total of 36 admissions. Housestaff wrote incorrect dosing intervals in 100% of the cases, incorrect dosages in 11% of cases, and incorrect food/fasting requirements in nine of 15 (60%) of applicable cases. Thus physician training and experience impacts the choice and management of antiretroviral regimens and therefore likely impacts treatment failure and patient survival.

Virologic response and clinical failure

In an early study of antiretroviral therapy, O'Brien and colleagues [67] studied 270 patients from the Veterans Affairs Cooperative Study on AIDS, and compared immediate *versus* deferred treatment with zidovudine monotherapy. In this study, 34 of 129 (26%) patients in the immediate-treatment group and 57 of 141 (40%) patients in the delayed-treatment group progressed to AIDS ($p = 0.03$), and progression was strongly correlated with baseline HIV RNA levels and CD4 cell counts. A response over the first 6 months of zidovudine treatment with a decrease in HIV RNA levels of at least 75% from baseline accounted for 59% of the treatment benefit, defined as delayed progression to AIDS, and adding a 10% increase in CD4 cell counts from baseline could explain a total of 79% of the treatment benefit. In a follow-up report, the same group showed that for each decrease of 0.5 $\log_{10}$ cpm of HIV RNA averaged over the 6 months after treatment randomization, the relative risk for progression to AIDS was 0.67 ($p > 0.001$) [68].

Hughes et al. [69] studied 198 HIV-infected, NRTI-experienced patients with <350 CD4 cells/mm^3 who received new combination antiretroviral regimens. In this study, using a multivariate model, the risk for disease progression was reduced by 56% ($p = 0.028$, 95% CI, 8–79%) for every 10-fold lower HIV RNA level at baseline, by 67% ($p < 0.001$, 95% CI, 42–81%) for every two-fold higher CD4 cell count at baseline, and by 52% ($p = 0.071$, 95% CI, 6–79%) for every 10-fold reduction in HIV RNA level at 8 weeks after starting antiretroviral treatment.

Virologic response and virologic failure

HIV RNA nadir

One of the first groups to examine the relationship between HIV RNA response to therapy and virologic outcome was Raboud and colleagues [70] reporting data from the INCAS study. In this randomized controlled study of combinations of zidovudine, didanosine and nevirapine they examined predictors of achieving and maintaining suppression of HIV RNA load levels. A total of 104 patients in this study had plasma viral load levels >500 cpm at baseline and then a nadir viral load level of <500 cpm while on treatment. Of these, 77 patients had an increase of viral load level to >500 cpm. The relative risk of a viral load level increased to >500 cpm was 0.11 for those with viral load nadir levels ≤20 cpm on treatment, compared with those with viral load levels >20 cpm ($p = 0.0001$). The relative risk of an increased viral load to >5000 cpm was 0.05 for those with viral load nadir levels ≤20 cpm on treatment and 0.37 for those with viral load nadir levels of 20–400 cpm, compared with those with nadir levels >400 copies/ml (Fig. 2). In addition, those with a nadir level <20 cpm were at a lower risk of virologic failure than those with nadir levels of 20–400 cpm ($p = 0.0001$).

A similar conclusion was reached by Kempf et al. [71]. In a retrospective analysis, these investigators assessed the durability of virologic response, which they defined as the time from initiation of therapy to the time when a sustained increase of 0.6 $\log_{10}$ cpm above the nadir HIV RNA level occurred. Two groups of subjects were analyzed: 29 subjects who received the PI, ritonavir (with or without other antiretroviral agents) and experienced virologic rebound with mutations conferring resistance to ritonavir and 102 subjects who initially responded to ritonavir (with or without zidovudine). In their examination for possible predictors of durability, they found a strong correla-

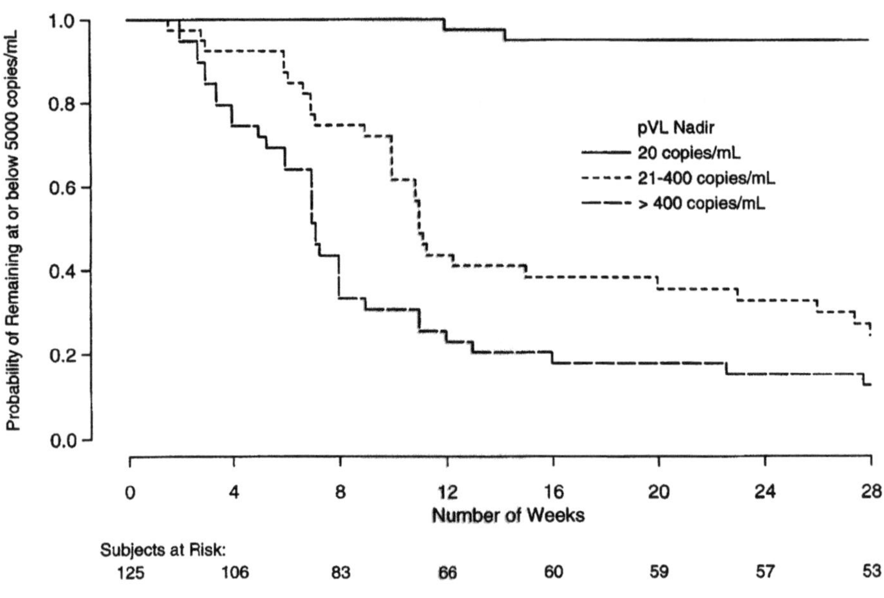

Figure 2. Probability of maintaining HIV RNA < 5000 cpm according to the treatment-induced HIV RNA nadir on the INCAS study (Raboud et al. [70]).

tion between the nadir viral load level on treatment and the durability of virologic response ($p < 0.01$).

Havlir and colleagues [72] did a retrospective analysis of 254 subjects enrolled in studies of combination antiretroviral therapy who had HIV RNA levels <200 cpm after at least 6 months of therapy to assess the clinical significance of intermittent low-level viremia (so-called viral load "blips"). The investigators defined intermittent viremia as a plasma HIV RNA level >50 cpm with a subsequent level of <50 cpm without evidence of virologic failure (two consecutive HIV RNA levels >200 cpm). In their analysis, intermittent viremia occurred in 96 (40%) of the 241 subjects from one of the studies and did not predict virologic failure: 10 (10%) of 96 patients with intermittent viremia *versus* 20 (14%) of 145 patients without intermittent viremia ultimately experienced virologic failure (RR 0.76, 95% CI, 0.29–1.72). They concluded that intermittent viremia occurred commonly but was not associated with virologic failure. Similar results were reported by Greub and colleagues [73] in their analysis of 1858 patients from the Swiss HIV Cohort Study and the Frankfurt HIV Clinical Cohort.

Perrin and colleagues [74] used a modified HIV RNA assay that has a limit of quantification of 3 cpm to study the relationship of the HIV RNA level to virologic rebound (defined as two consecutive HIV RNA >400 cpm following a level <400 cpm) in a group of subjects who took a lopinavir/ritonavir-based combination antiretroviral regimen. A total of 54 of 96 (54%) of patients on treatment had HIV RNA ≤3 cpm at least once and 32 of 96 (33%) demonstrated multiple levels ≤3 cpm through 72 weeks of follow-up. No greater risk of virologic failure through week 144 was seen in the patients who had HIV RNA ≤3 cpm, compared with those who did not suppress to ≤3 cpm ($p = 0.91$). Also, the risk of virologic failure over 144 weeks was similar in those who had multiple levels ≤3 cpm to those who had no level or one level ≤3 cpm over the first 72 weeks of treatment ($p = 0.59$). Similar trends were observed using a definition of virologic suppression to ≤50 copies/ml. The authors concluded that achieving or sustaining an HIV RNA level ≤3 copies/ml over 72 weeks of therapy did not predict the risk of virologic failure through 144 weeks of follow-up.

Early HIV RNA responses

Raboud and colleagues [75] extended their observations from the INCAS study by pooling data with the AVANTI-2 and –3 studies and comparing virologic responses in patients receiving two-drug *versus* three-drug therapy. Overall, at 52 weeks of follow-up, 59–79% of the patients receiving three-drug therapy and 16–43% receiving two-drug therapy had viral load decreases to <500 cpm. In this analysis, the investigators found that the week 16 virologic response could predict 68% and 74% of the virologic responses on triple and double therapy, respectively, at 1 year of follow-up.

Lepri and colleagues [76] analyzed a large clinical database that included 453 antiretroviral-naïve patients who had their HIV RNA levels monitored every 4 weeks and calculated the observed probability of a viral load level ≤500 cpm at 4 and 8 weeks after starting a potent antiretroviral regimen. They noted that 42% and 70% had viral load levels <500 cpm by week 4 and 8, respectively, and that both levels were highly predictive of week 24 response (log higher viral load level at week 4, OR 0.45, $p = 0.01$; $\log_{10}$ higher viral level at week 8, OR 0.42, $p = 0.0004$) and were not dependent on baseline HIV RNA levels. The week 4 viral load level also was the strongest predictor of virologic suppression to <50 cpm at week 24 ($\log_{10}$ higher viral load level at week 4, OR 0.39; $p = 0.03$). They further noted that in patients with viral load levels >10,000 cpm at week 8, only 42% achieved ≤500 cpm by week 24.

Polis and colleagues [77] assessed very early virologic responses in 124 patients beginning a PI-containing regimen. Using data from three cohorts of patients, they assessed viral load levels at days 1, 2, 3, and 6 and weeks 2, 4, 8, and 12 after starting therapy and defined those patients who had HIV RNA levels decreased by 1.5 $\log_{10}$ cpm or to less than the limit of detection as responders. They found that initial viral decay rates correlated significantly with HIV RNA levels at 4, 8, and 12 weeks ($p < 0.0001$) and correctly predicted 84% of the virologic responses. They further noted that a decrease in HIV RNA of <0.72 $\log_{10}$ cpm by day 6 predicted a poor 12-week virologic response in >99% of patients.

Huang and colleagues [78] assessed patterns of initial virologic response in 1518 HIV-infected subjects from four AIDS Clinical Trials Group (ACTG) studies and correlated them with virologic suppression. Responses at each follow-up visit were coded as "on track" or "off track" for a trajectory to virologic suppression below the limit of detection. They found only a minority of naïve or NRTI-experienced patients went off track within the first 2–4 weeks, whereas over half of heavily treatment-experienced patients did so by week 4. Interestingly, off-track profiles after week 4 were characterized by abrupt viral load rebounds that were not predictable from the prior response profile.

Durability of virologic response

Gulick and colleagues [79] reported the 3-year follow-up results of one of the first patient groups to receive three-drug therapy. Of 33 patients, nine experienced virologic failure (two consecutive HIV RNA levels >500 cpm): six in the first year and three in the third year. In a follow-up presentation of the 5-year results from this same cohort of patients, no additional episodes of virologic failure occurred [80].

Phillips and colleagues [81] studied 336 treatment-naïve patients from the Frankfurt HIV Cohort who started combination antiretroviral regimens and had HIV RNA levels <50 cpm by 24 weeks for over 3 years of follow-up. In this cohort, 25% experienced virologic rebound (two consecutive HIV RNA levels >500 cpm or a level >50 cpm that prompted a regimen change); however, for most of the patients, this was associated with complete interruption of therapy. Without interruption of therapy, only a 5% risk of virologic rebound occurred. Overall, the risk of virologic breakthrough decreased significantly with increased duration of virologic suppression ($p = 0.01$).

Virologic, immunologic, and clinical responses

Staszewski et al. [82] studied 154 patients from the Frankfurt AIDS cohort who started indinavir-containing regimens and continued for at least 72 weeks of follow-up. They found that the mean increase in CD4 cell counts, adjusted for baseline HIV RNA, CD4 cell count, and age, was strongly associated with the mean level of HIV RNA suppression ($p < 0.0001$). Specifically over 72 weeks, the CD4 increase was 161 cells/mm^3 over baseline in the group with a 2–3 $\log_{10}$ cpm decrease in HIV RNA, compared to a 314 cells/mm^3 increase over baseline in the group with >3 $\log_{10}$ cpm decrease in HIV RNA level ($p = 0.0004$).

Miller and colleagues [83] reported related findings from the same cohort. They investigated 558 patients, about half of whom were treatment naïve, had begun potent antiretroviral therapy and had HIV RNA <500 cpm and related the risk of virologic rebound to CD4 counts. In total, 306 patients experienced HIV RNA rebound to >500 cpm. Using a Kaplan-Meier method, they estimated that 42% of patients experienced virologic rebound by week 24 and 64% by week 84. The greatest predictor of the risk of virologic rebound was the treatment-induced change in CD4 cell counts from baseline. For example, compared to those patients with CD4 cell counts ≥500 cells/mm^3 on therapy, those with lower CD4 cell counts had a greater risk of virologic failure: CD4 < 20/mm^3 (RH 5.4); 20–100 (RH 2.4–2.5); 300–499 (RH 2.0) (overall $p < 0.001$).

Deeks et al. [4] studied 291 patients who had experienced virologic failure on a PI-containing regimen, defined as a persistent HIV RNA level >500 cpm, but had a CD4 cell count greater than their pre-therapy level. In this cohort, the median time to immunologic failure (defined as a return to the pre-therapy baseline CD4 cell counts) after the onset of virologic failure was 3.1 years. The treatment-induced change in HIV RNA level from baseline was inversely associated with the risk of immunologic failure. For example, for those patients with a change in HIV RNA <0.69 $\log_{10}$ cpm from baseline, the relative hazard of immunologic failure was 5.5 ($p = 0.04$). Discontinuing therapy was also associated with immunologic failure, even after adjustment for viral load level (HR 2.0, $p = 0.004$).

Piketty and colleagues [5] studied 162 patients with advanced HIV disease who began an indinavir-containing regimen. A total of 57% of patients responded to therapy with a >1 $\log_{10}$ cpm decrease from baseline HIV RNA levels and a >50 cell/mm^3 increase from baseline CD4 cell count; 17% did not respond, and 21% had discrepant responses, with half of these experiencing CD4 cell count increases without a virologic response and half experiencing a virologic response without a CD4 increase. A total of 16 (10%) patients had an AIDS-defining illness and nine (6%) died during the follow-up period. The incidence of AIDS-defining events was similar in those patients with a CD4 cell count response without a virologic response (1 of 17), (6%) and those with both virologic and immunologic responses (7 of 92), (8%), but higher in those without a virologic or immunlogic response (2 of 8), (25%) or those with a virological response without a CD4 cell response (3 of 17), (18%). Thus an immunologic response to therapy was associated with improved clinical outcome over a relatively short-term follow-up period, even in the absence of a virologic response.

D'Arminio Monforte and colleagues [84] followed 585 patients beginning combination antiretroviral regimens for the occurrence of clinical failure, defined as the occurrence of new or recurrent AIDS-defining illnesses or death. In this study, 55 (9%) patients experienced clinical failure, during a median follow-up of 483 days. Forty-five new AIDS-defining events occurred in 38 patients, six experienced recurrent AIDS-defining events and 11 patients died. A total of 24 (53%) of the new AIDS-defining events occurred during the first 3 months following institution of antiretroviral therapy. The mean CD4 cell counts were 144/mm^3 in patients with clinical failure *versus* 322/mm^3 in patients who did not experience clinical failure ($p < 0.0001$). In a multivariate analysis, after adjusting for baseline CD4 cell count and AIDS diagnosis at enrollment, only the CD4 cell count at 3 months was independently associated with clinical failure (hazard risk 4.8; 95% CI, 1.4–16.5). More than half of the AIDS defining-illnesses occurred during the first months after starting antiretroviral therapy, presumably due to incomplete immune reconstitution.

Ledergerber and colleagues [20] using data from the Swiss HIV Cohort Study, analyzed data in 2674 patients who started combination antiretroviral therapy in 1995–98. At 30 months of follow-up, AIDS or death occurred in 7% of patients with HIV RNA <400 cpm, 9% who had virologic rebound and 20% of those who have never suppressed their viral load levels to <400 copies/ml. The relative hazard of AIDS or death was 1.0 for the virologic rebound group and 2.4 for the never-suppressed group, compared with the suppressed group. The authors concluded that although virological failure rates were high, clinical progression and mortality were low in this cohort.

Grabar and colleagues [6] prospectively followed 2236 PI-naïve patients who began therapy with a PI-containing regimen between 7/96 and 3/97. On the basis of a virologic response (HIV RNA decreased by >1 $\log_{10}$ cpm or to <1000 cpm) and/or an immunologic response (CD4 increase of >50 cells/mm^3), patients were classified into four groups: virologic and immunologic response (48%), no response (16%), virologic response only (17%) and immunologic response only (19%). From 6 months to a median of 18 months of follow-up, 123 patients experienced a new AIDS-defining illness and 69 patients died. After adjustment, the difference in clinical progression rates was not significantly different in those with both virologic and immunologic responses and those with an immunologic response only, but rates were significantly higher in non-responders (RR 3.4) and those with a virologic response only (RR 2.0).

Demeter and colleagues [85] recently assessed outcomes in 489 patients receiving indinavir, zidovudine, and lamivudine with a CD4 cell count ≤200/mm^3 after 8 weeks of treatment as part of the ACTG 320 study. In this analysis, the HIV RNA level at 8 weeks was associated with the probability of suppression of HIV RNA to <500 or <50 cpm at 24 weeks. HIV RNA level at week 8 also predicted clinical outcome, with clinical progression experienced by 2% of those with an HIV RNA level <500 cpm at week 8, compared to 11% of those who had an HIV RNA level >500 cpm. In addition, after adjustment for HIV RNA level, patients with a CD4 count of ≥50 cells/mm^3 at week 8 had a decreased hazard for clinical progression compared with those with <50 cells/mm^3 at week 8 (HR 0.25, 95% CI 0.09, 0.67).

Adherence and treatment failure

It is clear that a patient's ability to take an antiretroviral regimen impacts treatment outcome. Properties of the treatment regimen, the patient, the health care provider, the patient-provider relationship, the stage of disease, and the clinical setting are important determinants of adherence. With the development of more potent antiretroviral drugs and combination regimens, groups have assessed the impact of adherence on virologic outcome.

One of the first clinical studies of potent antiretroviral therapy to assess adherence was the INCAS study [12]. In this blinded randomized study, 153 subjects were randomized to receive zidovudine together with didanosine, nevirapine, or both. Medication adherence was assessed by patient self-report. In this study, non-adherence was defined as failure to take any dose of at least one of the study medications for a total of 28 days over the 52-weeks of the study. The results of this study demonstrated that the rate of protocol-defined non-adherence was 75% (38 of 51 subjects) in the three-drug arm and overall was not different among the three treatment groups ($p = 0.77$). In this study, non-adherence was associated frequently with a failure to sustain maximum suppression of HIV RNA level.

Adherence on clinical studies may be quite different from adherence in the general population, in particular because clinical trials subjects are often a self-selected, highly-motivated group. Several groups have assessed adherence and virologic outcome in clinical cohorts. Haubrich and colleagues of the California Collaborative Treatment Group prospectively studied a group of patients who were asked to complete adherence questionnaires at baseline, 2 and 6 months after starting antiretroviral therapy [86]. Patients were classified into four groups based on their self-reported estimation of the proportion of prescribed medications taken within the prior 4 weeks (<80%, 80–95%, 95–99%, 100%) and HIV RNA levels were measured twice monthly. In this study, 41% of patients reported never missing a dose of antiretroviral therapy.

Overall, greater suppression of HIV RNA at 6 months was seen as adherence increased across the four groups ($p = 0.009$ for linear trend). In addition, the immunologic response to therapy at 6 months also correlated with adherence ($p = 0.029$ for linear trend).

Bangsberg and colleagues [87] assessed the relationship of adherence, HIV RNA, and drug resistance in a group of 34 homeless or marginally-housed HIV infected people in a cross-sectional manner. Adherence was assessed by periodic unannounced pill counts, electronic medication monitoring, and self-report over 6–10 weeks. Overall, median adherence was 89% (self-report), 73% (pill count), and 67% (electronic medication monitor), with 38% of the population demonstrating over 90% adherence by pill counts over the course of the study. A strong linear relationship was found between adherence and HIV RNA level, with electronic monitoring ($r = -0.81$, $p < 0.0001$) and pill count ($r = -0.67$, $p < 0.0001$) analyses correlating most closely. In a multivariate analysis, pill count adherence was a highly significant predictor of HIV RNA level ($r = 0.73$, $p = 0.001$) and every 10% decrease in adherence was associated with a doubling in the HIV RNA level.

Paterson and colleagues [88] prospectively studied 99 HIV-infected patients attending a Veterans Affairs or university medical center clinic who were prescribed a PI-containing regimen. Adherence was assessed using an electronic monitoring system, and the adherence rate was calculated as the number of doses taken divided by the number of doses prescribed. In the 81 evaluable patients, the overall adherence rate was 75% and the median duration of follow-up was 6 months. Rates of virologic failure (HIV RNA >400 cpm) were directly related to the degree of adherence: 22% of patients with ≥95% adherence experienced virologic failure, compared with 61% of patients with 80–94.9% adherence, and 80% in patients with <80% adherence (Fig. 3). In a Cox proportional hazard survival regression analysis, only the degree of adherence (relative risk 0.97; $p = 0.01$; 95% CI, 0.96, 0.98) and baseline viral load (relative risk 1.1; $p = 0.03$; 95% CI, 1.0, 1.3) were independent predictors of virologic failure. Additionally, changes in CD4 cell counts were also significantly associated with adherence ($r = 0.31$, $p = 0.006$). The degree of adherence was also significantly associated with the length of hospital stay during the study period ($p = 0.001$). The authors concluded that patients with ≥95% adherence to antiretroviral therapy had better virologic and immunologic responses and a lower hospitalization rate than those with lower rates of adherence.

The ICONA study showed that non-compliance was a common reason for discontinuation of potent combination antiretroviral regimens: Of 2002 treatment-naïve patients who started treatment, 857 (43%) discontinued therapy and 197 (23% of those who discontinued) did so for non-compliance [29].

Nonadherence also correlates with low drug concentrations. For example, Murri et al. [89] performed a cross-section analysis of 140 HIV-infected patients taking ritonavir- or indinavir-based regimens. By self-report, 12% missed at least one drug dose the prior day and 24% missed doses in the prior three days. A total of 14% had a plasma PI concentration below the assay limit

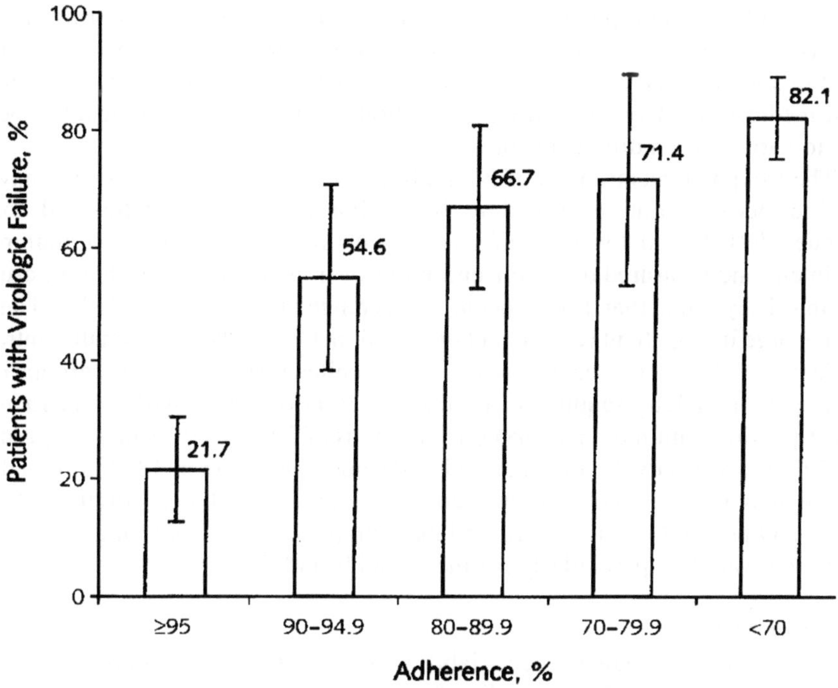

Figure 3. Adherence to antiretroviral therapy and the risk of virologic failure (Paterson et al. [88]).

of quantification. In a multivariate analysis, self-reported nonadherence and confusion were each associated with non-quantifiable PI levels (OR 15.8 and 9.9, respectively). Nieuwkerk and colleagues [90] found similar results in a study of 224 patients taking potent antiretroviral therapy who completed a questionnaire on adherence. Only 47% reported taking all drugs on time. Patients who reported deviation from their regimen had lower drug exposure and were less likely to have HIV RNA levels <500 cpm than fully adherent patients (adjusted OR 4.0, 95% CI 1.4–11.6).

Pharmacokinetics and treatment failure

Variability in the pharmacokinetics of antiretroviral drugs may lead to subtherapeutic drug concentrations and ultimately result in drug failure. While this may be seen with all three classes of antiretrovirals, it is particularly true for the NNRTIs and PIs [91]. Pharmacokinetics in an individual may be influenced by variability in drug absorption, intestinal and hepatic cytochrome P450 (CYP) metabolism, protein binding, cellular phosphorylation, the activity of p-glycoprotein, a cellular transport protein, drug elimination and drug-drug interactions. For example, the activity of the CYP enzyme system varies greatly in the general population, with the quantity of enzyme isoforms varying as much as 10–20 fold. Variability of these factors between patients may lead to a wide range of drug exposures in a population. In addition, concomitant medications may cause drug interactions with antiretroviral agents leading to variability in drug exposures.

The trough concentrations of NNRTIs and PIs between subjects may have coefficients of variation exceeding 100%. For example, Marzolini and colleagues [92] studied 130 HIV-infected patients receiving efavirenz-containing regimens and measured drug concentrations over an average of 14 h after drug dosing. They found that drug concentrations ranged from 125 to 15230 μg/l and noted large inter-patient (coefficient of variation, CV, 118%) with smaller intra-patient (CV 30%) differences. Schapiro and colleagues [93] studied groups of eight patients taking saquinavir at doses of 3600 or 7200 mg/day and found corresponding minimum plasma concentrations of 20 (CV 49%) and 123 μg/ml (CV 67%), respectively. Fletcher and colleagues [94] studied 18 children taking indinavir and found minimum concentrations of 0.01–1.28 mg/liter (CV 110%). Other studies have confirmed that intrapatient drug concentrations may be highly variable, particularly for the NNRTIs and PIs.

Pharmacogenetics

Genetics may contribute to the variability of pharmacokinetics among individuals. Allelic variants of genes coding for enzymes of the cytochrome P450 system or p-glycoprotein could account for some of this variability. Several groups have studied the effect of genetic polymorphisms on drug concentrations in patients both with and without HIV infection. For example, Fellay and colleagues [95] studied 123 HIV-infected patients and characterized the genes coding for isoenzymes of CYP3A4 as well as the multidrug-resistance transporter gene (MDR1) that codes for P-glycoprotein, a protein that regulates drug entry in a number of cells (e.g., intestinal, blood-brain and blood-testis barriers, lymphocytes). They correlated the genotypes with concentrations of efavirenz and nelfinavir, and with HIV RNA and CD4 cell counts. In this study, patients who were homozygous or heterozygous for a CYP2D6 allele associated with a poor-metabolizer phenotype had higher median plasma concentrations of both efavirenz and nelfinavir, than those with an extensive-metabolizer genotype. This was not seen with other isoenzymes (e.g., CYP2C19, CYP3A4). In addition, they found that MDR1 polymorphisms were associated with differing levels of MDR1 expression, different antiretroviral drug concentrations and differences in immune recovery, as measured by CD4 cell increases on therapy.

Correlation of drug levels with antiretroviral responses: nucleoside analogue reverse transcriptase inhibitors

Members of the NRTI class of drugs must undergo intracellular triphosphorylation to become virologically active. Thus, the plasma concentration of the nucleoside analogue may not correlate with either the intracellular concentration of the triphosphorylated form of the drug or the antiretroviral responses [96]. In one report, Fletcher and colleagues [97] studied 20 HIV-infected adults randomized to receive a standard dose of zidovudine *versus* a regimen of zidovudine targeted to achieve a plasma concentration of 0.7 µmol/l. At steady-state, the plasma concentration of zidovudine in the standard dose group was 0.62 µmol/l (coefficient of variation, CV, 32%) *versus* 0.76 µmol/l (CV 12%) in the adjusted group. Intracellular zidovudine triphosphate concentrations averaged 92 fmol/10^6 PBMCs for standard therapy and 160 fmol/10^6 PBMCs for adjusted therapy. CD4 cell count changes were –7% (standard therapy) *versus* +22% (adjusted therapy). In a small follow-up study, Fletcher and colleagues [98] assessed intracellular concentrations of zidovudine and lamivudine in eight patients taking a combination regimen and found a relationship between the intracellular concentrations of both drugs and both HIV RNA decreases and CD4 cell percentage changes.

Non-nucleoside analogue reverse transcriptase inhibitors (NNRTI)
Veldkamp and colleagues [99] studied patients participating in the INCAS study who took zidovudine, didanosine and nevirapine. They found that the random steady-state concentration of nevirapine was an independent predictor of the initial time of HIV RNA decrease to less than 20 copies/ml ($p = 0.011$) and that a nevirapine plasma concentration above the median at weeks 12, 24, 36, or 52 was significantly associated with HIV RNA <20 cpm at week 52 ($p < 0.03$ for each timepoint, OR 4.3–9.3). Similarly, Para and colleagues [100] performed a dose-ranging study of delavirdine monotherapy (ACTG 260) and found that higher delavirdine concentrations were associated with greater declines in HIV RNA ($p < 0.0001$). In the study of 130 patients taking efavirenz-based regimens, Marzolini and colleagues [92] determined the plasma level of efavirenz at mid-dosing interval after 3–18 months of dosing. They observed a significant inverse correlation between efavirenz levels and HIV RNA levels.

Protease inhibitors
A number of studies of HIV protease inhibitors have related drug concentrations to antiviral effect. For example, in their study of high-dose saquinavir monotherapy, Schapiro and colleagues [93] showed a correlation between 24-hour AUC determined at week 4 and the decrease in HIV RNA from baseline to week 4 ($r = 0.80$) in 16 subjects. At 24 weeks, 9 of 20 (45%) subjects receiving 3600 mg/day and 4 of 19 (21%) receiving 7200 mg/day of saquinavir developed substitutions at HIV protease positions 48 or 90 associated with saquinavir resistance ($p = 0.09$) suggesting a trend towards a relationship between drug dose and the development of drug resistance. In an early study of ritonavir monotherapy, Danner and colleagues [101], noted that HIV RNA and CD4 cell count responses were only sustained in subjects receiving doses of ritonavir that had measured mean trough concentrations exceeding the 90% effective concentration (~2.1 µg/ml).

Acosta et al. [102] studied 23 subjects who had started indinavir in combination with nucleoside analogues and related the pharmacokinetic and virologic parameters. They found significant intersubject variability in indinavir concentrations (8-h AUC, 5.4–68.0 µM-h) and that 5- and 8-h indinavir concentrations were significantly greater in patients with HIV RNA levels <500 cpm than those with HIV RNA levels >500 cpm ($p = 0.007$), (Fig. 4). Burger and colleagues studied 65 patients who started indinavir in a three-drug combination and found in a multivariate analysis that the plasma concentration of indinavir was an independent predictor of virologic failure, defined as an HIV RNA level >200 cpm after 24 weeks of treatment (OR 0.1) [103].

In an early study of nelfinavir monotherapy given twice or three times daily, Markowitz and colleagues [104] noted greater drug exposure in the three times daily dosing group. While HIV RNA levels decreased in all groups by day 14 in this study, they returned towards baseline in the twice daily dosing groups compared to sustained suppression over 28 days in the three times daily dosing groups. More recently, Kerr and colleagues [105] described the results of a pharmacokinetic analysis from a phase III study of nelfinavir as part of a combination antiretroviral regimen. In this study of subjects receiving nelfinavir and two nucleoside analogues, the probability of an HIV RNA level <50 cpm was dependent on baseline HIV RNA level and nelfinavir concentration, with the 2 h post-dose concentration being more predictive than an untimed trough concentration.

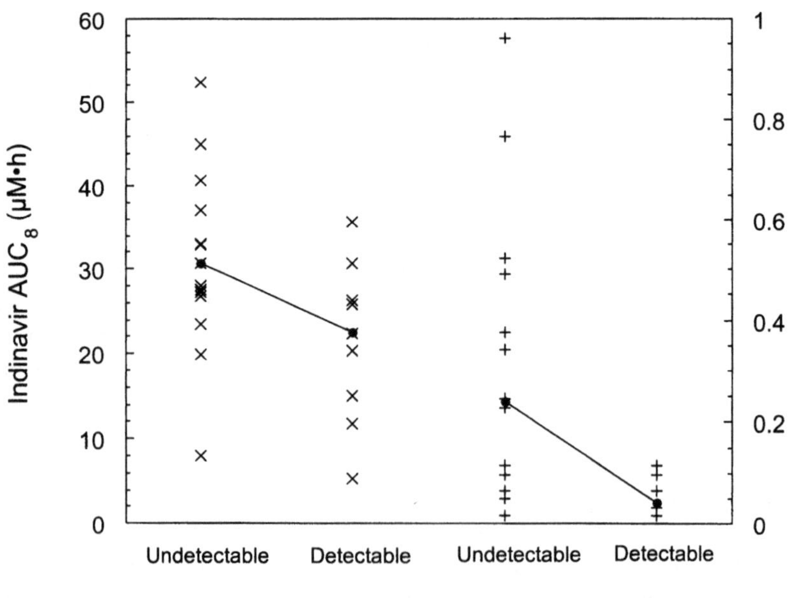

Figure 4. Differences in indinavir 8-h area under the curve (AUC_8) and steady-state concentration (C_{8h}) among undetectable (HIV RNA < 500 cpm) and detectable (HIV RNA ≥500 cpm) groups (Acosta et al. [102]).

Additional studies have shown similar associations between HIV protease inhibitor concentrations and virologic effects. Combinations of protease inhibitors have also demonstrated correlations between drug concentrations and virologic effects. For example, Hoetelmans and colleagues [106] described a group of 29 patients who received a four-drug regimen including the protease inhibitor combination of nelfinavir and saquinavir. Plasma drug concentrations and HIV RNA levels were determined at days 7, 14, 28 and 56. The investigators found a significant positive correlation with the plasma HIV RNA elimination rate and the median nelfinavir concentration ratio ($p = 0.011$) in a multivariate analysis.

In a substudy of the Viradapt study, Durant and colleagues [107] studied 81 patients who experienced virologic failure (HIV RNA level >10,000 cpm) taking a PI as part of a combination antiretroviral regimen. PI trough concentrations were performed at least three times over 48 weeks. A suboptimal concentration was defined as at least two plasma PI levels less than two times the 95% inhibitory concentration (IC95) of the drug. The investigators found that 26 of 81 (32%) had suboptimal concentrations and 55 of 81 (68%) had optimal concentrations of the HIV protease inhibitors. The HIV RNA change from baseline to week 48 was –1.28 $\log_{10}$ cpm in the optimal concentration group, compared with –0.36 log10 copies/ml in the suboptimal concentration group ($p = 0.0048$). In a multivariate analysis, higher drug concentration was an independent predictor of virologic response (odds ratio –2.0).

Cellular resistance

Nucleoside analogue HIV reverse transcriptase inhibitors require intracellular phosphorylation for antiretroviral activity. The nucleoside analogues have been divided into two groups based on their preferential phosphorylation in activated or resting cells [108]. Shirasaka et al. [109] found that the comparative anti-HIV activity in peripheral blood mononuclear cells was: zidovudine > zalcitabine > stavudine > didanosine in activated cells and zalcitabine > didanosine >> zidovudine and stavudine in resting cells. Van't Wout and colleagues [110] described selective inhibition of non-syncytium-inducing (NSI) HIV by zidovudine and syncytium-inducing (SI) HIV by didanosine, while the PI ritonavir inhibited both equally. Thus cell activation and phenotype may lead to differential antiretroviral activities among the nucleoside analogues, with the potential to lead to drug failure.

The MDR1 multidrug transporter, P-glycoprotein or P-gp, is a cell-surface phosphoglycoprotein that functions as an energy-dependent efflux pump for a wide variety of structurally-unrelated chemical compounds [111]. P-gp can be found on the surface of cells of tumors, gastrointestinal mucosa, liver, kidney, blood-brain barrier, blood-testis barrier, uterus, bone marrow, and peripheral blood monocytes and CD4+ lymphocytes [112]. Lee and colleagues showed in *in vitro* studies that the PIs indinavir, ritonavir and saquinavir were substrates for P-gp, and that inhibition of HIV replication by the PI was reduced in a cell line that expressed P-gp compared with one that did not express P-gp [113].

Furthermore, they showed reduced inhibition of HIV replication by the PIs could be reversed by adding agents known to inhibit P-gp activity. The authors concluded that P-gp has a role in reducing the antiretroviral effect of the HIV PI, and that cells that express P-gp will be relatively resistant.

Kim and colleagues [114] studied indinavir, nelfinavir and saquinavir and found that they were transported by P-gp in an *in vitro* system. In mice lacking the gene for P-gp, plasma concentrations of the drugs were elevated two to five-fold and brain concentrations seven to 36-fold compared to control mice. They concluded that P-gp limits both oral bioavailability and brain penetration of the PIs. Using an *in vitro* model, Profit and colleagues [115] found that the HIV PIs could modulate P-gp function, with the effect of ritonavir being the greatest, followed by saquinavir, nelfinavir, and indinavir, and they postulated that this inhibitory effect could explain some of the effects of dual PIs to increase drug concentrations. Van der Sandt and colleagues [116] extended these observations using amprenavir, indinavir and ritonavir in an *in vitro* model, but showed that when simultaneously administered, the PIs were not able to inhibit P-gp efflux in the brain.

Drug interactions

Patients with HIV commonly take a number of medications, including prescribed drugs, over-the-counter drugs, and herbal and other supplements. The potential for drug interactions among these multiple drugs is high through a number of different mechanisms [117], and some of these interactions may lead to reduced drug concentrations and result in virologic failure. Because both the HIV NNRTIs and PIs are metabolized by hepatic cytochrome P450 enzymes, inducers of these enzymes may lead to reductions in drug levels. One common example is the antituberculous drug rifampin that decreases the areas under the curve (AUCs) of the NNRTIs by 13–96% and the HIV PIs by 35–92%, prompting formal recommendations to avoid concomitant use [118]. However, updated recommendations specify that rifampin may be used with efavirenz, ritonavir or a dual protease inhibitor combination regimen [119].

Antiepileptics (e.g., carbamazepine, phenytoin, phenobarbital) are cytochrome P450 inducers that also would be expected to lower NNRTI and PI levels [117]. At least one well-documented case was reported of a patient with virologic suppression on an indinavir-containing regimen who started carbamezapine for post-herpetic neuralgia and experienced virologic rebound and lamivudine resistance in the setting of substantially reduced indinavir concentrations [120].

Antiretrovirals themselves may be cytochrome P450 inducers, e.g., ritonavir (which also acts as an inhibitor) and the NNRTIs, nevirapine and efavirenz. For example, Murphy et al. [121] studied 24 patients who started a combination regimen containing indinavir and nevirapine and found that nevirapine significantly reduced indinavir AUC by 27% and trough concentrations by 48%. Falloon and colleagues [122] reported an unfavorable drug interaction in which efavirenz lowered the AUC of amprenavir by 43% and correlated this with a blunted virologic response in one study of HIV salvage therapy.

Some non-prescription supplements also have been associated with reduced concentrations of antiretroviral drugs. For example, Piscitelli and associates [123] found that St. John's wort, a common herbal supplement, reduced the AUC of indinavir by 57% and the trough by 81% in healthy volunteers. The same group reported that garlic supplements reduced the AUC of saquinavir by 51%, the trough level by 49% and the peak level by 54% in healthy volunteers [124]. Reductions by these and other supplements may lead to virologic failure.

Multiple evaluations for treatment failure

Mueller and colleagues [125] studied 41 children taking ritonavir for 12 weeks, followed by three-drug combination therapy and evaluated predictors of 12-week (or longer) virological responses. The authors found that the virologic responses correlated with pretreatment HIV RNA levels ($p = 0.005$) and CD4 cell counts ($p = 0.02$), day 7 plasma ritonavir levels ($p = 0.004$), and the initial viral decay rate at days 1, 2, 3, 4, and 7 of therapy ($p < 0.001$). A combination of these parameters could predict 88% of the 12-week or longer virologic treatment effect, using the statistically weighted syndromes method, a sophisticated pattern recognition analysis.

Powderly and colleagues [126] also assessed predictors of 48-week virologic response in a phase III study of zidovudine, lamivudine and nelfinavir. A total of 196 patients received the three-drug antiretroviral combination regimen, with one of two doses of nelfinavir, and 196 (73%) responded with at least two consecutive HIV RNA levels <400 cpm. Of these, 42 experienced virologic rebound to >400 cpm over 12 months. In a step-wise regression analysis, the investigators evaluated factors that predicted HIV RNA levels <400 cpm or <50 cpm. The nelfinavir dose (750 *versus* 500 mg three times daily), the baseline HIV RNA level, and the 2-h concentration of nelfinavir were all predictive of virologic response.

In addition, the $\log_{10}$ cpm reduction in HIV RNA level from baseline to week 4 was highly correlated with virologic response (for each log decrease, the OR for reduction to <400 cpm was 4.6 and for reduction to <50 cpm was 4.2). Subjects with HIV RNA levels <50 cpm had significantly more durable responses than those with nadirs of 50–400 cpm ($p = 0.0001$). The majority of patients who did not reduce their viral load levels to <50 copies/ml had virologic rebounds. Also, the 63 patients with HIV RNA <50 cpm all had HIV RNA levels decreased to <400 cpm by week 16 and 59 of them had HIV RNA <50 cpm by week 24. Based on their data, the authors suggested that a long-term virological response was unlikely if there was not a measurable decrease in viral load level by 4–8 weeks, or a decrease to <400 cpm by 12–16 weeks.

As part of the Trilege study in treatment-naïve patients of induction therapy with indinavir, zidovudine, and lamivudine followed by continued three-drug therapy, zidovudine/lamivudine, or zidovudine/indinavir, Descamps and colleagues [127] evaluated multiple mechanisms of virologic failure using a case-control design. They identified 58 cases of patients with HIV RNA rebound to >500 cpm in two consecutive samples and randomly matched them to 58 control patients with sustained virologic suppression. In this study, virologic failure was not associated with the development of mutations associated with zidovudine or indinavir resistance, though genotypic evidence of lamivudine resistance occurred commonly. They found that plasma indinavir levels were undetectable in all cases of virologic failure. Drug adherence rates were lower for cases than controls for zidovudine ($p = 0.05$) and indinavir ($p = 0.05$). They differentiated responses in the three-drug group in whom low indinavir levels, low adherence rates, and the pattern of virologic rebound compared to baseline levels suggested adherence as the principal cause of failure in contrast to the zidovudine/lamivudine group in whom adherence rates were not significantly different from controls and virologic rebound was low-level, suggesting insufficient drug potency as the principal cause of drug failure.

Similarly, Havlir et al. [128] reported their assessment of virologic rebound in patients on the ACTG 343 study of induction therapy with indinavir, zidovudine and lamivudine followed by maintenance therapy with three-drugs, indinavir monotherapy, or zidovudine/lamivudine. A total of 26 subjects who experienced confirmed HIV RNA ≥200 cpm were compared with 10 control subjects with sustained virologic suppression. Once again, mutations associated with indinavir resistance were not seen, though genotypic evidence of lamivudine resistance occurred commonly. Low plasma indinavir levels were more common in these cases than the control group ($p = 0.03$). The authors suggested poor adherence with resultant low drug concentrations and suboptimal antiretroviral potency contributed significantly to virologic failure.

Conclusions

HIV treatment failure must be assessed in terms of virologic, immunologic and clinical responses and discordant responses may occur. Virologic failure occurs commonly in patients taking antiretroviral regimens followed in clinical studies and clinical cohorts. Treatment discontinuation also occurs commonly but may not signify treatment failure since patients may be able to benefit from other antiretroviral regimens. There are multiple potential causes of treatment failure including patient factors (e.g., baseline parameters, adherence), viral factors (e.g, drug resistance, virologic response), provider factors (e.g., experience treating HIV, antiretroviral regimen choice), and treatment regimen factors (e.g., potency, toxicity, pharmacokinetics, drug interactions). One of the great challenges in assessing treatment failure in a particular patient is that any, some, or all of these factors may play a role. Future progress in reducing antiretroviral treatment failure will result from a multi-tiered approach, e.g., developing simpler and less toxic antiretroviral regimens to promote adherence, assessing more potent antiretroviral regimens to improve antiviral activity and durability, developing newer antiretroviral drugs with activity against drug-resistant viral variants, and providing ongoing education for health providers).

References

1 Panel on Clinical Practices for Treatment of HIV Infection. Guidelines for the use of antiretroviral agents in HIV-infected adults and adolescents. Accessed February 4, 2002. Available at http://www.hivatis.org

2 Carpenter CC, Cooper DA, Fischl MA, Gatell JM, Gazzard BG, Hammer SM, Hirsch MS, Jacobsen DM, Katzenstein DA, Montaner JS et al. (2000) Antiretroviral therapy in adults: updated recommendations of the International AIDS Society -USA Panel. *JAMA* 283: 381–390

3 BHIVA Writing Committee on Behalf of the BHIVA Executive Committee (2001) British HIV Association (BHIVA) guidelines for the treatment of HIV-infected adults with antiretroviral therapy. *HIV Med* 2: 276–313

4 Deeks SG, Barbour JD, Grant RM, Martin JN (2002) Duration and predictors of CD4 T-cell gains in patients who continue combination therapy despite detectable plasma viremia. *AIDS* 16: 201–207

5 Piketty C, Castiel P, Belec L, Batisse D; Si Mohamed A, Gilquin J, Gonzalez-Canali G, Jayle D, Karmochkine M, Weiss L et al. (1998) Discrepant responses to triple combination antiretroviral therapy in advanced HIV disease. *AIDS* 12: 745–750

6 Grabar S, LeMoing V, Goujard C, Leport C, Kazatchkine MD, Costagliola D, Weiss L (2000) Clinical outcome of patients with HIV-1 infection according to immunlogic and virologic response after 6 months of highly active antiretroviral therapy. *Ann Intern Med* 133: 401–410

7 Jacobson MA, French M (1998) Altered natural history of AIDS-related opportunistic infections in the era of potent combination antiretroviral therapy. *AIDS* 12: S157–S163

8 Eron JJ, Benoit SL, Jemsek M, MacArthur RD, Santana J, Quinn JB, Kuritzkes DR, Fallon MA, Rubin M (1995) Treatment with lamivudine, zidovudine, or both in HIV-positive patients with 200 to 500 CD4+ cell per cubic millimeter. North American HIV Working Party. *N Engl J Med* 333: 1662–1669

9 Hammer SM, Katzenstein DA, Hughes MD, Gundacker H, Schooley RT, Haubrich RH, Henry WK, Lederman MM, Phair JP, Niu M et al. (1996) A trial comparing nucleoside monotherapy with combination therapy in HIV-infected adults with CD4 cell counts from 200 to 500 per cubic millimeter. AIDS Clinical Trials Group Study 175 Study Team. *N Engl J Med* 335: 1081–1090

10 Hammer SM, Squires KE, Hughes MD, Grimes JM, Demeter LM, Currier JS, Eron JJ Jr, Feinberg JE, Balfour JJ Jr, Deyton LR et al. (1997) A controlled trial of two nucleoside analogues plus indinavir in persons with human immunodeficiency virus infection and CD4 cell counts of 200 per cubic millimeter or less. AIDS Clinical Trial Group 320 Study Team. *N Engl J Med* 337: 724–733

11 Gulick RM, Mellors JW, Havlir D, Eron JJ, Gonzalez C, McMahon D, Richman DD, Valentine FT, Jonas L, Meibohm A, Emini EA et al. (1997) Treatment with indinavir, zidovudine, and lamivudine in adults with human immunodeficiency virus infection and prior antiretroviral therapy. *N Engl J Med* 337: 734–739

12 Montaner JSG, Reiss P, Cooper D, Vella S, Harris M, Conway B, Wainberg MA, Smith D, Robinson P, Hall D et al. (1998) A randomized, double-blind trial comparing combinations of nevirapine, didanosine, and zidovudine for HIV-infected patients: the INCAS trial, Italy, The Netherlands, Canada and Australia Study. *JAMA* 279: 930–937

13 AVANTI Study Group (2000) AVANTI 2: Randomized, double-blind trial to evaluate the efficacy and safety of zidovudine plus lamivudine *versus* zidovudine plus lamivudine plus indinavir in HIV-infected antiretroviral-naïve patients. *AIDS* 14: 367–374

14 Gartland M, AVANTI Study Group (2001) AVANTI 3: A randomized, double-blind trial to compare the efficacy and safety of lamivudine plus zidovudine *versus* lamivudine plus zidovudine plus nelfinavir in HIV-1-infected antiretroviral-naïve patients. *Antivir Ther* 6: 127–134

15 Saag MS, Tebas P, Sension M, Conant M, Myers R, Chapman SK, Anderson R, Clendeninn N, Viracept Collaborative Study Group (2001) Randomized, double-blind comparison of two nelfinavir doses plus nucleosides in HIV-infected patients (Agouron study 511). *AIDS* 15: 1971–1978

16 Bartlett JA, DeMasi R, Quinn J, Moxham C, Rousseau F (2001) Overview of the effectiveness of triple combination therapy in antiretroviral-naïve HIV-1 infected adults. *AIDS* 15: 1369–1377
17 Fatkenheuer G, Theisen A, Rockstroh J, Grabow T, Wicke C, Becker K, Wieland U, Pfister H, Reiser M, Hegener P et al. (1997) Virological treatment failure of protease inhibitor therapy in an unselected cohort of HIV-infected patients. *AIDS* 11: F113–F116
18 Casado JL, Perez-Elias MJ, Antela A, Sabido R, Marti-Belda P, Dronda F, Blazquez J, Quereda C (1998) Predictors of long-term response to protease inhibitor therapy in a cohort of HIV-infected patients. *AIDS* 12: F131–F135
19 D'Arminio Monforte A, Testa L, Adorni F, Chiesa E, Bini T, Moscatelli GC, Abeli C, Rusconi S, Sollima S, Balotta C et al. (1998) Clinical outcome and predictive factors of failure of highly active antiretroviral therapy in antiretroviral-experienced patients in advanced stages of HIV-1 infection. *AIDS* 12: 1631–1637
20 Ledergerber B, Egger M, Opravil M, Telenti A, Hirschel B, Battegay M, Vernazza P, Sudre P, Flepp M, Furrer H et al. (1999) Clinical progression and virological failure on highly active antiretroviral therapy in HIV-1 patients: a prospective cohort study. Swiss HIV Cohort Study. *Lancet* 353: 863–868
21 Lucas GM, Chaisson RE, Moore RD (1999) Highly active antiretroviral therapy in a large urban clinic: risk factors for virologic failure and adverse drug reactions. *Ann Intern Med* 131: 81–87
22 Staszewski S, Miller V, Sabin C, Carlebach A, Berger AM, Weidmann E, Helm EB, Hill A, Phillips A (1999) Virological response to protease inhibitor therapy in an HIV clinic cohort. *AIDS* 13: 367–373
23 Wit FW, Van Leeuwen R, Weverling GJ, Jurriaans S, Nauta K, Steingrover R, Schuijtemaker J, Eyssen X, Fortuin D, Weeda M et al. (1999) Outcome and predictors of failure of highly active antiretroviral therapy: one-year follow-up of a cohort of human immunodeficiency virus type 1-infected persons. *J Infect Dis* 179: 790–798
24 Valdez H, Lederman MM, Woolley I, Walker CJ, Vernon LT, Hise A, Gripshover BM (1999) Human immunodeficiency virus 1 protease inhibitors in clinical practice: predictors of virological outcome. *Arch Intern Med* 159: 1771–1776
25 Deeks SG, Hecht FM, Swanson M, Elbeik T, Loftus R, Cohen PT, Grant RM (1999) HIV RNA and CD4 cell count response to protease inhibitor therapy in an urban AIDS clinic: response to both initial and salvage therapy. *AIDS* 13: F35–F43
26 Paredes R, Mocroft A, Kirk O, Lazzarin A, Barton SE, van Lunzen J, Katzenstein TL, Antunes F, Lundgren JD, Clotet B (2000) Predictors of virological success and ensuing failure in HIV-positive patients starting highly active antiretroviral therapy in Europe: results from the EuroSIDA study. *Arch Intern Med* 160: 1123–1132
27 Le Moing V, Chene G, Carrieri MP, Alioum A, Brun-Vezinet F, Piroth L, Cassuto JP, Moatti JP, Raffi F, Leport C et al. (2002) Predictors of virological rebound in HIV-1-infected patients initiating a protease inhibitor-containing regimen. *AIDS* 16: 21–29
28 Fellay J, Boubaker K, Ledergerber B, Bernasconi E, Furrer H, Battegay M, Hirschel B, Vernazza P, Francioli P, Greub G et al. (2001) Prevalence of adverse events associated with potent antiretroviral treatment: Swiss HIV Cohort Study. *Lancet* 358: 1322–1327
29 D'Arminio Monforte A, Lepri AC, Rezza G, Pezzotti P, Antinori A, Phillips AN, Angarano G, Colangeli V, De Luca A, Ippolito G et al. (2000) Insights into the reasons for discontinuation of the first highly active antiretroviral therapy (HAART) regimen in a cohort of antiretroviral naïve patients. I.C.O.N.A. Study Group. Italian Cohort of Antiretroviral-Naïve Patients. *AIDS* 14: 499–507
30 Dorrucci M, Pezzotti P, Grisorio B, Minardi C, Muro MS, Vullo V, Monforte A, Italian Cohort of Antiretroviral–Naïve Patients Study Group (2001) Time to discontinuation of the first highly active antiretroviral therapy regimen: a comparison between protease inhibitor- and non-nucleoside reverse transcriptase inhibitor-containing regimens. *AIDS* 15: 1733–1736
31 Mocroft A, Youle M, Moore A, Sabin CA, Madge S, Lepri AC, Tyrer M, Chaloner C, Wilson D, Loveday C et al. (2001) Reasons for modification and discontinuation of antiretrovirals: results from a single treatment centre. *AIDS* 15: 185–194
32 Crystal S, Sambamoorthi U, Moynihan P, McSpiritt E (2001) Initiation and continuation of newer antiretroviral treatments among medicaid recipients with AIDS. *J Gen Intern Med* 16: 850–859
33 Ahdieh Grant L, Silverberg MJ, Palacio H, Minkoff H, Anastos K, Young MA, Nowicki M, Kovacs A, Cohen M, Munoz A (2001) Discontinuation of potent antiretroviral therapy: predictive value of and impact on CD4 cell counts and HIV RNA levels. *AIDS* 15: 2101–2108

34 Bini T, Testa L, Chiesa E, Adorni F, Abeli C, Casternuovo B, Melzi S, Sollima S, Bongiovanni M, d'Arminio Monforte A (2000) Outcome of a second-line protease inhibitor-containing regimen in patients failing or intolerant of a first highly active antiretroviral therapy. *J Acquir Immune Defic Syndr* 24: 115–122
35 Miller V, Mocroft A, Reiss P, Katlama C, Papadopoulos AI, Katzenstein T, van Lunzen J, Antunes F, Phillips AN, Lundgren JD (1999) Relations among CD4 lymphocyte count nadir, antiretroviral therapy, and HIV-1 disease progression: results from the EuroSIDA study. *Ann Intern Med* 130: 570–577
36 Skowron G, Street JC, Obee EM (2001) Baseline CD4(+) cell count, not viral load, correlates with virologic suppression induced by potent antiretroviral therapy. *J Acquir Immune Defic Syndr* 28: 313–319
37 Hecht FM, Grant RM, Petropoulos CJ, Dillon B, Chesney MA, Tian H, Hellmann NS, Bandrapalli NI, Digilio L, Branson B et al. (1998) Sexual transmission of an HIV-1 variant resistant to multiple reverse-transcriptase and protease inhibitors. *N Engl J Med* 337: 307–311
38 Boden D, Hurley A, Zhang L, Cao Y, Guo Y, Jones E, Tsay J, Ip J, Farthing C, Limoli K et al. (1999) HIV-1 drug resistance in newly infected individuals. *JAMA* 282: 1135–1141
39 Little SJ, Daar ES, D'Aquila RT, Keiser PH, Connick E, Whitcomb JM, Hellmann NS, Petropoulos CJ, Sutton L, Pitt JA (1999) Reduced antiretroviral drug susceptibility among patients with primary HIV infection. *JAMA* 282: 1142–1149
40 UK Collaborative Group on Monitoring the Transmission of HIV Drug Resistance (2001) Analysis of prevalence of HIV-1 drug resistance to primary infections in the United Kingdom. *Br Med J* 322: 1087–1088
41 Little SJ, Routy JP, Daar ES, Markowitz M, Collier AC, Koup RA, Conway B, Saag MS, Connick E, Holte S et al. (2001) Antiretroviral drug susceptibility and response to initial therapy among recently HIV infected subjects in North America. *Abstracts of the 8th Conference on Retroviruses and Opportunistic Infections*, Chicago, IL, USA, 4–8 February 2001, Abstract #756
42 Gulick RM, Mellors JW, Havlir D, Eron JJ, Gonzalez C, McMahon D, Jonas L, Meibohm A, Holder D, Schleif WA et al. (1998) Simlutaneous vs sequential intitiation of therapy with indinavir, zidovudine, and lamivudine for HIV-1 infection: 100 week follow-up. *JAMA* 280: 35–41
43 Havlir DV, Marschner IC, Hirsch MS, Collier AC, Tebas P, Bassett RL, Ioannidis JP, Holohan MK, Leavitt R, Boone G et al. (1998) Maintenance antiretroviral therapies in HIV infected patients with undetectable plasma HIV RNA after triple-drug therapy. AIDS Clinical Trials Group Study 343 Team. *N Engl J Med* 339: 1261–1268
44 Pialoux G, Raffi F, Brun-Vezinet F, Meiffredy V, Flandre P, Gastaut JA, Dellamonica P, Yeni P, Delfraissy JF, Aboulker JP (1998) A randomized trial of three maintenance regimens given after three months of induction therapy with zidovudine, lamivudine, and indinavir in previously untreated HIV-1-infected patients. Trilege (Agence Nationale de Recherches sur le SIDA 072) Study Team. *N Engl J Med* 339: 1269–1276
45 Reijers MH, Weverling GJ, Jurriaans S, Wit FW, Weigel HM, Ten Kate RW, Mulder JW, Frissen PH, van Leeuwen R, Reiss P et al. (1998) Maintenance therapy after quadruple induction therapy in HIV-1 infected individuals: Amsterdam Duration of Antiretroviral Medication (ADAM) study. *Lancet* 352: 185–190
46 Reijers MH, Weverling GJ, Juriaans S, Roos MT, Wit FW, Weigel HM, Ten Kate RW, Mulder JW, Richter C, Ter Hofstede JH et al. (2001) The ADAM study continued: maintenance therapy after 50 weeks of induction therapy. *AIDS* 15: 129–131
47 Staszewski S, Morales-Ramirez J, Tashima KT, Rachlis A, Skiest D, Stanford J, Stryker R, Johnson P, Labriola DF, Farina D et al. (1999) Efavirenz plus zidovudine and lamivudine, efavirenz plus indinavir, and indinavir plus zidovudine and lamivudine in the treatment of HIV-1 infection in adults. Study 006 Team. *N Engl J Med* 341: 1865–1873
48 Walmsley S, Badley A, Beall G, Cameron W, Johnson D, Johnson M, Ruane P, Rubio R, Stryker R, King M et al. (2000) Efficacy of ABT-378/r vs nelfinavir (NFV) in antiretroviral (ARV)-naïve subjects: results of a phase III blinded randomized clinical trial. *Abstracts of the 40th Interscience Conference on Antimicrobial Agents and Chemotherapy*, Toronto, Canada, 17–20 September 2000, Abstract 693
49 Matthews GV, Sabin CA, Mandalia S, Lampe D, Phillips AN, Nelson MR, Bower M, Johnson MA, Gazzard BG (2002) Virological suppression at 6 months is related to choice of initial regimen in antiretroviral-naïve patients: a cohort study. *AIDS* 16: 53–61
50 Barreiro P, Soriano V, Casas E, Gonzalez-Lahoz J. (2002) Different degree of immune recovery

using antiretroviral regimens with protease inhibitors or non-nucleosides. *AIDS* 16: 245–249
51 Gulick RM, Hu XJ, Fiscus SA, Fletcher CV, Haubrich R, Cheng H, Acosta E, Lagakos SW, Swanstrom R, Freimuth W et al. (2000) Randomized study of saquinavir with ritonavir or nelfinavir together with delavirdine, adefovir, or both in human immunodeficiency virus-infected adults with virologic failure on indinavir: AIDS Clinical Trials Group Study 359. *J Infect Dis* 182: 1375–1384
52 Eron J, Falloon J, Masur H et al. (1998) Activity of combination abacavir/amprenavir/efavirenz therapy in HIV-1 infected subjects failing their current protease inhibitor containing regimen. *AIDS* 12: OP5.2
53 Mellors J, Hammer S, Vaida F et al. (2000) A randomized, placebo-controlled trial of saquinavir soft gel (SQV), indinavir (IDV) or nelfinavir (NFV) in combination with amprenavir (APV), abacavir (ABC), efavirenz (EFZ) and adefovir (ADV) in patients with protease inhibitor (PI) failure. *Abstracts of the Third International Workshop on Salvage Therapy for HIV Infection*, Chicago, IL, 12–14 April 2000, Abstract #21
54 Tebas P, Patick AK, Kane EM, Klebert MK, Simpson JH, Erice A, Powderly WG, Henry K (1999) Virologic responses to a ritonavir-saquinavir-containing regimen in patients who had previously failed nelfinavir. *AIDS* 13: F23–F28
55 Gulick RM, Smeaton LM, D'Aquila RT, Eron JJ, Currier JS, Gerber JG, Acosta E, Sommadossi JP, Tung R, Snyder S et al. (2001) Indinavir, nevirapine, stavudine and lamivudine for human immunodeficiency virus-infected, amprenavir-experienced subjects: AIDS Clinical Trials Group Protocol 373. *J Infect Dis* 183: 715–721
56 Zhang L, Ramratnam B, Tenner-Racz K, Vesanen M, Lewin S, Talal A, Racz P, Perelson AS, Korber BT et al. (1999) Quantifying residual HIV-1 replication in patients receiving combination antiretroviral therapy. *N Engl J Med* 340: 1605–1613
57 Furtado MR, Callaway DS, Phair JP, Kunstman KJ, Stanton JL, Macken CA, Perelson AS, Wolinsky SM (1999) Persistence of HIV-1 transcription in peripheral-blood mononuclear cells in patients receiving potent antiretroviral therapy. *N Engl J Med* 340: 1614–1622
58 Ramratnam B, Mittler JE, Zhang L, Boden D, Hurley A, Fang F, Macken CA, Perelson AS, Markowitz M, Ho DD (2000) The decay of the latent reservoir of replication-competent HIV-1 is inversely correlated with the extent of residual viral replication during prolonged anti-retroviral therapy. *Nat Med* 6: 82–85
59 Martinez-Picado J, DePasquale MP, Kartsonis N, Hanna GJ, Wong J, Finzi D, Rosenberg E, Gunthard HF, Sutton L, Savara A et al. (2000) Antiretroviral resistance during successful therapy of HIV type 1 infection. *Proc Natl Acad Sci USA* 97: 10948–10953
60 Weverling GJ, Lange JM, Jurriaans S, Prins JM, Lukashov VV, Notermans DW, Roos M, Schuitemaker H, Hoetelmans RM, Danner SA et al. (1998) Alternative multidrug regimen provides improved suppression of HIV-1 replication over triple therapy. *AIDS* 12: F117–F122
61 Eron J, Junod P, Thompson M, Arduino R, Walmsley S, Pierce A. et al. (2000) NZT 4002: 64 week analysis of Combivir (COM)-based triple and quadruple therapy in antiretroviral-naïve, HIV-1 infected subjects. *Abstracts of the XIII International AIDS Conference*, Durban, South Africa, 9–14 July 2000, Abstract WeOrB608
62 Fischl H, Ribaudo H, Collier A, Rice A, Giuliano M, Dehlinger M, Eron JJ, Saag M, Hammer S, Vella S et al. (2002) A randomized trial comparing 2 4-drug antiretroviral regimens with a 3-drug regimen in advanced HIV disease. *Abstracts of the 9th Conference on Retroviruses and Opportunistic Infections*, Seattle, Washington, 24–28 February 2002, Abstract #41
63 Kitahata MM, Koepsell TD, Deyo RA, Maxwell CL, Dodge WT, Wagner EH. (1996) Physicians' experience with the acquired immunodeficiency syndrome as a factor in patients' survival. *N Engl J Med* 334: 701–706
64 Brosgart CL, Mitchell TF, Coleman RL, Dyner T, Stephenson KE, Abrams DI (1999) Clinical experience and choice of drug therapy for human immunodeficiency virus disease. *Clin Infect Dis* 28: 14–22
65 Stone VE, Mansourati FF, Poses RM, Mayer KH (2001) Relation of physician specialty and HIV/AIDS experience to choice of guideline-recommended antiretroviral therapy. *J Gen Intern Med* 16: 360–368
66 Edelstein H, Wilson M (2001) Antiretroviral medication errors were universal in hospitalized HIV-seropositive patients at a teaching hospital. *J Acquir Immune Defic Syndr* 28: 496
67 O'Brien WA, Hartigan PM, Martin D, Esinhart J, Hill A, Benoit S, Rubin M, Simberkoff MS, Hamilton JD (1996) Changes in plasma HIV-1 RNA and CD4+ lymphocyte counts and the risk of

progression to AIDS. Veterans Affairs Cooperative Study Group on AIDS. *N Engl J Med* 334: 426–431
68 O'Brien WA, Hartigan PM, Daar ES, Simberkoff MS, Hamilton JD (1997) Changes in plasma HIV RNA levels and CD4+ lymphocyte counts predict both response to antiretroviral therapy and therapeutic failure. *Ann Intern Med* 126: 939–945
69 Hughes MD, Johnson VA, Hirsch MS, Bremer JW, Elbeik T, Erice A, Kuritzkes DR, Scott WA, Spector SA, Basgoz N et al. (1997) Monitoring plasma HIV-1 RNA levels in addition to CD4+ lymphocyte count improves assessment of antiretroviral therapeutic response. ACTG 241 Protocol Virology Substudy Team. *Ann Intern Med* 126: 929–938
70 Raboud JM, Montaner JSG, Conway B, Rae S, Reiss P, Vella S, Cooper D, Lange J, Harris M, Wainberg MA et al. (1998) Suppression of plasma viral load below 20 copies/ml is required to achieve a long-term response to therapy. *AIDS* 12: 1619–1624
71 Kempf DJ, Rode RA, Xu Y, Sun E, Heath-Chiozzi ME, Valdes J, Japour AJ, Danner S, Boucher C, Molla A et al. (1998) The duration of viral suppression during protease inhibitor therapy for HIV-1 infection is predicted by plasma HIV-1 RNA at the nadir. *AIDS* 12: F9–F14
72 Havlir DV, Bassett R, Levitan D, Gilbert P, Tebas P, Collier AC, Hirsch MS, Ignacio C, Condra J, Gunthard HF et al. (2001) Prevalence and predictive value of intermittent viremia with combination HIV therapy. *JAMA* 286: 171–179
73 Greub G, Cozzi Lepri A, Ledergerber B, Staszewski S, Perrin L, Miller V, Francioli P, Furrer H, Battegay M, Vernazza P et al. (2001) Low-level HIV viral rebound and blips in patients receiving potent antiretroviral therapy. *Abstracts of the 8th Conference on Retroviruses and Opportunistic Infections*, Chicago, IL, 4–8 February 2001, Abstract 522
74 Perrin L, King M, Brun S et al. (2001) Failure to achieve HIV RNA ≤3 copies/ml does not predict loss of virologic response to Kaletra (lopinavir/ritonavir) therapy through 3 years. *Abstracts of the 41st Interscience Conference of Antimicrobial Agents and Chemotherapy*, Chicago, IL, 16–19 December 2001, Abstract #1927
75 Raboud JM, Rae S, Montaner JSG (2000) Predicting HIV RNA virologic outcome at 52-weeks follow-up in antiretroviral clinical trials. The INCAS and AVANTI Study Groups. *J Acquir Immune Defic Syndr* 24: 433–439
76 Lepri AC, Miller V, Phillips AN, Rabenau H, Sabin CA, Staszewski S (2001) The virological response to highly active antiretroviral therapy over the first 24 weeks of therapy according to the pre-therapy viral load and the weeks 4–8 viral load. *AIDS* 15: 47–54
77 Polis MA, Sidorov IA, Yoder C, Jankelevich S, Metcalf J, Mueller BU, Dimitrov MA, Pizzo P, Yarchoan R, Dimitrov DS (2001) Correlation between reduction in plasma HIV-1 RNA concentration 1 week after start of antiretroviral treatment and longer-term efficacy. *Lancet* 358: 1760–1765
78 Huang W, De Gruttola V, Fischl M, Hammer S, Richman D, Havlir D, Gulick R, Squires K, Mellors J (2001) Patterns of plasma human immunodeficiency virus type 1 RNA response to antiretroviral therapy. *J Infect Dis* 183: 1455–1465
79 Gulick RM, Mellors JW, Havlir D, Eron JJ, Meibohm A, Condra JH, Valentine FT, McMahon D, Gonzalez C, Jonas L et al. (2000) 3-year suppression of HIV viremia with indinavir, zidovudine, and lamivudine. *Ann Intern Med* 133: 35–39
80 Gulick RM, Mellors JW, Havlir D et al. (2001) Indinavir (IDV), zidovudine (ZDV), and lamivudine (3TC): 5-year follow-up. *Abstracts of the 1st IAS Conference on HIV Pathogenesis and Treatment*, Buenos Aires, Argentina, 8–11 July 2001, abstract 215
81 Phillips AN, Miller V, Sabin C, Cozzi Lepri A, Klauke S, Bickel M, Doerr HW, Hill A, Staszewski S (2001) Durability of HIV-1 viral suppression over 3.3 years with multi-drug antiretroviral therapy in previously drug-naïve individuals. *AIDS* 15: 2379–2384
82 Staszewski S, Miller V, Sabin C, Schlecht C, Gute P, Stamm S, Leder T, Berger A, Weidemann E, Hill A et al. (1999) Determinants of sustainable CD4 lymphocyte count increases in response to antiretroviral therapy. *AIDS* 13: 951–956
83 Miller V, Staszewski S, Sabin C, Carlebach A, Rottmann C, Weidmann E, Rabenau H, Hill A, Lepri AC, Phillips AN (1999) CD4 lymphocyte count as a predictor of the duration of highly active antiretroviral therapy-induced suppression of human immunodeficiency virus load. *J Infect Dis* 180: 530–533
84 D'Arminio Monforte A, Testori V, Adorni F, Castelnuovo B, Bini T, Testa L, Moscatelli G, Chiesa E, Rusconi S, Abeli C et al. (1999) CD4 cell counts at the third month of HAART may predict clinical failure. *AIDS* 13: 1669–1676

85 Demeter LM, Hughes MD, Coombs RW, Jackson JB, Grimes JM, Bosch RJ, Fiscus SA, Spector SA, Squires KE, Fischl MA et al. (2001) Predictors of virologic and clinical outcomes in HIV-1-infected patients receiving concurrent treatment with indinavir, zidovudine, and lamivudine. AIDS Clinical Trials Group Protocol 320. *Ann Intern Med* 135: 954–964
86 Haubrich RH, Little SJ, Currier JS, Forthal DN, Kemper CA, Beall GN, Johnson D, Dube MP, Hwang JY, McCutchan JA (1999) The value of patient-reported adherence to antiretroviral therapy in predicting virologic and immunologic response. California Collaborative Treatment Group. *AIDS* 13: 1099–1107
87 Bangsberg DR, Hecht FM, Charlebois ED, Zolopa AR, Holodniy M, Sheiner L, Bamberger JD, Chesney MA, Moss A (2000) Adherence to protease inhibitors, HIV-1 viral load and development of drug resistance in an indigent population. *AIDS* 14: 357–366
88 Paterson DL, Swindells S, Mohr J, Brester M, Vergis EN, Squier C, Wagener MM, Singh N (2000) Adhrence to protease inhibitor therapy and outcomes in patients with HIV infection. *Ann Intern Med* 133: 21–30
89 Murri R, Ammassari A, Gallicano K, De Luca A, Cingolani A, Jacobson D, Wu AW, Antinori A (2000) Patient-reported nonadherence to HAART is related to protease inhibitor levels. *J Acquir Immune Defic Syndr* 24: 123–128
90 Nieuwkerk PT, Sprangers MAG, Burger DM, Hoetelmans RM, Hugen PW, Danner SA, van Der Ende ME, Schneider MM, Schrey G, Meenhorst PL et al. (2001) Limited patient adherence to highly active antiretroviral therapy for HIV-1 infection in an observational cohort study. *Arch Intern Med* 161: 1962–1968
91 Acosta EP, Kakuda TN, Brundage RC, Anderson PL, Fletcher CV. (2000) Pharmacodynamics of human immunodeficiency virus type 1 protease inhibitors. *Clin Infect Dis* 30: S151–S159
92 Marzolini C, Telenti A, Decosterd LA, Greub G, Biollaz J, Buclin T (2001) Efavirenz plasma levels can predict treatment failure and central nervous system side-effects in HIV-1-infected patients. *AIDS* 15: 71–75
93 Schapiro JM, Winters MA, Stewart F, Efron B, Norris J, Kozal MJ, Merigan TC (1996) The effect of high-dose saquinavir on viral load and CD4+ T-cell counts in HIV-infected patients. *Ann Intern Med* 124: 1039–1050
94 Fletcher CV, Brundage RC, Remmel RP, Page LM, Weller D, Calles NR, Simon C, Kline MW (2000) Pharmacologic characteristics of indinavir, didanosine, and stavudine in human immunodeficiency virus-infected children receiving combination therapy. *Antimicrob Agents Chemother* 44: 1029–1034
95 Fellay J, Marzolini C, Meaden ER, Back DJ, Buclin T, Chave JP, Decosterd LA, Furrer H, Opravil M, Pantaleo G et al. (2002) Response to antiretroviral treatment in HIV-1-infected invididuals with allelic variants of the multidrug resistance transporter 1: a pharmacogenetics study. *Lancet* 359: 30–36
96 Peter K, Gambertoglio JG (1998) Intracellular phophorylation of zidovudine (ZDV) and other nucleoside reverse transcriptase inhibitors (RTI) used for human immunodeficiency virus (HIV) infection. *Pharm Res* 15: 819–825
97 Fletcher CV, Acosta EP, Henry K, Page LM, Gross CR, Kawle SP, Remmel RP, Erice A, Balfour HH Jr et al. (1998) Concentration-controlled zidovudine therapy. *Clin. Pharmacol Ther* 64: 331–338
98 Fletcher CV, Kawle SP, Kakuda TN, Anderson PL, Weller D, Bushman LR, Brundage RC, Remmel RP (2000) Zidovudine triphosphate and lamivudine triphosphate concentration-response relationships in HIV-infected persons. *AIDS* 14: 2137–2144
99 Veldkamp AI, Weverling GJ, Lange JM, Montaner JS, Reiss P, Cooper DA, Vella S, Hall D, Beijnen JH, Hoetelmans RM (2001) High exposure to nevirapine in plasma is associated with an improved virological response in HIV-1-infected invidivduals. *AIDS* 15: 1089–1095
100 Para MF, Meehan P, Holden-Wiltse J, Fischl M, Morse G, Shafer R, Demeter LM, Wood K, Nevin T, Virani-Ketter N (1999) ACTG 260: a randomized, phase I-II, dose-ranging trial of the anti-human immunodeficiency virus activity of delavirdine monotherapy. The AIDS Clinical Trials Group Protocol 260 Team. *Antimicrob Agents Chemother* 43: 1373–1378
101 Danner SA, Carr A, Leonard JM, Lehman LM, Gudiol F, Gonzalez J, Raventos A, Rubio R, Bouza E, Pintado V et al. (1995) A short-term study of the safety, pharmacokinetics, and efficacy of ritonavir, an inhibitor of HIV-1 protease. European-Australian Collaborative Ritonavir Study Group. *N Engl J Med* 333: 1528–1533
102 Acosta EP, Henry K, Baken L, Page LM, Fletcher CV (1999) Indinavir concentrations and antivi-

ral effect. *Pharmacotherapy* 19: 708–712
103 Burger DM, Hoetelmans RM, Hugen PW, Mulder JW, Meenhorst PL, Koopmans PP, Brinkman K, Keuter M, Dolmans W, Hekster YA (1998) Low plasma concentrations of indinavir are related to virological treatment failure in HIV-1 infected patients on indinavir-containing triple therapy. *Antiviral Ther* 3: 215–220
104 Markowitz M, Conant M, Hurley A, Schluger R, Duran M, Peterkin J, Chapman S, Patick A, Hendricks A, Yuen GJ et al. (1998) A preliminary evaluation of nelfinavir mesylate, an inhibitor of human immunodeficiency virus (HIV)-1 protease, to treat HIV infection. *J Infect Dis* 177: 1533–1540
105 Kerr B, Pithavala Y, Zhange M, Knowles M, Daniels R, Anderson R. (1998) Virologic response-plasma drug concentration relationship in phase III study of nelfinavir mesylate (Viracept). *Abstracts of the 12th World AIDS Conference*, Geneva, Switzerland, 28 June – 3 July 1998, Abstract #12304
106 Hoetelmans, RMW, Reijers MHE, Weverling GJ, Ten Kate RW, Wit FW, Mulder JW, Weigel HM, Frissen PH, Roos M, Jurriaans S et al. (1998) The effect of plasma drug concentrations on HIV-1 clearance rate during quadruple drug therapy. *AIDS* 12: F111–F115
107 Durant J, Clevenbergh P, Garraffo R, Halfon P, Icard S, Del Giudice P, Montagne N, Schapiro JM, Dellamonica P (2000) Importance of protease inhibitor plasma levels in HIV-infected patients treated with genotypic-guided therapy: pharmacological data from the Viradapt Study. *AIDS* 14: 1333–1339
108 Gao WY, Agbaria R, Driscoll JS, Mitsuya H (1994) Divergent anti-human immunodeficiency virus activity and anabolic phosphorylation of 2',3'-dideoxynucleoside analogs in resting and activated human cells. *J Biol Chem* 269: 12633–12638
109 Shirasaka T, Chokekijchai S, Yamada A, Gosselin G, Imbach J-L, Mitsuya H (1995) Comparative analysis of anti-human immunodeficiency virus type 1 activities of dideoxynucleoside analogs in resting and activated peripheral blood mononuclear cells. *Antimicrob Agents Chemother* 39: 2555–2559
110 Van't Wout AB, Ran LJ, de Jong MD, van Leeuwen R, Notermans DW, Loeliger AE, de Wolf F, Danner SA, Reiss P et al. (1997) Selective inhibition of syncytium-inducing and nonsyncytium-inducing HIV-1 variants in individuals receiving didanosine or zidovudine, respectively. *J Clin Invest* 100: 2325–2332
111 Gottesman MM, Hrycyna CA, Schoenlein PV, Germann UA, Pastan I. (1995) Genetic analysis of the multidrug transporter. *Annu Rev Genet* 29: 607–649
112 Lucia MB, Cauda R, Landay AL, Malorni W, Donelli G, Ortona L (1995) Transmembrane P-glycoprotein (P-gp/P-170) in HIV infection: analysis of lymphocyte surface expression and drug-unrelated function. *AIDS Res Hum Retroviruses* 11: 893–901
113 Lee CGL, Gottesman MM, Cardarelli CO, Ramachandra M, Jeang KT, Ambudkar SV, Pastan I, Dey S (1998) HIV-1 protease inhibitors are substrates for the MDR1 multidrug transporter. *Biochemistry* 37: 3594–3601
114 Kim RB, Fromm MF, Wandel C, Leake B, Wood AJ, Roden DM, Wilkinson GR (1998) The drug transporter P-glycoprotein limits oral absorption and brain entry of HIV-1 protease inhibitors. *J Clin Invest* 101: 289–294
115 Profit L, Eagling VA, Back DJ (1999) Modulation of P-glycoprotein function in human lymphocytes and Caco-2 cell monolayers by HIV-1 protease inhibitors. *AIDS* 13: 1623–1627
116 Van der Sandt IC, Vos CM, Nabulsi L, Blom-Roosemalen MC, Voorwinden HH, de Boer AG, Breimer DD (2001) Assessment of active transport of HIV protease inhibitors in various cell lines and the *in vitro* blood-brain barrier. *AIDS* 15: 483–491
117 Piscitelli SC, Gallicano KD (2001) Interactions among drugs for HIV and opportunistic infections. *N Engl J Med* 344: 984–996
118 Centers for Disease Control and Prevention. (1998) Prevention and treatment of tuberculosis among patients infected with human immunodeficiency virus: principles of therapy and revised recommendations. *MMWR* 47: 1–58
119 Centers for Disease Control and Prevention. (2000) Updated guidelines for the use of rifabutin or rifampin for the treatment and prevention of tuberculosis among HIV-infected patients taking protease inhibitors or nonnucleoside reverse transcriptase inhibitors. *MMWR* 49: 185–189
120 Hugen PW, Burger DM, Brinkman K, ter Hofstede HJ, Schuurman R, Koopmans PP, Hekster YA (2000) Carbamezapine-indinavir interaction causes antiretroviral therapy failure. *Ann Pharmacother* 34: 465–470

121 Murphy RL, Sommadossi JP, Lamson M, Hall DB, Myers M, Dusek A (1999) Antiviral effect and pharmacokinetic interaction between nevirapine and indinavir in persons infected with human immunodeficiency virus type 1. *J Infect Dis* 179: 1116–1123
122 Falloon J, Piscitelli S, Vogel S, Sadler B, Mitsuya H, Kavlick MF, Yoshimura K, Rogers M, LaFon S, Manion DJ et al. (2000) Combination therapy with amprenavir, abacavir, and efavirenz in human immunodeficiency virus (HIV)-infected patients failing a protease-inhibitor regimen: pharmacokinetic interactions and antiviral activity. *Clin Infect Dis* 30: 313–318
123 Piscitelli SC, Burstein AH, Chaitt D, Alfaro RM, Falloon J (2000) Indinavir concentrations and St. John's wort. *Lancet* 355: 547–548
124 Piscitelli SC, Burstein AH, Welden N, Gallicano KD, Falloon J. (2002) The effect of garlic supplements on the pharmacokinetics of saquinavir. *Clin Infect Dis* 34: 234–238
125 Mueller BU, Zeichner SL, Kuznetsov VA, Heath-Chiozzi M, Pizzo PA, Dimitrov DS (1998) Individual prognoses of long-term responses to antiretroviral treatment based on virological, immunological and pharmacological parameters measured during the first week under therapy. *AIDS* 12: F191–F196
126 Powderly WG, Saag MS, Chapman S, Yu G, Quart B, Clendennin NJ (1999) Predictors of optimal virological response to potent antiretroviral therapy. *AIDS* 13: 1873–1880
127 Descamps D, Flandre F, Calvez V, Peytavin G, Meiffredy V, Collin G, Delaugerre C, Robert-Delmas S, Bazin B, Albouker JP et al. (2000) Mechanisms of virologic failure in previously untreated HIV-infected patients from a trial of induction-maintenance therapy. Trilege (Agence Nationale de Recherches sur le SIDA 072) Study Team. *JAMA* 283: 205–211
128 Havlir DV, Hellmann NS, Petropoulos CJ, Whitcomb JM, Collier AC, Hirsch MS, Tebas P, Sommadossi JP, Richman DD (2000) Drug susceptibility in HIV infection after HIV rebound in patients receiving indinavir-containing regimens. *JAMA* 283: 229–234

Combination Therapy of AIDS
Edited by E.D.A. De Clercq and A.-M.I. Vandamme

Economic implication of HIV-1 resistance testing in overall clinical care

Christopher Holtzer[1] and Mike Youle[2]

[1] *University of California, San Francisco, Gilead Sciences Inc., Foster City, CA, USA*
[2] *Director of HIV Clinical Research, Royal Free Centre for HIV Medicine, Pond Street, London NW3 2QG, UK*

Introduction

Currently, there are two methodologies for assessing the sensitivity of HIV-1 to antiretroviral (ARV) medications: genotyping and phenotyping. Genotyping is generally done either by direct sequencing or with line-probe assays that look for specific mutations in the reverse transcriptase (RT), protease (PR) and other domains that might be associated with viral resistance to medications. Phenotyping involves either the direct culturing of HIV from infected patients, or the incorporation of relevant sections of the patients viral genomic RNA into laboratory constructs of HIV-1 to determine fold changes in susceptibility. Both tests yield data that can then be used to examine the sensitivity of the viral strain within an as yet untreated individual with HIV or to modify a failing ARV regimen to improve the chance of a virologic response or viral suppression. These diagnostic tools therefore have a potential impact on the total costs of HIV care [1].

The cost of care for HIV-infected individuals as for all chronically ill patients is of increasing concern to healthcare providers and payers due to increasingly limited healthcare resources and the rising costs of medical care. The current costs for resistance based testing are (in 2002 amounts): for genotype testing 350–600 Euro and for phenotype testing approximately 1,000–1,500 Euro. Resistance testing is currently being used approximately zero to four times per year in each patient (0–6,000 Euro/Year), but this varies immensely depending on treatment strategy and response. At the other end of the cost spectrum, antiretroviral (ARV) drug therapy may cost in excess of 15,000 Euro per year for current standard of care therapy. Interestingly, evidence exists that total cost-of-care declines with the use of ARV therapy due to decreased hospitalizations and clinic utilization [2–4]. Mostly these costs have been determined in aggregate for patient care and are only now being analyzed separately for the various stages of the disease in an attempt to determine the range of actual costs per patient for individual treatment modalities or strategies. Within this frame the majority of the direct costs are for drugs and in-patient care, with diagnostics accounting for <5% of total costs (even in a clinical trial setting [5]).

The majority of patients in clinical practice have HIV successfully suppressed below the current limit of detection. However, a variable proportion have unsatisfactory virologic responses and/or virologically rebound under their ARV regimens within the 18 months following initiation of therapy, especially if they have been experienced to mono- or dual therapy in the past. Retrospective research from patient databases has shown that there is additional cost to successful treatment following virologic failure in the range of, on average, 250 Euro per month throughout all lines of therapy [6]. These data add a cost incentive to the clinical incentive of achievement and maintenance of virologic control of HIV. Knowledge of viral resistance patterns has been shown, in randomised studies, to improve the percentage virologic control seen in experienced patient populations [7–11]. However, the declining mortality rate in HIV disease means that the total cost of care for an HIV infected population is rising rapidly due mainly to the extended use of expensive ARV medications. It is therefore important to assess, in addition, the impact of resistance testing on overall costs of therapy in order to make informed decisions on the use of these tests in HIV disease treatment.

Studies completed to date

There is little argument anymore that the advent of highly active antiretroviral therapy (HAART) in 1996 has had a profound change upon the treatment of HIV-infection and the associated costs. Data from the HIV Cost and Services Utilization Study, which is at the moment the only representative survey of HIV-infected persons under treatment in the USA, found that the costs of pharmaceuticals were already, at that time, $9,460 per person per year. The amount spent on all other components of outpatient care, within this patient population, was $3,300 per person per year – even before the advent of ARV [2]. In France costs relating to hospitalizations for HIV-infected patients have been falling since 1995 (a reduction of 24% between 1995–1998), but during the same period medication costs increased from $600 in 1995 to $3,200 per person per year by 1998. Note that these figures cannot be directly compared to those from the US survey, since that survey only concerned patients receiving care, whereas the French figures are an average of total drug expenditures divided by the total number of HIV-infected persons, whether or not they received care). Universally, centers are reporting increases in pharmaceutical related costs that are directly attributable to these perceived to be expensive therapies [3, 4, 12, 13]. Clinically, all reports to date during this period in time have indicated a decrease in the morbidity and mortality associated with HIV and AIDS [13–16]. Some studies have found that this decrease in clinical severity has offset the increased costs seen with the use of these medications resulting in either cost savings or cost neutrality from previously examined time periods [4, 13, 17–19].

Although other studies have argued that ARV therapy may not result in cost savings over the long term (>1 year), a recent review found that existing studies "strongly" suggest that ARV is a cost-effective use of medical resources [18–20]. A current definition of what is cost-effective in medical care has been defined as an increase of less than $25,000 per QALY [6, 21]. Since the median efficacy of many ARV regimens has not yet been reached due to short follow-up then the median cost effectiveness of these interventions are not particularly clear.

If ARV therapy is cost-effective without genotypic analysis, what is the potential impact of a test costing $350–500 on the total cost of care of an HIV-infected patient? The stated goal of HIV-1 resistance testing is to target HIV therapy effectively with the minimum amount of and dosage of antiretroviral medications. This should result in better clinical outcomes with a lower incidence of virologic failure and potentially lower cost of care.

The cost of ARV failure in HIV patients has been recently explored and quantified [6]. Seven hundred and forty-three US patients had the total cost of their care and ARV status determined between January 1996 and December 1998. ARV therapy failure was defined as two successive viral loads >1000 copies/ml, ≥2 weeks apart. Cost of care encompassing hospitalizations, pharmacy, laboratory, skilled nursing care, home health as well as professional and infusion care costs was assessed. The cost of care for patients who had never experienced an episode of ARV failure was significantly lower than those who had experienced a single ARV failure ($1,195 per month *versus* $1,433 per month; $p = 0.05$). Patients who had experienced a single ARV failure had significantly lower costs than patients who had experienced > one episode of ARV failure ($1433 *versus* $1863; $p = 0.014$). The conclusions of this research have potential implications for ARV therapy and genotypic testing. The authors suggested that failure of ARV therapy results in an increase in costs of at least $250 per month with costs rising as a patient experiences multiple episodes of virologic failure. Therefore, they concluded that there is an economic as well as therapeutic premium in the prevention of ARV failure.

There are some data that have explored the possible cost of resistance testing in clinical practice. Most studies to date, however, have explored this using cost models with only a few studies actually analyzing prospective datasets from clinical trials.

Genotypic data

To date, there has only been one study examining the economics of resistance testing that has utilised prospectively collected data [22]. This study examined costs within the VIRADAPT study; a prospective, open-label, randomized trial comparing patients assigned to standard of care ($n = 43$), *versus* a genotypic-guided new treatment regimen ($n = 64$) for 6 months with a open-label phase extending to 1 year [9]. Hospitalizations number and costs of tests taken, drugs prescribed and clinic visit data were retrieved from the VIRADAPT study case report forms and database. The costs in this study were computed from the viewpoint of the healthcare system and were estimated from the actual, not billed, costs of the corresponding Diagnosis Related Groups derived from the French national cost database. The unit costs of tests and clinic visits were determined using the French National Social Security reimbursement price. The prices of prescribed medications were based upon purchase price by either retail pharmacies or hospitals. The cost of the genotyping assay in the trial was estimated at $500. This cost is relatively high by current standards, with standard costs being approximately $350 per test in 2002.

The authors concluded that the mean (standard deviation) yearly cost per patient was $20,412 (±10,129) in the standard of care group and $18,484 (±9,652) in the genotyping group ($p = 0.35$). Drug costs represented 55% of total costs in both groups. While this difference did not yield a statistically significance difference between the two groups in overall cost savings, there was a trend ($p = 0.07$) toward a decrease in drug costs in the genotyping arm as compared to the standard of care group. Sub-analysis of the data found that the greatest reduction in cost were from the decreased use of protease inhibitors in the genotyping arm and a corresponding increase in non-nucleoside reverse transcriptase inhibitor (NNRTI) use. The authors concluded that the additional expense of genotyping appeared to be offset by the savings obtained in drug costs. While not reaching statistical significance, this study was suggestive of streamlined drug utilization possibly leading to decreased costs. This preliminary analysis was hampered by its sample size (107 patients) and brief period of follow-up (6 months of randomization in the study, 12 months overall). Also, there was and is a wide variability in the relative costs of antiretroviral agents, which is unrelated to potency and this may influence the outcome and the significance of data analyzed on a local cost basis in a different country or region. In addition, in the analysis, the relatively high assumed cost of genotyping may have hampered the generation of significant results.

A retrospective study undertaken at the US Veterans Affair medical center in Palo Alto, CA, compared total yearly expenditures for HIV-infected patients in 1993, 1995 and 1997 [23]. In 1997, drug resistance genotyping and highly active antiretroviral therapy (HAART) were widely used. Despite the advent of the use of these two costly modalities, when the data from 1993 and 1997, representing the eras of pre and post-HAART introduction and wide acceptance, were compared directly, total yearly expenditures were reduced from $4,999 to $3,806. During this 4-year time-period genotyping costs increased from $51 to $280 per patient per year. Of note, the costs of genotyping represented 7% of the total costs. While the direct effects of genotyping on the cost of care cannot be separated from the effects of the introduction of HAART, the primary modality that has been shown to reduce health-care expenditures in the setting of HIV infection, they are suggestive of possible benefits from its use.

A number of smaller model simulations have shown potential cost savings for genotyping in specific clinical circumstances. A group of researchers in Canada reported that the use of genotyping was associated with a global reduction in healthcare costs of Ca$595 when the assay was used together with a 1-month drug pause as compared to immediate switching to ritonavir/saquinavir salvage therapy. [24]. A second modeling trial, based on retrospective clinical data from 36 patients, demonstrated that genotyping would result in a cost savings of Ca$202–803 when used in protease inhibitor-experienced patients [25].

A cost-effectiveness analysis that compared the treatment options in patients failing first-line ARV therapy using a Monte Carlo simulation was recently reported in abstract form. Genotyping was assumed to cost $400. When compared to either no second-line therapy or second-line therapy guided by clinical judgement only, second-line treatment guided by drug-resistance genotyping appeared to be cost-effective at US$26,000 per additional QALY (Quality Adjusted Life Year). This is well within the accepted Cost-effectiveness ration (CER) limit of US$50,000/QALY for the introduction of technologies and treatment into the most developed healthcare markets, but outside more recent cost-effectiveness guidelines [26].

In a more in-depth study [21], the authors looked at the cost-effectiveness in more detail in regard to specific studies that had demonstrated the clinical effectiveness of genotyping. When they applied their Monte Carlo simulation based on the results of the GART trial (showing an approximate 25% improvement in ARV therapy effectiveness out to 16 weeks) a cost-effectiveness ratio of US$17,900 per quality adjusted life year (QALY) and a 2.2 month increase in life-expectancy was shown. Using the outcomes of the VIRADAPT trial (38% improvement in ARV therapy at 24 weeks) the authors demonstrated a US$16,300 cost benefit ratio per QALY and a 3.7 month increase in life-expectancy. Data on these models in ARV experienced patients was extended to both a 5% and a 50% improvement in ARV therapy effectiveness. A 5% improvement yielded a cost benefit ratio of US$24,000 per QALY with a 0.7 month increase in QALY and a 50% benefit yielded a US$16,100 ratio per QALY and an 5.7 month increase in QALY. The 5% improvement level was felt to be the likely improvement that could be suggested in ARV-naïve patients. When the cost of the genotyping test was extended to US$1,000 per test, the approximate current cost of phenotyping tests, it was found that cost-effectiveness was achieved at a 10% improvement in ARV effectiveness (US$25,000 per QALY) and a 25% increase yielded a ratio of US$13,200 per QALY. As this data approximated what has been demonstrated in clinical trials, this result could be extended to show that phenotyping is cost-effective [7].

From these studies and others, a table comparing the cost effectiveness of resistance testing as compared to other HIV treatment modalities has been developed and it illustrates quite well the relative position of resistance testing

Table 1 Cost-effectiveness ratios for interventions in HIV disease

Intervention/Comparator	Cost effectiveness US$ 1998	Source
1 Low dose TMP/SMX *versus* no prophylaxis	8,700/QALY	Goldie SMDM abstract
2 3 ARV drugs *versus* 2 ARV drugs mean CD4 500 at baseline	13,000/QALY	Freedberg SMDM abstract
3 Genotyping for 2nd line HAART *versus* clinical judgement	16,500/QALY	Weinstein SMDM abstract
4 3 ARV drugs *versus* 2 ARV drugs mean CD4 87 at baseline	18,000/QALY	Freedberg SMDM abstract
5 Dapsone for TMP/SMX sensitive *versus* no prophylaxis for PCP	18,600/QALY	Goldie SMDM abstract
6 Genotyping for 1st HAART in naïve patients * *versus* clinical judgment	20,500/QALY	Weinstein SMDM slide presentation

in the setting of HIV infection [27] (Tab. 1). These modeling trials have also produced cost-effectiveness ratios for the use of genotypic resistance testing from between US$16,500 to US$20,500 per quality adjusted life year (QALY). However, before these analyses can serve as definitive proof for the conclusions drawn, their modeling assumptions need to be validated in clinical practice.

Phenotypic data

To date there have only been few modeling studies that have addressed the issue of HIV-1 phenotyping.

Sheriff et al. [28] took a prospective look at medical costs of first ARV failure patients who either had their salvage therapies directed by phenotype (Phenosense, Virologic) or who had received standard of care. The modeling they used also attempted at predicting virologic suppression rates in each arm. Costs were determined from the perspective of the payer and were based on year 2000 treatment guidelines and published literature. Over a 2-year time period, they concluded that the standard of care arm would have a 29% virologic success rate, with costs totaling US$40,724. For the phenotyping arm, they estimated a virologic suppression rate of 55% and a total cost of US$38,703. This implied a cost saving of US$2021 every 2 years with the use of phenotyping. The authors concluded that phenotyping should improve virologic outcomes and decrease the cost of care [28].

The costs and benefits of genotyping and phenotyping have been estimated in models that considered a 90-day time horizon. We found that using drug resistance genotyping before the initiation of treatment in patients failing first-line HAART was beneficial and saved US$202–803 per patient, while phenotyping increase the total costs of treatment by US$266–615 per patient. This difference appeared to be explained by the higher cost of phenotyping in the modeling study by Anis et al. [24, 25].

However, the previously shown Weinstein data on genotyping with a cost of US$1000 per test and a 25% increase in effectiveness potentially contradicts this data, developed within the framework of a genotyping model [21]. The parameters are within the costs and what has been shown for phenotyping test in the setting of first-line failure. Before this can be shown as truly applicable, the same model should be run using data from the VIRA 3001 trial (phenotyping compared to standard of care in the setting of first ARV failure), and until this occurs it can only be used as suggestive rather than direct proof [7].

Issues with defining the economic costs of resistance testing in HIV therapy

A critical issue in all economic analyses is measuring the costs and effectiveness of the intervention [29]. Cost-effectiveness analyses of the use resistance testing in the treatment of HIV disease faces some particular challenges in measuring both costs and effectiveness.

The nature of HIV therapy presents specific challenges in the measurement of the cost-effectiveness of HIV-1 resistance testing. One challenge is that HIV therapies are changing so rapidly that it is difficult to determine their cost-effectiveness without significant confounding data. In addition, such analyses require the measurement of treatments and outcomes over a long period of time, which is difficult to do for reasons of the changing nature of HIV therapy. Although observational rather than clinical trial data may be the only feasible approach to obtaining the large sample sizes and long follow-up necessary to accurately track costs and outcomes, such data are hampered by potential selection bias in who obtains resistance testing and what patient populations are studied.

Another important challenge for economic evaluations, particularly of HIV-1 resistance testing, is the number of outcomes that need to be considered. HIV-1 resistance testing is of value both as a preventive measure and as a diagnostic measure. In order to fully measure the benefits of resistance testing, therefore, analyses need to deal with challenges that are similar to any economic evaluation of preventive service [30]. In addition, there are potentially quality-of-life or "intangible" costs and benefits of resistance testing. Ultimately, economic evaluations will be of most relevance to practice and policy to the extent that they consider the broader issues that arise from the use of resistance testing.

In the case of HIV-1 resistance testing, a unique ethical issue is the role of adherence. Adherence to ARV therapies is a major concern because patients have to adhere to complicated regimens at levels unprecedented in other long-term disease states [31, 32]. Patients who are unable to adhere to their regimens are more likely to develop drug resistance and require the use of resistance testing to adjust their regimens, leading to a potential cycle of non-adherence and efforts to rectify the results.

The broader policy implications of resistance testing also need to be considered. In this regard, an important distinction can be made between resistance testing, which is focused on the patient and individual clinical decisions, compared to pharmacogenomics, which also considers the broader societal implications of developing new drugs based on resistance testing results [30, 33]. An example would be the tailoring of drug treatment that may lead to the streamlining of regulatory requirements for new agent approvals and thereby facilitate quicker access to new drugs. In addition, the use of resistance testing may lead to increased interest from drug companies in developing new drugs that are tailored to resistance testing, or, conversely, it may decrease the interest of drug companies in developing drugs that are suitable for a smaller population.

Overall evidence of cost-effectiveness

In summary, the economic studies performed to-date indicate that HIV-1 resistance testing may be reasonably cost-effective and have the potential for recognizing short-term reductions in expenditures due to the avoidance of hospitalizations and drug costs. This is particularly relevant in light of recent recommendations in Europe that have validated resistance testing as a major part of HIV care [34]. The data are not currently complete, but the initial data are encouraging, and further research is required. It is important to recognize and understand that the cost-savings and benefits from resistance testing are likely to be seen in cost shifting from one budget to another (i.e., pharmacy to laboratory). This may produce an accompanying global reduction in costs that might not be seen by all in the setting of a fragmented healthcare budget [35].

Access to new interventions such as HIV resistance testing vary across Europe and is currently limited by two major issues: each country has a specific system of healthcare that is structured and reimbursed differently and access to all aspects of HIV healthcare varies across Europe [36]. This affects the ability of the treating physician and patient to utilize these diagnostic tools to their best advantage in each country and administrative region. In addition, the expertise and experience of physicians across Europe is affected by HIV epidemiology that is evidenced by the markedly varying HIV prevalence rates across Europe [37]. Models of treatment and care are developed through research, usually at larger treating institutions and disseminated to smaller healthcare settings to hopefully set up even more cost-effective uses for these tests. Further research is imperative to more clearly define the cost-effectiveness of both genotypic and phenotypic resistance testing and to clarify the optimal mechanisms for reimbursement, however, current research might lead one to be optimistic about the results of this future research.

References

1 Chaix-Couturier C, Holtzer C, Phillips KA, Durand-Zaleski I, Stansell J (2000) HIV-1 drug resistance genotyping. A review of clinical and economic issues. *Pharmacoeconomics* 18: 425–433

2 Bozzette SA, Berry SH, Duan N, Frankel MR, Leibowitz AA, Lefkowitz D, Emmons CA, Senterfitt JW, Berk ML, Morton SC et al. (1998) The care of HIV-infected adults in the United States. HIV Cost and Services Utilization Study Consortium. *N Engl J Med* 339: 1897–1904

3 Perdue BE, Weidle PJ, Everson-Mays RE, Bozek PS (1998) Evaluating the cost of medications for ambulatory HIV-infected persons in association with landmark changes in antiretroviral therapy. *J Acquir Immune Defic Syndr Hum Retrovirol* 17: 354–360

4 Lapins D, Barrett J, Hamel E, Urdaneta M, Markson LE (1998) Trends in costs of care for patients with HIV. 38th Interscience Conference on Antimicrobial Agents and Chemotherapy, San Diego, CA, USA, 24–27 September 1998. Abstract 460

5 Lacey L, Youle M, Trueman P, Staszewski S, Shrappe M, Behrens M (1999) A prospective evaluation of the cost effectiveness of adding lamivudine to zidovudine-containing antiretroviral treatment regimens in HIV infection. *Pharmacoeconomics* 15 Suppl 1: 39–53

6 Stansell J, Barrett J, de Guzman C, Holtzer C, Lapins D (2000) Incremental costs of HIV suppression in HIV therapeutic failure. 7th Conference on Retroviruses and Opportunistic Infections, San Francisco, CA, USA, 30 January 30–2 February 2000. Abstract 761

7 Cohen CJ, Hunt S, Sension M, CFarthing C, Conant M, Jacobson S, Nadler J, Verbiest W, Hertogs K, Ames M et al. (2002) A Randomized trial assessing the impact of phenotypic resistance testing on antiretroviral therapy. *AIDS* 16: 579–588

8 Clevenbergh P, Durant J, Halfon P, del Giudice P, Mondain V, Montagne N, Schapiro JM, Boucher CA, Dellamonica P (2000) Persisting long-term benefit of genotype-guided treatment for HIV-infected patients failing HAART. The Viradapt study: week 48 follow-up. *Antiviral Ther* 5: 65–70

9 Durant J, Clevenbergh P, Halfon P, del Giudice P, Porsin S, Simonet P, Montagne N, Boucher CA, Schapiro JM, Dellamonica P (1999) Drug-resistance genotyping in HIV-1 therapy: the VIRADAPT randomised controlled trial. *Lancet* 353: 2195–2199

10 Baxter JD, Mayers DL, Wentworth DN, Neaton JD, Merigan TC (1999) A pilot study of the short-term effects of antiretroviral management based on plasma genotypic antiretroviral resistance testing (GART) in patients failing antiretroviral therapy. 6th Conference on Retroviruses and Opportunistic Infections, Chicago, IL, USA, 31 January–4 February 1999. Abstract LB8

11 Tural C, Ruiz L, Holtzer C, Schapiro J, Viciana P, Gonzalez J, Domingo P, Boucher C, Rey-Joly C, Clotet B (2002) Clinical utility of HIV-1 genotyping and expert advice: the Havana Trial. *AIDS* 16: 209–218

12 Anis AH, Hogg RS, Yip B, Wang XH, Montaner JS, O'Shaughnessy MV, Schechter MT (1999) Average annual drug cost and its determinants in a population based cohort of HIV-positive adult men and women. *Pharmacoeconomics* 13: 327–336

13 Shapiro MF, Morton SC, McCaffrey DF, Senterfitt JW, Fleishman JA, Perlman JF, Athey LA, Keesey JW, Goldman DP, Berry SH et al. (1999) Variations in the care of HIV-infected adults in the United States: results from the HIV Cost and Services Utilization Study. *JAMA* 281: 2305–2315

14 Holtzer CD, Jacobson MA, Hadley WK, Huang L, Stanley HD, Montanti R, Wong MK, Stansell JD (1998) Decline in the rate of specific opportunistic infections at San Francisco General Hospital: 1994–1997. *AIDS* 12: 1931–1932

15 Brettle RP, Wilson A, Povey S, Morris S, Morgan R, Leen CL, Hutchinson S, Lewis S, Gore S (1998) Combination therapy for HIV: the effect on inpatient activity, morbidity and mortality of a cohort of patients. *Int JSTD AIDS* 9: 80–87

16 Palella FJ Jr, Delaney KM, Moorman AC, Loveless MO, Fuhrer J, Satten GA, Aschman DJ, Holmberg SD (1998) Declining morbidity and mortality among patients with advanced human immunodeficiency virus infection. HIV Outpatient Study Investigators. *N Engl J Med* 338: 853–860

17 Rabeneck L, Menke T, Wray NP (1999) How good are US studies of HIV costs of care? *Med Care* 37: 748–759

18 Moore RD, Gebo K, Bartlett JG, Chaisson RE (1999) Health Care Costs for 1995–1998 in AIDS. 6th Conference on Retroviruses and Opportunistic Infections, Chicago IL, USA, 31 January–4 February 1999. Abstract 102

19 Holtzer C, Deeks SG (1998) Impact of HIV-1 protease inhibitors on the cost of treating HIV/AIDS patients. *Drug Benefit Trends* 10: 27–31
20 Saag, MS (2001) HIV resistance testing in clinical practie: a QALY-fied success. *Ann Intern Med* 134: 475–477
21 Weinstein MC, Goldie SJ, Losina E, Cohen CJ, Baxter JD, Zhang H, Kimmel AD, Freedberg KA (2001) Use of genotypic resistance testing to guide HIV therapy: clinical impact and cost-effectiveness. *Ann Intern Med* 134: 440–450
22 Chaix C, Grenier-Sennelier C, Clevenbergh P, Durant J, Schapiro JM, Dellamonica P, Durand-Zaleski I (2000) Economic evaluation of drug resistance genotyping for the adaptation of treatment in HIV-infected patients in the VIRADAPT study. *J Acquir Immune Defic Syndr* 24: 227–231
23 Mole L, Ockrim K, Holodniy M (1999) Decreased medical expenditures for care of HIV-seropositive patients. The impact of highly active antiretroviral therapy at a US Veterans Affairs Medical Center. *Pharmacoeconmics* 16: 307–315
24 Anis AH, Wang X, Harrigan R, Hogg RS, Yin B, O'Shaughnessy MV, Schlechter MT, Montaner JSG (1999) The cost-effectiveness of immediate ritonavir-saquinavir therapy *versus* resistance testing and a drug holiday in HIV patients failing protease inclusive regimens. Abstract. Society for Medical Decision Making, Reno
25 Anis AH, Wang X, Harrigan R, Hogg RS, Yin B, O'Shaghnessy MV, Schlechter MT, Montaner JSG (1999) Optimizing drug treatment: a cost-effectiveness analysis of HIV/AIDS drug resistance testing. Abstract. Society for Medical Decision Making, Reno
26 Weinstein MC, Goldie SJ, Cohen C, Losina H, Zhang HJ, Kimmel AD (1999) Resistance testing to guide the choice of second-line antiretroviral therapy in HIV: clinical impact and cost-effectiveness. Abstract. Society for Medical Decision Making, Reno
27 Simpson KN (2000) Budget impacts of the integration of genotypic antiretroviral resistance testing (GART) into HIV treatment guidelines. 7th Conference on Retroviruses and Opportunistic Infections, San Francisco, CA, USA, 30 January– February 2000. Abstract 789
28 Sheriff SK et al. (2000) 38th Annual Meeting of the Infectious Diseases Society of America, New Orleans, LA, USA, 7–10 September 2000. Abstract 449
29 Gold MR, Siegel JE, Russell LB, Weinstein MC (1996) *Cost-Effectiveness in Health and Medicine*. Oxford University Press, New York
30 Phillips KA, Hotlgrave DR (1997) Using cost-effectiveness/cost-benefit analysis to allocate health resources: a level playing field for prevention? *Am J Prev Med* 13: 18–25
31 DeMasi RA, Graham NM, Tolson JM, Pham SV, Capuano GA, Fisher RL, Shaefer MS, Pakes GE, Sawyerr GA, Eron JJ Jr, (2001) Correlation between self-reported adherence to highly active antiretroviral therapy (HAART) and virologic outcome. *Adv Ther* 18: 163–173
32 McNabb JC, Ross JW, Abriola K, Turley C, Nightingale CH, Nicolau DP (2001) Adherence to highly active antiretroviral therapy predicts virologic outcome at an inner-city human immunodeficiency virus clinic. *Clin Infect Dis* 33: 700–705
33 Sadee W (1998) Genomics and drugs: finding the optimal drug for the right patient. *Pharm Res* 15: 959–963
34 The EuroGuidelines Group for HIV Resistance (2001) Clinical and laboratory guidelines for the use of HIV-1 drug resistance testing as part of treatment management: recommendations for the European setting. *AIDS* 15: 309–320
35 Klein DB, Hurley L (1999) HAART-related cost saving from decreased utilization of diagnositc microbiology lab. Abstract 465. 38th Interscience Conference on Antimicrobial Agents and Chemotherapy, San Francisco, USA
36 Lundgren J, Katlama C, Johnson A, Pradier C, Antunes F, Mulcahy F et al. (1999) The changing spectrum of AIDS across Europe 1994–1999. The Eurosida Study. 7th European Conference on Clinical Aspects of Treatment of HIV Infection, Lisbon, Portugal, 23–27 October 1999. Abstract 244
37 Dreezen C, Barrata C, Coppieters Y, Caldeira L, Liotta G, Colebunders R (1999) Utilization of Health services for HIV+ patients in Europe. 7th European Conference on Clinical Aspects of Treatment of HIV Infection, Lisbon, Portugal, 23–27 October 1999. Abstract 240

Combination Therapy of AIDS
Edited by E.D.A. De Clercq and A.-M.I. Vandamme

Guidelines for antiretroviral therapy

Charles C.J. Carpenter

Department of Medicine, The Miriam Hospital, 164 Summit Avenue, Providence, RI 02906, USA

Progress in the treatment of HIV infection over the past decade has resulted in dramatic reductions in morbidity, mortality and health care utilization as more effective treatment regimens have been developed and simpler schedules of drug administration have been devised [1, 2]. As more effective regimens have provided durable suppression of viral replication, it has become clear that reversal of disease progression can be achieved even in persons with advanced HIV infection [3]. It is now possible to achieve at least transient interruption of disease progression in almost all patients who have received no prior antiretroviral therapy, and who are therefore responsive to most available antiretroviral drugs [4].

Despite the great reductions in morbidity and mortality, two major problems still limit the effectiveness of antiretroviral therapy, even for patients with ready access to appropriate HIV medications. The first is the fact that resistance of the HIV-1 virus to antiretroviral drugs, although not inevitable, does eventually develop in the majority of persons who receive antiretroviral medications, requiring sequential changes in the regimen [5]. The second is the frequent development of undesirable side-effects of the medications; although metabolic alterations and fat distribution changes are perhaps the most well known of these, no antiretroviral medications are totally free of side-effects, and the cumulative side-effects of long-term antiretroviral therapy present a major challenge [6].

This chapter will deal sequentially with: when antiretroviral therapy should be initiated in individuals with established HIV infection; which regimens are most effective for treatment of HIV infection; how antiretroviral therapy, once undertaken, should be monitored; and when antiretroviral regimens need be changed in individual patients.

Currently available three- and four-drug antiretroviral regimens, if taken on appropriate schedules, will almost always rapidly reduce the plasma HIV RNA levels in antiretroviral-naïve individuals with HIV infection [4, 7]. This reduction in plasma HIV RNA levels is usually followed by a more gradual increase in the CD4 cell counts, in some cases to normal levels, but in most cases to levels adequate to prevent more serious opportunistic infections [8]. An initial increase in memory CD4 cells occurs within weeks after initiation of effective therapy, resulting from redistribution of CD4 cells from lymphoid tissues to the circulation [9]. Naïve CD4 cells, essential for effective responses to new antigenic challenges, are restored more gradually, associated with continued effective suppression of viral replication over months [10]. CD4 cell counts are restored to a safe range more rapidly in persons with higher CD4 counts at the time treatment is initiated [4]. Total CD4 cell counts are restored to levels at which patients are no longer generally susceptible to serious opportunistic infections in the great majority of patients. Since currently available antiretroviral regimens will not eradicate HIV-1 [11, 12], a prolonged commitment to antiretroviral therapy is essential, with the long-term goal of durable inhibition of viral replication so that the patient can maintain an effective immune response to most potential microbial pathogens.

When to initiate antiretroviral therapy (ART)

Retrospective cohort data have provided strong support for the CD4 cell count as the major determinant of when therapy should be initiated [3, 4, 13, 14]. It is now certain that ART should be initiated before a threshold of 200 cells/μl has been reached [3, 4, 13, 14]. Most of the life threatening opportunistic infections occur only after the CD4 count is less than 200 cells/μl, and the immune reconstitution syndrome with its considerable morbidity, is largely limited to individuals who initiate therapy with very low CD4 counts, usually below 100 cells/μl [15]. Laboratory markers also demonstrate less favorable responses when antiretroviral therapy is delayed until the CD4 counts are below 200 cells/μl [4, 16]; these include a less rapid increase in CD4 cell counts, and a lesser ability to reduce plasma HIV RNA levels below the limit of detection. Certain serious infections, including bacteremic pneumococcal pneumonia and pulmonary tuberculosis, do begin to occur with increased frequency in persons with less severe immunosuppression (e.g., with CD4 counts in the 200 to 350 cells/μl range) suggesting that it may be helpful to initiate antiretroviral therapy at a threshold greater than 200 cells/μl [17–19]. There are, however, now no sound data with clinical endpoints that define precisely where the threshold for initiation of antiretroviral therapy should lie.

In persons with CD4 cell counts higher than 350 cells/μl, the risk of near term (3 year) clinical progression is very small [20], and concerns about the impact of antiretroviral drugs on quality of life, adherence to regimens, emergence of viral resistance, risks of serious adverse effects of the drugs and limitations on future treatment options generally outweigh the benefits of durable viral suppression [8].

Thus physicians and patients must thoroughly weigh the known risks and benefits of starting antiretroviral therapy for persons with CD4 counts in the 200 to 350 cells/μl range and make individualized informed decisions about exactly when to initiate therapy (Fig. 1). The strength of a recommendation to initiate therapy should depend not only on the patient's immunologic status but also on the rapidity of the fall of CD4 lymphocyte count and even more impor-

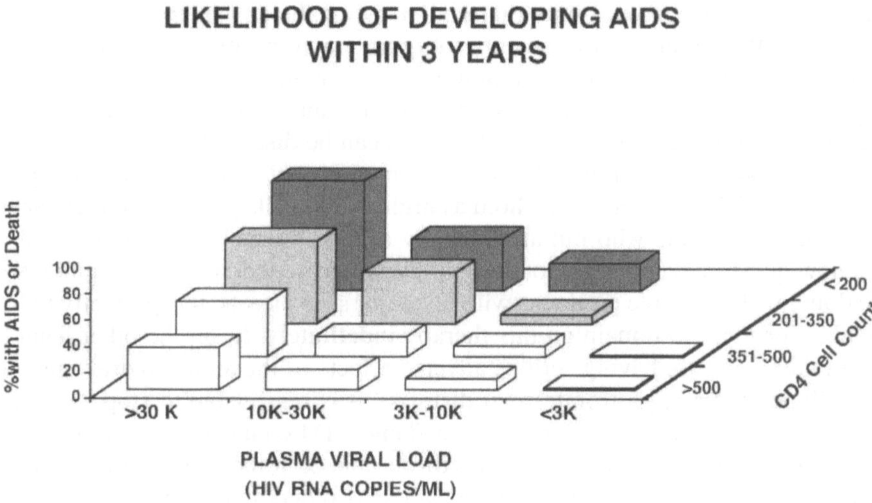

Figure 1. Recommendations for initiation of antiretroviral therapy in relation to the likelihood of developing symptomatic AIDS or death within three years in the absence of antiretroviral therapy. (Adapted from Mellors et al. (20), with modifications). It is strongly recommended that HAART be initiated before the patient's biological markers fall in the black zone. Current data indicate that known risks of adverse consequences of HAART outweigh potential benefits when biological markers are in the white zone. Therefore, HAART should be initiated when biological markers fall in the gray zone. Data do not permit recommendation of a precise CD4 count within the gray zone which is optimal for initiation of HAART. Within this zone, treatment must be individualized: a rapid rate of CD4 decline (<100 cells/µL/annum), a high HIV-RNA level (>50,000 to 100,000 copies/mL) and/or strong desire of the patient to initiate therapy would all point to somewhat earlier initiation of treatment.

tantly on his or her understanding of the long-term commitment to an antiretroviral regimen. If a patient initiates a regimen to which he or she is not fully committed, the patient is more likely to have poor adherence, and thus be at greater risk for emergence of resistance to antiretroviral drugs which will clearly compromise future treatment options.

Although available data do not support a plasma HIV-RNA threshold as an independent indicator for initiating antiretroviral therapy, patients with CD4 cell counts at any level who have >100,000 HIV RNA copies/mL of plasma should be followed closely, probably at 2-month intervals, because the CD4 cell count predictably decreases more rapidly in untreated patients with very high HIV RNA levels [20] (Fig. 1).

Although some observational studies have indicated that HIV RNA levels early in the course of HIV disease are lower in women than in men [21, 22], there are no documented sex differences in the relation of the CD4 cell count to the risk of opportunistic infections. Recommendations as to when to initiate ART are therefore the same in women as in men.

Treatment is recommended in all persons with symptomatic established HIV infection. However, immediate treatment of a serious opportunistic infection may take precedence over initiating antiretroviral therapy in these instances. Whenever the potential for adverse drug interactions exists (for example between rifampin and protease inhibitors in the treatment of *Mycobacterium tuberculosis*), it is often reasonable to delay the antiretroviral treatment until the specific antibiotic causing the interaction can be discontinued.

Because earlier recommendations for initiation of antiretroviral therapy suggested a CD4 cell count threshold as high as 500 cells/µl [23, 24], there are a number of patients who initiated therapy at this threshold, have done well, and now question whether or not they should interrupt therapy. If such patients are doing well on stable regimens with no major side-effects, it is perfectly reasonable for them to remain on this therapy indefinitely. If, on the other hand, such individuals are having serious adverse effects of the antiretroviral agents (e.g., lipodystrophy, gastrointestinal distress), it is reasonable to stop therapy with close monitoring of plasma viral load and CD4 counts over the next several months. Some of such patients will be able to remain off antiretroviral therapy for periods of months to years, whereas others may show a relatively rapid rise in plasma viral load, associated with a fall in CD4 count indicative of need to resume therapy.

Effective antiretroviral regimens for initial therapy

There are now several highly potent three- or four-drug antiretroviral regimens which can effectively be used for initiation of therapy in persons living with HIV (Tab. 1). No one regimen is optimal for all patients, and each regimen should be individualized so that it fits best with the lifestyle, dietary habits and concomitant medications taken by the individual patient.

Durability of several effective highly active antiretroviral therapy (HAART) regimens has been demonstrated over periods of 3 or more years [25, 26]. Major factors which contribute to the durability of the regimen include ease of adherence, tolerability and convenience of the regimen. Drugs with long half-lives which impose no dietary restrictions have clear-cut advantages in producing durable suppression of viral replication and related increase in CD4 counts [27]. Even minor side-effects may interfere with adherence over the long-term. Twice daily regimens are associated with better adherence than regimens taken more frequently. Once daily regimens permit directly observed therapy which appears to be very effective in dealing with the management of certain hard-to-reach patients [28].

Certain initial durable regimens have proved to be effective for several years in over half of individuals who have volunteered for controlled clinical trials [25, 26]. Most patients, however, do ultimately experience treatment failure of the initial regimen. Because of intraclass cross resistance (which is almost complete between available NNRTIs, but is less extensive between NRTIs and protease inhibitors), it is of tremendous importance that the choice of the initial regimen results in effective long-term suppression of viral replication.

Table 1. Selected antiretroviral regimens for starting therapy

This table provides a guide to the use of available treatment regimens for individuals with no prior experience on HIV therapy. Priority is given to regimens in which clinical trials data suggest the following: sustained suppression of HIV plasma RNA (particularly in patients with high baseline viral load) and sustained increase in CD4 + T-cell count (in most cases over 48 weeks), and favorable clinical outcome (i.e., delayed progression to AIDS and death) for preferred regimens. Emphasis is given to regimens that have been compared directly with other effective regimens; for "preferred" regimens, additional consideration is given to the regimen's pill burden, dosing frequency, food requirements, convenience, toxicity, and drug interaction profile compared with other regimens.

All antiretroviral agents have potential serious toxic and adverse events associated with their use. The reader is strongly encouraged to consult definitive sources [6, 8] for detailed discussion of potential toxicities of available antiretroviral drugs.

NNRTI-based regimens		# of pills/day
Regimens preferred by author	Efavirenz[1] (or Nevirapine) + Lamivudine + (Zidovudine or Tenofovir)	3–5
Alternative regimens	Efavirenz[1] (or Nevirapine) + Lamivudine + (Stavudine or Didanosine)	3–5
PI-based regimens		**# of pills/day**
Regimens preferred by author	Lopinavir + Ritonavir + Lamivudine + (Zidovudine or Stavudine)	8–10
Alternative regimens	Amprenavir + Lamivudine + (Zidovudine or Stavudine)	12–14
	Indinavir + Lamivudine + (Zidovudine or Stavudine)	8–10
	Indinavir + Ritonavir + Lamivudine + (Zidovudine or Stavudine)	8–12
	Nelfinavir + Lamivudine + (Zidovudine or Stavudine)	12–14
	Saquinavir + Ritonavir + Lamivudine + (Zidovudine or Stavudine)	14–16
Triple NRTI-based regimens		**# of pills/day**
Alternative regimen	Abacavir + Lamivudine + Zidovudine	2

[1]Except for pregnant women or women with childbearing potential

For initial therapy three initial types of regimens can be recommended: (Tab. 1)

1. A protease inhibitor (with or without "boosting" by low-dose ritonavir) and two nucleoside reverse transcriptase inhibitors (NRTIs);
2. A nonnucleoside reverse transcriptase inhibitor (NNRTI) and two NRTIs;
3. Three NRTIs.

Two other potential regimens (described below under Alternative Regimens) may be used in special circumstances, but are not recommended for initial therapy.

Protease inhibitor-based regimens

Each of the six currently available protease inhibitors, when combined with two nucleoside reverse transcriptase inhibitors, can be effective in durably reducing HIV RNA levels in antiretroviral-naïve patients, with reduction of plasma HIV RNA to levels <50 copies/mL in a majority of patients receiving the regimens. Major limitations of protease inhibitor-based regimens fall into two categories:

1. Metabolic abnormalities (hyperlipidemia and insulin resistance) and/or alterations in body fat distribution occur in 30 to 50 per cent of patients taking protease inhibitor-containing HAART regimens [6].
2. Regimens containing a single protease inhibitor are difficult to take because of three-times a day schedules and/or food constraints (e.g., indinavir, nelfinavir) [27]. Also, many patients have been unable to tolerate the side-effects of the protease inhibitors, or have had only borderline effective antiviral blood levels, resulting in failure of the regimen within relatively short periods of time.

The metabolic consequences of protease inhibitor-containing regimens remain a major concern. Although the adverse metabolic effects can be anticipated and minimized by appropriate treatment (e.g., statins for hypercholesterolemia) they cannot entirely be avoided.

The problems with difficult dosing requirements can, however, be at least partially alleviated by the use of low-dose ritonavir (either 100 or 200 mg daily) to pharmacokinetically enhance (or "boost") the area under the curve of saquinavir, indinavir, amprenavir or lopinavir. (Lopinavir is commercially available only in a co-formulated tablet with ritonavir). This boosting allows each of these four protease inhibitors to be administered on a twice-daily regimen, in doses which less commonly cause gastrointestinal side-effects. The ritonavir boosting also eliminates the need for dietary constraints with indinavir [8].

The benefits of utilizing certain protease inhibitors in combination with low-dose ritonavir have been most encouraging. Ritonavir inhibits enzymes of the cytochrome P450 system: it may act early on absorption in first-pass metabolism, increasing peak plasma concentrations of the co-administered protease inhibitor (e.g., indinavir), or it may inhibit subsequent metabolism and extend the half-life of the second protease inhibitor by increasing the trough level of the second drug (e.g., lopinavir, saquinavir). In addition to the advantage of once or twice daily dosing and reduction of specific food requirements provided by ritonavir-enhanced regimens, the increase in trough levels of the second protease inhibitor greatly improves the likelihood that a clinically effective level of the drug will be present throughout the 24-h period, minimizing the likelihood of the development of resistance.

NNRTI-based regimens

A triple drug regimen containing one NNRTI and two NRTI's is also recommended for initial therapy of established infection. In head-to-head studies, efavirenz and two NRTIs are at least equally as effective as indinavir and two NRTIs in reducing plasma HIV RNA to undetectable levels, with corresponding rises in CD4 cell counts, for a period of at least 24 months [29]; there was less attrition of patients because of adverse side-effects with the efavirenz-containing regimens. Regimens containing an NNRTI and 2 NRTIs are less often associated with the blood lipid abnormalities and alterations in fat distribution which often occur with protease inhibitor-containing regimens.

Of the three available NNRTIs, nevirapine, efavirenz and delavirdine, both efavirenz and nevirapine can be administered on a once daily basis, which makes them generally more desirable than delavirdine, which must be administered on a three-times daily regimen with a relatively high pill burden. Efavirenz may cause especially vivid dreams, mild dizziness on standing, and rarely hallucinations. These side-effects, however, usually disappear after 2 or 3 weeks of treatment; if patients are forewarned, the side-effects seldom cause discontinuation of the regimen. In studies in pregnant primates, efavirenz has caused CNS malformations; therefore, this medication must be avoided in women who intend to become pregnant.

The potency of nevirapine when combined with two NRTIs appeared to be comparable to that of nelfinavir and of indinavir when combined with two NRTIs in two separate studies [30, 31]; these studies were not powered adequately to provide definitive conclusions. Nevirapine is more often associated with rashes, although the rashes are usually mild and generally resolve spontaneously. They may rarely be severe, involve the mucous membranes and require discontinuation of the medication. Hepatic toxicity, manifested by jaundice and elevation of transaminase enzymes, occurs in up to 2 per cent of patients, and necessitates withdrawal of nevirapine [8]. As with efavirenz, nevirapine has been associated with improvement in plasma lipid profiles when substituted for protease inhibitors in HAART regimens.

Triple NRTI regimens

Triple NRTI regimens which include abacavir may also be recommended for initial therapy. Abacavir, zidovudine, and lamivudine regimens produced a decrease in plasma HIV RNA level at 48 weeks which was generally equivalent to that produced by zidovudine/lamivudine/indinavir in a randomized equivalence trial [32]. However, reduction of the plasma HIV RNA to undetectable levels was less likely with the triple NRTI regimen in patients who had a pretreatment viral load >100,000 copies/mL. Triple NRTI combinations are therefore not routinely recommended as initial therapy for patients with viral loads in excess of 100,000 copies/mL. To date, adverse side-effects of twice daily triple NRTI regimens have been less than that observed with protease inhibitor-containing regimens, and appear to be comparable to those of NNRTI-containing regimens. Abacavir/zidovudine/lamivudine was the first triple drug combination to be available in a single fixed dose combination and this appears to have facilitated adherence. However, hypersensitivity reactions to abacavir, necessitating immediate cessation of the drug, occur in roughly 3 per cent of patients who receive abacavir (8); this is an especially serious consideration for groups of patients who do not have ready access to follow-up medical care.

Alternative regimens

Two other regimens have been used effectively in exceptional circumstances, but neither is generally recommended for initial therapy.

1. In patients who have experienced moderate to severe lactic acidosis as a complication of NRTI therapy, a regimen containing a protease inhibitor enhanced with low dose ritonavir and an NNRTI has been used successfully in a number of instances, but there are no available data on the long-term effectiveness of such regimens.
2. A regimen containing drugs from all three classes (a protease inhibitor, an NNRTI and a NRTI), has been utilized for patients with high near term mortality risk because of very low CD4 counts, but there are no long-term data which support the effectiveness of such a regimen. Regimens utilizing all three classes of available antiretroviral drugs are generally avoided because of the danger, especially in patients who have problems with adherence, of development of resistance to all three antiretroviral drug classes.

Monitoring antiretroviral therapy

Routine monitoring of antiretroviral therapy utilizes two biological markers, the CD4 cell count and the plasma HIV RNA level. In the asymptomatic individual, who has had no previous antiretroviral therapy, both a decrease in the HIV RNA level and an increase in the CD4 cell count can be anticipated after initiation of HAART [2–4, 8]. The plasma HIV RNA level, indicating the rate of virus replication, usually decreases rapidly before changes in CD4 cell counts occur. The CD4 cell counts begin to increase shortly after the initial rapid fall in plasma HIV RNA. Effective antiretroviral therapy generally reduces the plasma HIV RNA by more than 90 per cent (1 $\log^{10}$) within 2 weeks of treatment; failure to observe such a response most commonly results from poor adherence, but may result from either antecedent viral resistance, or inadequate exposure to the drug. After initiation of therapy, follow-up measurements of HIV RNA levels should therefore be carried out at 2 and 4 weeks.

If anticipated reductions in plasma HIV RNA level are not achieved during this period, careful discussion with the patient about adequacy of adherence must be initiated; if there is no indication of a problem with adherence, it is reasonable to initiate antiretroviral genotypic resistance testing, and to make changes in the regimen if indicated. Once an acceptable decrease in plasma HIV RNA level has been achieved (a decrease in viral load to a level below the limits of detection on two sequential measurements), monitoring can usually be carried out at 12-week intervals [8]. It is important to continue to emphasize the need for close adherence at the time of each monitoring visit.

After initiation of therapy, the increase in CD4 count generally follows the decrease in plasma HIV RNA level. If therapy has been started at a CD4 count >200 cells/µl, an increase in CD4 count cell count of roughly 50 cells/µl can be anticipated within 8 weeks after initiation of successful antiretroviral therapy. This is usually followed by a more gradual increase of 50 to 80 cells/µl per year; the mean rate of increase varies with the CD4 cell count at the time of initiation of therapy. The CD4 count is generally measured at the same intervals as indicated above for the plasma viral load, but the measurement of plasma CD4 count during the first 8 weeks after of initiation therapy is less important than the measurement of HIV RNA; the CD4 cell count response is less helpful in therapeutic decisions during the critical initial 2 to 4 weeks after initiation of antiretroviral therapy.

In patients who begin therapy with CD4 cell counts <200 cell/µl, once the CD4 cell count has exceeded 200 cells/µl for 3 to 6 months with effective therapy, the risks of *Pneumocystis carinii* pneumonia and most other serious opportunistic infections are greatly reduced, and prophylaxis against most opportunistic pathogens can be stopped [33].

In some instances the CD4 cell counts do not increase, and may rarely decrease despite successful suppression of viral replication to <50 copies/µl. This phenomenon is most often observed in persons with underlying hepatitis C infection associated with splenomegaly. There is no specific intervention which has been shown to be helpful in such instances of a discordant response, and, in general, patients do well as long as their HIV RNA levels remain below the limits of detection.

Although the virologic goal of effective therapy is to maintain the plasma HIV RNA levels below limits of detection (generally <50 copies/mL), occasional small increases in HIV RNA to the level of 50–200 copies/mL may occur despite good adherence to the regimen [34]. There is no evidence that such occasional modest increases in viral load are predictive of subsequent treatment failure.

Antiretroviral resistance testing

Randomized prospective trials have generally noted a significant clinical benefit of drug resistance testing in the management of treatment failure [35].

Although there are no definitive data as to what method of resistance testing is superior, viral genotype testing has been more widely used because results are obtained more rapidly (and are therefore of greater utility) and have been more carefully standardized. Resistance testing is demonstrably helpful in determining the choice of what drugs to exclude from a new regimen following treatment failure; testing must be performed when the selective pressure of the failing regimen is still present, and testing in most laboratories can be carried out only when the HIV RNA level is >1,000 copies/mL. Although the utility of genotypic resistance testing has thus far been demonstrated only in patients who are changing failing regimens, such testing should be seriously considered in patients who are likely to have been recently infected with resistant viral strains, and in individuals who have an inadequate initial response to an antiretroviral regimen, in the face of documented excellent adherence [36].

Therapeutic drug concentration monitoring

Most patients with no previous antiretroviral drug experience develop adequate plasma levels of recommended antiretroviral agents with standard recommended doses. There are, however, substantial interpatient variations in pharmacokinetics, and some patients may not respond adequately to standard doses. Therapeutic drug concentration monitoring may be helpful in such individuals. Such monitoring appears to be more helpful with protease inhibitors [37] than with NRTIs; the latter require intracellular activation, and since intracellular concentrations are impossible to measure on a routine basis, therapeutic drug monitoring is less helpful.

Monitoring of drug levels should take place only after careful assessment of adherence. Monitoring must be carried out at a defined period of time following the last dose of medication. Ideally a drug steady-state should have been reached before the plasma concentration is measured, and steady-states are obtained at different time intervals after ingestion of the various antiretroviral agents. Although therapeutic drug concentration monitoring is commercially available in a number of regions, the techniques are not well standardized, and few studies have demonstrated the clinical utility of drug monitoring to date.

Treatment interruptions

Treatment interruptions, either strategic, structured, or supervised (all frequently described as STI's), have been studied in three distinct situations: after effective therapy of the acute HIV syndrome, following effective suppressive therapy of established infection, and in the hope of improving salvage after failures of several antiretroviral regimens [38].

The goal of strategic treatment interruptions following effective treatment of acute HIV infection is to further boost HIV specific immune responses, with the objective that the heightened responses will decrease viral replication sufficiently to obviate the need for further antiretroviral treatment. Although preliminary uncontrolled data from small numbers of patients provided hopeful information [39], definitive data are not available, and all patients who would like to undertake such strategic therapeutic interventions should be encouraged to enter a controlled clinical trial of this approach.

Structured treatment interruptions, with predetermined HIV-RNA and/or CD4 cell count thresholds for resumption of therapy, have been directed toward reducing long-term toxicity and cost by limiting drug exposure in patients receiving effective therapy of established infection. The boosting of HIV-specific immune responses, initially thought to be a potential goal, has seldom been observed in this setting. Responses to interruption of therapy in persons who have had effective reduction of HIV RNA levels for 2 or more years have been quite variable. In some instances there have been recurrences of acute retroviral syndromes associated with rapid rises in HIV RNA levels and sharp falls in CD4 cell counts [40, 41]. At the other extreme, occasional patients have showed gradual increases in HIV RNA levels over periods of 6 to 9 months or longer, with only modest decreases in CD4 cell counts [42]. Since there is no known means of predicting the response in an individual patient, structured treatment interruptions in this setting should be carried out only with careful monitoring of the biological markers of HIV disease. With careful monitoring, few undesirable consequences have resulted from such interruptions, provided all drugs are stopped simultaneously, and reinitiated simultaneously when a previously agreed upon threshold of CD4 count or HIV RNA level has been reached. Although such structured treatment interruptions cannot be recommended, they may be helpful in instances in which undesirable side-effects of medications can be eliminated for a variable period of time.

Supervised treatment interruptions following failure of successive antiretroviral regimens have been studied as a possible approach to enhance emergence of wild type virus, which might theoretically increase antiretroviral drug options of subsequent regimens. This approach is not recommended, because in such individuals precipitous reductions in CD4 cell counts may occur and subsequent increases in CD4 count with resumption of therapy cannot be predicted [43, 44].

Changing therapy during treatment

First and second failures

A change in therapy during established infection is indicated by an increased level of plasma HIV RNA. This may result from drug toxicity, poor adherence or virological failure. When the HIV RNA level rises in an individual who has earlier achieved suppression to levels below the limit of detection, an in-depth discussion should occur between patient and physician to determine whether the failure of therapy may be related to incomplete adherence, adverse effects or intolerance. If it appears that adverse effects or intolerance is causing the patient to skip doses of therapy, it is often possible, after discussion of the problem with the patient, to determine that a single agent is responsible for the adverse effects. Examples of these would include protracted diarrhea secondary to nelfinavir, or pain and paraesthesias in the feet secondary to stavudine. In such instances a single substitution for an individual component of the regimen is usually effective. Examples would be substitution of zidovudine for stavudine or nevirapine for nelfinavir. In settings in which the responsible antiretroviral agent cannot be identified, treatment can be completely interrupted while acute adverse effects subside, prior to initiation of an alternative regimen which is unlikely to cause the same side-effects.

When the adverse side-effects are increased serum lipid levels or unwanted changes in body fat distribution secondary to protease inhibitor-containing regimens, efavirenz, nevirapine, or abacavir can often be substituted for the protease inhibitor component. Improvement in serum lipid abnormalities can be expected rapidly following such changes; although improvement in body fat distribution does not occur in most instances, no further progression of the fat distribution abnormalities is likely to occur after the change.

Though failure often occurs within the first year of therapy in treatment-naïve patients, it is important to recognize that, with good adherence, a substantial proportion of patients should be able to maintain an undetectable virus for a period of several years [25]. The fact that this does not always occur in clinical practice is, in large part, related to inadequate adherence to the regimen. A large proportion of therapeutic failures occur during the first year of HAART; lower, but finite, rates of treatment failure occur in each of the following years.

Only after the possibility of inadequate adherence to the regimen, caused by any one of a number of problems (these may include simple forgetfulness, occupational demands, alcohol or drug use, in addition to drug intolerance) has been ruled out, can virological failure be ascribed to the treatment regimen. The working definition of treatment failure varies during the course of the treatment of the individual patient. Virologic failure of the first regimen is generally defined as an inability to maintain levels of HIV RNA below the assay detection limits (generally less than 50 copies/ml) on two successive determinations over a period of 2 to 3 months [8]. A second regimen at this point is relatively easy to construct, utilizing three new antiretroviral agents to which the patient has had no previous exposure. Although it would be ideal to do resistance testing at this time in order to identify, when possible, a single drug to which the patient's virus has become resistant, this can only be done when the serum HIV RNA level rises above 1,000 copies/ml; most physicians would prefer to make the initial change in regimen before the HIV RNA has increased to that level.

Subsequent failures of antiretroviral therapy

With failure of a second or subsequent regimen, it can be anticipated that both the level and the breadth of virus resistance is increased, and it will be more difficult to construct a third effective antiretroviral treatment regimen. With successive failures, it becomes increasingly difficult to find a regimen which will achieve an undetectable viral load, and the goal of therapy becomes that of preventing clinical progression; therefore the failure of such third or fourth regimens can not be simply defined by an increase in HIV RNA, but rather by an associated fall in CD4 count and/or the development of a major opportunistic infection.

Following the third and subsequent failures of antiretroviral therapy, results of resistance testing as well as treatment histories often indicate that it will be impossible to administer three new drugs to which the virus appears to be sensitive. Subsequent regimens may not, therefore, be able to achieve undetectable plasma HIV RNA levels.

Evidence of low rates of clinical progression in patients who continue antiretroviral therapy despite moderate levels of viral replication, indicated by plasma HIV RNA levels in the 3,000 to 30,000 area, argues strongly for continuing treatment in the face of the demonstrated virologic failure [43, 45]. Some, but not all, viruses with multi-drug resistant mutations have reduced replicative capacity relative to wild type viruses. Presumably for this reason, stabilization of CD4 cell numbers, associated with lack of clinical progression, often occurs as long as the plasma HIV RNA level is sustained at a level at least 0.5 $\log^{10}$ below the patient's pretreatment value. If all drugs in a regimen are stopped in this situation, there may be a reversion to wild type virus within a few weeks; this may be associated with rapid increase in plasma viral load and a marked decrease in CD4 cell counts [43]. In such situations, the CD4 cell counts often do not increase when antiretroviral therapy is resumed. Therefore continuing drug regimens that maintain selective pressure on the virus is preferable to discontinuing all antiretroviral therapy [45]. This is clearly true in settings in which the CD4 counts are stable despite a rebound in viral replication, but appears also to be preferable in such situations in which the CD4 counts are showing a gradual fall.

If the CD4 count falls to levels associated with serious opportunistic infections, a change in regimen must be considered, but there are no good guidelines as to how to choose an alternative therapeutic regimen in this situation. The alternative regimen will, by necessity, often include drugs recycled from previously failed regimens. Multi-drug rescue therapy, utilizing six or more antiretroviral drugs, may exhibit an effective antiretroviral effect, but at the cost of multiple adverse side-effects, unanticipated drug interactions, and increased costs [46]. In this setting, the CD4 cell responses are not predictable.

Adjuvant immunologic enhancing therapy

Although the concept of improving immune response to HIV-1 infection is an exciting one, current data do not provide strong support for any specific immune-enhancing regimen [8]. A number of studies have indicated that interleukin-2 (IL-2) consistently increases the peripheral blood CD4 cell counts in a dose-dependent fashion; these responses occur most dependably in persons with CD4 cell counts which are >200 cells/mm^3 at baseline, and it has not been possible to demonstrate that such therapy improves the outcome of persons living with HIV infection [47, 48]. Additionally, with IL-2 doses which dependably increases CD4 cell counts in persons with baseline CD4 cells >200 cells/mm^3, adverse side-effects including fever, myalgias and weakness, usually occur [49]. No data with clinical endpoints support the use of IL-2 in current clinical practice.

Efforts to develop "therapeutic vaccines" which may boost the host immune response to HIV are underway. Although multiple studies have demonstrated that viral subunits, or inactivated viral particles, can generate HIV-specific immune responses in some individuals, there is no evidence of clinical benefit to the patient [50, 51]. Further studies of the potential value of therapeutic vaccines are currently underway.

Acute HIV infection

Over half of patients acutely infected with HIV will experience at least some symptoms of the acute retroviral syndrome (ARS) [52, 53]. However, acute HIV infection is often not recognized in the primary care setting because of the similarity of the symptom complex with those of more common acute viral illnesses. Healthcare providers must maintain a high level of suspicion for HIV infection in all patients presenting with a compatible clinical syndrome and should obtain appropriate laboratory testing. Evidence for acute HIV infection includes detectable HIV RNA in plasma using sensitive PCR or bDNA assays, together with a negative or indeterminate HIV antibody test. There may be low titer (<10,000 copies/ml) false-positive tests for HIV RNA levels; so patients diagnosed with HIV infection by HIV RNA testing should have subsequent confirmatory ELISA testing performed.

Data indicate that treatment of primary HIV infection with combination therapy has a beneficial effect on biological markers of disease progression [54, 55]. The theoretical rationale for antiretroviral treatment of acute HIV infection include: a decrease in the severity of acute disease; an alteration of the initial viral "setpoint," which may ultimately affect the rate of disease progression; a reduction in the rate of viral mutation due to the suppression of viral replication; and preservation of immune function [8]. However, the potential disadvantages of initiating therapy for acute HIV infection include additional lengthy exposure to antiretroviral therapy without a known clinical benefit; this may result in significant toxicities, development of antiretroviral drug resistance, and adverse effect on quality of life. Ongoing clinical trials are now addressing the question of the long-term clinical benefit of potent treatment regimens.

In addition to patients with symptomatic primary HIV infection, many experts would also consider therapy for patients in whom seroconversion has been documented to have occurred within the previous 6 months. Although the initial burst of viremia in adults has usually resolved by 2 months, treatment during the 2 to 6 month period after infection is based on the likelihood that virus replication in lymphoid tissue is still not maximally contained by the immune system during this time [56]. Although therapy may be offered to patients who test antibody positive and who believe the infection occurred within the previous 6 months, there are no clinical outcome data to support such a decision.

When the clinician and patient decide to utilize antiretroviral therapy for primary HIV infection, treatment should be directed toward suppressing plasma HIV RNA levels to below detectable levels. There are insufficient data to make firm recommendations regarding specific drug therapy; potential combinations of agents available are the same as those used in established infection [57]. Patients must fully understand the possibility of developing drug resistance that may limit future options; this will occur if virus replication is not adequately suppressed or if the patient has been infected with a viral strain that is already resistant to one or more agents. Since the goal of therapy is suppression of viral replication to below the level of detection, any regimen that is not expected to maximally suppress viral replication is not appropriate for treating the acutely HIV-infected individual.

If HAART is initiated in this setting, testing for plasma HIV RNA levels and CD4+ T-cell count and toxicity monitoring should be performed as described under Monitoring above. The optimal duration of therapy in this setting is unknown; ongoing clinical trials should provide data relevant to these issues.

Conclusion

Dramatic improvement in HIV-associated morbidity and mortality has been demonstrated throughout the industrialized world over the past 6 years as a result of more effective antiretroviral therapy. The problems posed by adherence, development of viral resistance, and long-term toxicities have been addressed, with variable effectiveness. The availability of simplified once or twice daily regimens with no dietary restrictions has greatly improved adherence and should result in further improvement in clinical outcomes. Increasing knowledge of toxicities related to specific antiretroviral agents have allowed tailoring of antiretroviral drugs in a manner which minimizes toxicity in individual patients. There has been only modest progress toward developing drugs in the three currently available classes of antiretroviral agents, which do not share cross-class resistance with earlier agents. Recent data demonstrate that HIV-1 entry inhibitors provide an effective fourth class of antiretroviral agents, which can be administered in combination with existing drugs [58, 59]. Furthermore, a fifth class of antiretroviral agents, integrase inhibitors, are now entering Phase I-II clinical trials.

The greatest challenge in this field over the next decade will be to provide the known benefits of antiretroviral therapy to persons in the developing world, where over 90% of persons living with HIV now live.

References

1 Palella FJ Jr, Delaney KM, Moorman AC, Loveless MO, Fuhrer J, Satten GA, Aschman DJ, Holmberg SD (1998) Declining morbidity and mortality among patients with advanced human immunodeficiency virus infection. *N Engl J Med* 338: 853–860

2 Hogg RS, Heath KV, Yib B, Craib KJ, O'Shaughnessy MV, Schechter MT, Montaner JS (1998) Improved survival among HIV-infected individuals following initiation of antiretroviral therapy. *JAMA* 279: 450–454

3 Hogg RS, Yip B, Chen KJ, Wood E, Craib KJ, O'Shaughnessy MV, Montaner JS (2001) Rates of disease progression by baseline CD4 cell count and viral load after initiating triple-drug therapy. *JAMA* 286: 2568–2577

4 Phillips AN, Staszewski S, Weber R, Kirk O, Francioli P, Miller V, Vernazza P, Lundgren JD, Ledergerber B (2001) HIV viral load response to antiretroviral therapy according to the baseline CD4 cell count and viral load. *JAMA* 286: 2560–2567

5 Hirsch MS, Conway B, D'Aquila RT, Johnson VA, Brun-Vezinet F, Clotet B, Demeter LM, Hammer SM, Jacobsen DM, Kuritzkes DR et al. (1998) Antiretroviral drug resistance testing in adults with HIV infection: implications for clinical management. International AIDS Society-USA panel. *JAMA* 279: 1984–1991

6 Carr A, Cooper DA (2000) Adverse effects of antiretroviral therapy. *Lancet* 356: 1423–1430

7 Rizzardi GP, de Boer RJ, Hoover S, Tambussi G, Chapuis A, Halkic N, Bart PA, Miller V, Staszewski S, Notermans DW et al. (2000) Predicting the duration of antiviral treatment needed to suppress plasma HIV-1 RNA. *J Clin Invest* 105: 777–782

8 Department of Health and Human Services. Guidelines for the use of antiretroviral agents in HIV-infected adults and adolescents. April 23, 2001. Available at: http://hivatis.org. Accessed February 4, 2002

9 Carcelain G, Li TS, Renaud M, Blanc C, Tubiana R, Calvez V, Leibowitch J, Debre P, Agut H, Katlama C et al. (1999) Antiretroviral therapy and immune reconstitution. *J Soc Biol* 193: 5–10

10 Tortajada C, Garcia F, Plana M, Gallart T, Maleno MJ, Miro JM, Gatell JM (2000) Comparison of T-cell subsets' reconstitution after 12 months of highly active antiretroviral therapy initiated during early *versus* advanced states of HIV disease. *J Acquir Immune Defic Syndr* 25: 296–305

11 Hockett RD, Kilby JM, Derdeyn CA, Saag MS, Sillers M, Squires K, Chiz S, Nowak MA, Shaw GM, Bucy RP (1999) Constant mean viral copy number per infected cell in tissues regardless of high, low, or undetectable plasma HIV RNA. *J Exp Med* 189: 1545–1559

12 Finzi D, Blankson J, Siliciano JD, Margolick JB, Chadwick K, Pierson T, Smith K, Lisziewicz J, Lori F, Flexner C et al. (1999) Latent infection of CD4+ T cells provides a mechanism for lifelong persistence of HIV-1, even in patients on effective combination therapy. *Nat Med* 5: 512–517

13 Cozzi Lepri A, Phillips AN, d'Arminio Monforte A, Castelli F, Antinori A, de Luca A, Pezzotti P, Alberici F, Cargnel A, Grima P et al. (2001) When to start highly active antiretroviral therapy in chronically HIV-infected patients: evidence from the ICONA study. *AIDS* 15: 983–990

14 Kaplan J, Hanson D, Karon J, Cohn D, Thompson M, Buskin S, Fleming P, Dworkin M (2001) Late initiation of antiretroviral therapy (at CD4+ lymphocyte count <200 cells/µL) is associated with increased risk of death. 8th Conference on Retroviruses and Opportunistic Infections, Chicago, IL, USA, 4–8 February 2001. Abstract 520

15 DeSimone JA, Pomerantz RJ, Babinchak TJ (2000) Inflammatory reactions in HIV-1-infected persons after initiation of highly active antiretroviral therapy. *Ann Int Med* 133: 447–454

16 Kaufmann GR, Bloch M, Finlayson R, Zaunders J, Smith D, Cooper DA (2002) The extent of HIV-1-related immunodeficiency and age predict the long-term CD4 T lymphocyte response to potent antiretroviral therapy. *AIDS* 16: 359–367
17 Chaisson RE, Schecter GF, Theuer CP, Rutherford GW, Echenberg DF, Hopewell PC (1987) Tuberculosis in patients with the acquired immunodeficiency syndrome. Clinical features, response to therapy, and survival. *Am Rev Respir Dis* 136: 570–574
18 Hirschtick RE, Glassroth J, Jordan MC, Wilcosky TC, Wallace JM, Kvale PA, Markowitz N, Rosen MJ, Mangura BT, Hopewell PC (1995) Bacterial pneumonia in persons infected with the human immunodeficiency virus. Pulmonary Complications of HIV Infection Study Group. *N Engl J Med* 333: 845–851
19 Crowe SM, Carlin JB, Stewart KI, Lucas CR, Hoy JF (1991) Predictive value of CD4 lymphocyte numbers for the development of opportunistic infections and malignancies in HIV-infected persons. *J Acquir Immune Defic Syndr* 4: 770–776
20 Mellors JW, Munoz A, Giorgi JV, Margolick JB, Tassoni CJ, Gupta P, Kingsley LA, Todd JA, Saah AJ, Detels R et al. (1997) Plasma viral load and CD4+ lymphocytes as prognostic markers of HIV-1 infection. *Ann Intern Med* 126: 946–954
21 Farzadegan H, Hoover DR, Astemborski J, Lyles CM, Margolick JB, Markham RB, Quinn TC, Vlahov D (1998) Sex differences in HIV-1 viral load and progression to AIDS. *Lancet* 352: 1510–1514
22 Junghans C, Ledergerber B, Chan P, Weber R, Egger M (1999) Sex differences in HIV-1 viral load and progression to AIDS. Swiss HIV Cohort Study [letter]. *Lancet* 353: 589–591
23 Carpenter CC, Fischl MA, Hammer SM, Hirsch MS, Jacobsen DM, Katzenstein DA, Montaner JS, Richman DD, Saag MS, Schooley RT et al. (1996) Antiretroviral therapy for HIV infection in 1996. Recommendations of an international panel. International AIDS Society-USA. *JAMA* 276: 146–154
24 Feinberg MB, Carpenter C, Fauci AS, Stanley SK, Cohen O, Bartlett JG, Kaplan JE, Abrutyn E (1998) Report of the NIH panel to define principles of therapy of HIV infection and guidelines for the use of antiretroviral agents in HIV-infected adults and adolescents. *Ann Intern Med* 128: 1057–1100
25 Gulick RM, Mellors JW, Havlir D, Eron JJ, Meibohm A, Condra JH, Valentine FT, McMahon D, Gonzalez C, Jonas L et al. (2000) Three-year suppression of HIV viremia with indinavir, zidovudine and lamivudine. *Ann Int Med* 133: 35–39
26 Hirsch M, Steigbigel R, Staszewski S, Mellors J, Scerpella E, Hirschel B, Lange J, Squires K, Rawlins S, Meibohm A et al. (1999) A randomized, controlled trial of indinavir, zidovudine and lamivudine in adults with advanced human immunodeficiency virus type 1 infection and prior antiretroviral therapy. *J Infect Dis* 180: 659–665
27 Stone VE, Hogan JW, Schuman P, Rompalo AM, Howard AA, Korkontzelou C, Smith DK (2001) Antiretroviral regimen complexity, self-reported adherence, and HIV patients' understanding of their regimens: survey of women in the HER study. *J Acquir Immune Defic Syndr* 28: 124–131
28 Mitty JA, Stone VE, Sands M, Macalino G, Flanigan T (2002) Directly observed therapy for the treatment of people with human immunodeficiency virus infection: a work in progress. *Clin Infect Dis* 34: 984–990
29 Staszewski S, Morales-Ramirez J, Tashima KT, Rachlis A, Skiest D, Stanford J, Stryker R, Johnson P, Labriola DF, Farina D et al. (1999) Efavirenz plus zidovudine and lamivudine, efavirenz plus indinavir, and indinavir plus zidovudine and lamivudine in the treatment of HIV-1 infection in adults. *N Eng J Med* 341: 1865–1873
30 Murphy RL, Katlama C, Johnson V et al. (1999) The Atlantic study: a randomized, open-label trial comparing two protease inhibitor (PI)-sparing antiretroviral strategies *versus* a standard PI-containing regimen, 48 week data. 39th Annual Meeting of the Interscience Conference on Antimicrobial Agents and Chemotherapy, San Francisco, CA, USA, 26–29 September 1999. Abstract LB-22
31 Podzamczer D, Ferrer E, Consiglio E, Gatell JM, Perez P, Perez JL, Luna E, Gonzalez A, Pedrol E, Lozano L et al. (2001) Final 12-month results from the combine study: a randomized, open, multicenter trial comparing combivir plus nelfinavir or nevirapine in naïve patients. 1st International AIDS Society Conference on HIV Pathogenesis and Treatment, Buenos Aires, Argentina, 8–11 July 2001. Abstract 7
32 Staszewski S, Kaiser P, Montaner J, Raffi F, Gathe J, Brotas V, Hicks C, Hammer SM, Cooper D, Johnson M et al. (2001) Abacavir-lamivudine-zidovudine vs indinavir-lamivudine-zidovudine in

antiretroviral-naïve HIV-infected adults: a randomized equivalence trial. *JAMA* 285: 1155–1163
33 USPHS/IDSA Prevention of Opportunistic Infections Working Group. 2001 USPHS/IDSA guidelines for the prevention of opportunistic infections in persons infected with human immunodeficiency virus. Updated November 28, 2001. Available at: http://www.hivatis.org. Accessed December 3
34 Havlir DV, Bassett R, Levitan D, Gilbert P, Tebas P, Collier AC, Hirsch MS, Ignacio C, Condra J, Gunthard HF et al. (2001) Prevalence and predictive value of intermittent viremia with combination HIV therapy. *JAMA* 286: 171–179
35 Hirsch MS, Brun-Vezinet F, D'Aquila RT, Hammer SM, Johnson VA, Kuritzkes DR, Loveday C, Mellors JW, Clotet B, Conway B et al. (2000) Antiretroviral drug resistance testing in adult HIV-1 infection: recommendations of the International AIDS Society-USA panel. *JAMA* 283: 2417–2426
36 International AIDS Society-USA Resistance Mutations Project Panel (2001) Update on drug resistance mutations in HIV-1. *Topics in HIV Medicine* 9: 21–23
37 Flexner C (1998) HIV-protease inhibitors. *N Engl J Med* 338: 1281–1292
38 Haubrich RH (2000) Structured treatment interruptions: review of current data. *Topics in HIV Medicine* 8: 9–13
39 Rosenberg ES, Altfeld M, Poon SH, Philips MN, Wilkes BM, Eldridge RL, Robbins GK, D'Aquila RT, Goulder PJ, Walker BD (2000) Immune control of HIV-1 after early treatment of acute infection. *Nature* 407: 523–526
40 Kilby JM, Goepfert PA, Miller AP, Gnann JW Jr, Sillers M, Saag MS, Bucy RP (2000) Recurrence of the acute HIV syndrome after interruption of antiretroviral therapy in a patient with chronic HIV infection: a case report. *Ann Intern Med* 133: 435–438
41 Colven R, Harrington RD, Spach DH, Cohen CJ, Hooton TM (2000) Retroviral rebound syndrome after cessation of suppressive antiretroviral therapy in three patients with chronic HIV infection. *Ann Intern Med* 133: 430–434
42 Ruiz L, Martinez-Picado J, Romeu J, Paredes R, Zayat MK, Marfil S, Negredo E, Sirera G, Tural C, Clotet B (2000) Structured treatment interruption in chronically HIV-1 infected patients after long-term viral suppression. *AIDS* 14: 397–403
43 Deeks SG, Wrin T, Liegler T, Hoh R, Hayden M, Barbour JD, Hellman NS, Petropoulos CJ, McCune JM, Hellerstein MK et al. (2001) Virologic and immunologic consequences of discontinuing combination antiretroviral-drug therapy in HIV-infected patients with detectable viremia. *N Engl J Med* 344: 472–480
44 Miller V, Sabin C, Hertogs K, Bloor S, Martinez-Picado J, D'Aquila R, Larder B, Lutz T, Gute P, Weidmann E et al. (2000) Virological and immunological effects of treatment interruptions in HIV-1 infected patients with treatment failure. *AIDS* 14: 2857–2867
45 Kaufman D, Pantaleo G, Sudre P, Telenti A (1998) CD4-cell count in HIV-1-infected individuals remaining viraemic with highly active antiretroviral therapy (HAART). Swiss HIV Cohort Study. *Lancet* 351: 723–724
46 Montaner JSG, Harrington PR, Jahnke N, Raboud J, Castillo E, Hogg RS, Yip B, Harris M, Montessori V, O'Shaughnessy MV (1998) Multiple drug rescue therapy for HIV-infected individuals with prior virologic failure to multiple regimens. *AIDS* 15: 61–69
47 Kovacs JA, Baseler M, Dewar RJ, Vogel S, Davey RT Jr, Falloon J, Polis MA, Walker RE, Stevens R, Salzman NP et al. (1995) Increases in CD4 T lymphocytes with intermittent courses of interleukin-2 in patients with human immunodeficiency virus infection. *N Eng J Med* 332: 567–574
48 Davey RT, Murphy RL, Graziano FM, Boswell SL, Pavia AT, Cancio M, Nadler JP, Chaitt DG, Dewar RL, Sahner DK et al. (2000) Immunologic and virologic effects of subcutaneous interleukin-2 in combination with antiretroviral therapy: a randomized controlled trial. *JAMA* 284: 183–189
49 Davey RT Jr, Chaitt DG, Albert JM, Piscitelli SC, Kovacs JA, Walker RE, Falloon J, Polis MA, Metcalf JA, Masur H et al. (1999) A randomized trial of high- *versus* low-dose subcutaneous interleukin-2 outpatient therapy for early human immunodeficiency virus type 1 infection. *J Infect Dis* 179: 849–858
50 Birx DL, Loomis-Price LD, Aronson N, Brundage J, Davis C, Deyton L, Garner R, Gordin F, Henry D, Holloway W et al. (2000) Efficacy testing of recombinant human immunodeficiency virus (HIV) gp160 as a therapeutic vaccine in early-stage HIV-1-infected volunteers. rgp 160 Phase II Vaccine Investigators. *J Infect Dis* 181: 881–889
51 Kahn JO, Cherng DW, Mayer K, Murray H, Lagakos S (2000) Evaluation of HIV-1 immunogen,

an immunologic modifier, administered to patients infected with HIV having 300 to 549 × 10(6)/L CD4 cell counts: a randomized controlled trial. *JAMA* 284: 2193–2202
52 Kinloch-de Loes S, de Saussure P, Saurat JH, Stalder H, Hirschel B, Perrin LH (1993) Symptomatic primary infection due to human immunodeficiency virus type 1: review of 31 cases. *Clin Infect Dis* 17: 59–65
53 Schacker T, Collier AC, Hughes J, Shea T, Corey L (1996) Clinical and epidemiologic features of primary HIV infection. *Ann Intern Med* 125: 257–264
54 Lillo FB, Ciuffreda D, Veglia F, Capiluppi B, Mastrorilli E, Vergani B, Tambussi G, Lazzarin A (1999) Viral load and burden modification following early antiretroviral therapy of primary HIV-1 infection. *AIDS* 13: 791–796
55 Malhotra U, Berrey MM, Huang Y, Markee J, Brown DJ, Ap S, Musey L, Schacker T, Corey L, McElrath MJ (2000) Effect of combination antiretroviral therapy on T-cell immunity in acute human immunodeficiency virus type-1 infection. *J Infect Dis* 181: 121–131
56 Pantaleo G, Cohen OJ, Schacker T, Vaccarezza M, Graziosi C, Rizzardi GP, Kahn J, Fox CH, Schnittman SM, Schwartz DH et al. (1998) Evolutionary pattern of human immunodeficiency virus (HIV) replication and distribution in lymph nodes following primary infection: implications for antiviral therapy. *Nat Med* 4: 341–345
57 Lafeuillade A, Poggi C, Tamalet C, Profizi N, Tourres C, Costes O (1997) Effects of a combination of zidovudine, didanosine, and lamivudine on primary human immunodeficiency virus type-1 infection. *J Infect Dis* 175: 1051–1055
58 Lazezari JP, Henry K, O'Hearn M, Montaner JSG, Piliero PJ, Trottier B, Walmsley S, Cohen C, Kuritzkes DR, Eron JJ et al. (2003) Enfuvirtide, an HIV-1 fusion inhibitor, for drug-resistant HIV infection in North and South America. *N Engl J Med* 348: 2175–2185
59 Lazzarin A, Bonaventura C, Cooper D, Reynes J, Arasteh K, Nelson M, Katlama C, Stellbrink H, Delfraissy JF, Lange J et al. (2003) Efficacy of enfuvirtide in patients infected with drug-resistant HIV-1 in Europe and Australia. *N Engl J Med* 348: 2186–2195

Combination Therapy of AIDS
Edited by E.D.A. De Clercq and A.-M.I. Vandamme

Visions for the future of antiretroviral therapy

Ume L. Abbas and John W. Mellors

Division of Infectious Diseases, School of Medicine, University of Pittsburgh, Scaife Hall, Suite 818, 3550 Terrace Street, Pittsburgh PA 15261, USA

Introduction

The spreading epidemic of human immunodeficiency virus type 1 (HIV-1) infection and acquired immunodeficiency syndrome (AIDS) is poised to become the greatest scourge of humankind in recorded history. As efforts to control HIV-1 continue globally, attitudes about this formidable opponent have changed from pathos of initial despair, and after a brief heady euphoria, have settled into modern-day cautious optimism.

At the beginning of the twenty-first century, the utopian goals of universal protection from HIV-1 infection and eradication of HIV-1 in infected persons were exchanged for more realistic ones – a partially effective vaccine that protects against disease rather than infection, and improved duration and quality of life for the infected individuals through daily administration of antiretroviral therapy. This shift in focus was forced by several sobering realizations: i) a safe and effective HIV-1 vaccine that imparts sterilizing immunity is beyond our current understanding of immune protection [1]; ii) existing antiretroviral therapy is not potent enough to completely suppress HIV-1 replication and thus could never eradicate HIV-1 infection [2–7]; and iii) HIV-1 infection may persist indefinitely in cellular reservoirs including resting CD4+ memory T-cells and can serve as a source of recrudescent viremia when therapy is stopped [8–10].

Despite its limitations, antiretroviral therapy has taken center stage in the fight to reduce worldwide morbidity and mortality from HIV. Recent emphasis has been placed not only on the development of new therapies, but also on optimizing existing ones for both industrialized and developing nations. The optimal timing, duration, composition and sequencing of antiretroviral treatment regimens are being better defined to optimize patient outcome. Improved monitoring and decision support tools are likely to improve the efficacy of antiretroviral therapy in the future. This chapter will provide a glimpse into the future of antiretroviral therapy including i) promising new antiretrovirals; ii) expected enhancements of existing therapies; and iii) novel treatment strategies including strategic treatment interruptions, therapeutic vaccines and immunomodulators.

Promising new antiretrovirals

Many efforts are underway to discover new antiretroviral agents including those from existing drug classes (nucleoside reverse transcriptase inhibitors [NRTIs], non-nucleoside reverse transcriptase inhibitors [NNRTIs] and protease inhibitors [PIs]), as well as those from new classes (e.g., entry and integrase inhibitors).) Although the path from inhibitor discovery to approval for use is fraught with pitfalls, several of the agents described below provide optimism that the armament of antiretrovirals will continue to expand.

Nucleoside reverse transcriptase inhibitors

DAPD (amdoxovir)

DAPD (β-D-2, 6-diaminopurine dioxolane) is a guanosine analog that is converted *in vivo* by adenosine deaminase to dioxolane guanine (DXG), which has activity against both HIV-1 and HBV. DXG retains activity against recombinant and clinical isolates of HIV-1 that are resistant to NRTIs including those encoding the 69 insert multi-nucleoside resistance complex in reverse transcriptase (RT). Only isolates containing the K65R or L74V mutation, or the Q151M complex of mutations in RT show reduced susceptibility to DXG. In Phase I/II clinical trials, DAPD has dose-dependent antiviral activity and no major short-term toxicity. In treatment naïve subjects given 300 or 500 mg twice daily, reductions in plasma HIV-1 RNA from baseline after 2 weeks of therapy were 1.45 and 1.46 $\log_{10}$ copies/ml, respectively [11–12]. Phase I/II trials in treatment-experienced patients have also shown that DAPD has antiviral activity, but the impact of baseline NRTI resistance mutations on drug activity has not been clearly defined. The clinical development of DAPD has been complicated by the findings of lens abnormalities, diabetes mellitus and obstructive uropathy in long-term animal toxicology studies. These toxicities have required intensive ophthalmologic and laboratory monitoring in current clinical trials of DADP and have slowed its development.

Emtricitabine (FTC)

FTC [(–)-(2R,5S)-5-fluoro-1-[2(hydroxymethyl)-1,3-oxathiolane-5-yl]-cytosine] is a cytidine analog similar to lamivudine (3TC) in structure (L-nucleoside), but shows a several-fold greater potency against HIV-1 *in vitro* [13] and can be dosed once daily. The resistance profile of FTC is similar to 3TC with the M184V mutation in RT conferring >10-fold resistance. Three randomized, controlled Phase III trials of FTC 200 mg once-daily compared with 3TC 150 mg twice-daily as part of triple drug regimens in antiretroviral-naïve patients show equivalent antiretroviral activity and safety of FTC and 3TC [14]. These studies have led to Food and Drug Administration (FDA) approval of FTC in the United States (US). An important unanswered question is whether the greater *in vitro* potency of FTC will translate into antiretroviral activity in treatment-experienced patients with 3TC-resistant virus.

D-D4FC (DPC 817)
D-D4FC (β-D-2',3'-didehydro-2',3'-dideoxy-5-fluorocytidine) is a cytosine analog that is active against both HIV-1 and HBV. It is converted *in vivo* to the active metabolite D-D4FC-triphosphate that has an intracellular half-life of 13–17 h. D-D4FC is promising because of its resistance profile against zidovudine- (AZT) or 3TC-resistant virus and its lack of inhibition of human bone marrow progenitor cell growth or mitochondrial DNA synthesis *in vitro* [15, 16]. Single-dose pharmacokinetic studies in normal volunteers and HIV-infected persons suggest that D-D4FC can be dosed once daily and has substantial antiretroviral activity. Multi-dose trials of D-D4FC in treatment-naïve and treatment-experienced HIV-infected persons are in progress.

FLT (alovudine/miv-310)
FLT (3'deoxy-3'-fluorothymidine) is a fluorinated thymidine analog for which development was stopped earlier due to excess bone marrow toxicity and lack of advantage over AZT. However, a lower dose of 7.5 mg once daily for 4 weeks [17] added to the regimen of 15 patients with median plasma HIV-1 RNA of 3.93 $\log_{10}$ copies/ml and more than two thymidine analogue mutations (TAMs), produced a median plasma HIV-1 RNA reduction of 1.13 $\log_{10}$ copies/ml. A greater plasma HIV-1 RNA reduction (1.88 $\log_{10}$ *versus* 0.57) in patients not receiving stavudine (d4T) suggests a possible negative interaction between FLT and d4T. The drug has activity against multinucleoside-resistant viruses [18], including those carrying multiple TAMs, the Q151M mutation, or the 69 insertions. FLT shows promise for salvage therapy, but the potential for bone marrow suppression may ultimately limit its usefulness.

Non-nucleoside reverse transcriptase inhibitors

Identifying "second" generation NNRTIs with favorable toxicity, pharmacokinetic and activity profiles against viruses that are resistant to approved NNRTIs (nevirapine, delavirdine, efavirenz) has proved exceedingly difficult because of extensive within class cross-resistance. Two candidates that remain in development are described below.

Capravirine
Capravirine has potent *in vitro* antiviral activity [19] against HIV-1 variants with NNRTI resistance mutations including K103N, which confers cross-resistance to all approved NNRTIs. In a recent clinical trial of capravirine *versus* placebo (plus nelfinavir and two new NRTIs) in 75 patients failing an NNRTI-containing regimen [20], more than 50% of patients had plasma HIV-1

RNA < 400 copies/ml after 12 weeks, suggesting that capravirine has activity against NNRTI-resistant viruses. Despite these encouraging findings, clinical development of capravirine was temporarily put on hold by the US FDA because of vasculitis observed in long-term toxicology studies in dogs. This hold has been lifted and capravirine is entering larger Phase II studies in patients who have failed approved NNRTIs.

TMC 125

TMC 125 has potent *in vitro* activity against both wild-type HIV-1 and mutants exhibiting high-level resistance to approved NNRTIs. In a randomized, double-blind, placebo-controlled study [21], 900 mg of TMC125 BID as monotherapy for 7 days in therapy-naïve HIV-infected patients was safe and reduced plasma HIV-1 RNA by a mean of 1.99 $\log_{10}$ copies/ml. In a pilot study of TMC 125 (900 mg BID) for 7 days in 16 patients failing therapy with an NNRTI-based regimen [22], the mean reduction in HIV-1 RNA from baseline was 0.9 $\log_{10}$ copies/ml and 44% of patients had a decrease of >1.0 $\log_{10}$ copies/ml. These results clearly demonstrate promising *in vivo* activity of TMC 125 against NNRTI-resistant virus. Longer-term dosing studies are needed to determine the durability of the antiretroviral effect and whether resistance to TMC 125 develops.

Protease inhibitors

Atazanavir

Atazanavir is a potent azapeptide HIV-1 protease inhibitor with *in vitro* EC_{50} in the 2.6–5.3 nM range. It is active against both wild-type HIV-1 and variants resistant to single PIs including nelfinavir, saquinavir and amprenavir, but indinavir- or ritonavir-resistant variants exhibit 6- to 9-fold resistance [23], and viruses that are resistant to multiple PIs are usually cross-resistant to atazanavir. Favorable pharmacokinetics permit once daily dosing of atazanavir. In a study comparing atazanavir with nelfinavir in therapy-naïve patients beginning combination therapy, atazanavir was well tolerated and had similar antiretroviral activity as nelfinavir, but with no elevation in serum cholesterol or triglycerides [24]. In patients without prior PI experience, atazanivir selects a unique I50L mutation in protease, which decreases sensitivity to atazanavir, but not to other protease inhibitors. Atazanavir was recently approved (June 2003) for clinical use in the US. It is expected that atazanavir will become the first-line protease inhibitor of choice because of its once daily dosing, absence of lipid elevations and lack of cross-resistance to other PIs from the I50L mutation.

Tipranavir

Tipranavir is a nonpeptidic PI which is active against most HIV-1 viral strains with >10-fold resistance to three to four approved PIs [25]. Exposure to tipranavir is markedly enhanced by co-administration with ritonavir (200 mg BID). A recent Phase II study comparing three different doses of tipranavir with ritonavir identified the optimal dose to be tipranavir/ritonavir 500 mg/200 mg BID [26]. This dose is currently being tested in two pivotal Phase III trials in PI-experienced patients. Tipranavir's potential lies in its activity against HIV-1 variants resistant to multiple PIs.

TMC 114

TMC 114 is a bis-furan-sulphonamide with potent *in vitro* activity against HIV-1 variants resistant to multiple PIs. As for tipranavir, exposure to TMC 114 is markedly increased by co-administration with ritonavir (100 mg QD or BID). Arasteh et al. [27] recently reported the results of a Phase II pilot trial of three doses of TMC 114/ritonavir administered for 14 days to patients with multiple prior PI-experience and multiple PI-resistance. TMC 114/ritonavir was safe and showed potent antiretroviral activity, reducing plasma HIV-1 RNA by a median of 1.35 $\log_{10}$ copies/ml. These promising initial results have led to 24-week Phase IIB safety and dose-finding trials.

Novel classes and new targets

Fusion inhibitors

T-20 (Enfuvirtide/Fuzeon). T-20 is a 36 amino acid synthetic peptide administered by subcutaneous injection that binds to the helical domains of gp41 preventing their association, and thus disrupting the conformational change required for fusion of HIV-1 with the target cell membrane. Twenty-four week data from two, randomized international Phase III clinical trials, TORO 1 and TORO 2 have led to approval of T20 in the US [28, 29]. Both trials had a similar design; patients were randomized 2:1 to T-20 (90 mg SC BID) plus optimized background (three to five antiretrovirals) or optimized background alone (no T-20). Study patients were heavily therapy-experienced with an average of 12 prior antiretrovirals. At 24 weeks, there was significantly greater reduction in plasma HIV-1 RNA in the T-20 plus optimized background arm of 0.934 $\log_{10}$ copies/ml (TORO 1) and 0.78 $\log_{10}$ copies/ml (TORO 2). Forty-eight week data are pending. By week 24, only 20% and 12% of patients receiving T-20 achieved plasma HIV-1 RNA less than 50 copies/ml in TORO 1 and TORO 2, respectively, and the durability of virologic response was directly related to the number of active antiretrovirals in the optimized background regimen. These results indicate that T-20 is a potent drug with potential use in salvage therapy, but it should probably be reserved for use in combination with at least two other active antiretrovirals. In addition, the need for refrigeration and twice-daily injections, injection site reactions (observed in 98% of study patients), and costs may create barriers to its widespread use.

T-1249. This 39 amino acid peptide has greater potency than T-20 *in vitro* and is active against T-20 resistant variants [30]. In a 14-day pilot study of

T-1249, administered subcutaneously in doses ranging from 6.25 mg to 50 mg/day, median plasma HIV RNA responses in the different dose groups ranged from 0.1–1.40 $\log_{10}$ copies/ml [31]. In a pilot trial in 25 patients with virologic failure on T-20 [32], a switch to T-1249 (192 mg/day SC) resulted in a median HIV RNA reduction of 1.12 $\log_{10}$ copies/ml after 10 days. A failure on T-20 of <48 weeks duration was associated with a better virologic outcome. Hence, it is expected that T-1249 will provide an important treatment option for patients in whom resistance to T-20 has developed, but as for T-20, its use will likely be limited by costs and the requirement for subcutaneous injections.

Integrase inhibitors

L-870,810 and L-870,812. Potent diketo acid inhibitors of HIV-1 integrase that block the DNA strand transfer reaction essential for integration were described in 2000 [33], but this chemical class was not developed because of unfavorable pharmacokinetics and extensive protein binding. Structurally related naphthyridine carboxamides have been identified that are potent inhibitors of strand transfer and that have favorable pharmacokinetic and resistance profiles. One of these, L-870,812, has been shown to markedly reduce viremia in macaques acutely infected with SHIV. L-870,810 was chosen as the lead clinical candidate of this class and is being evaluated in short-term pilot trials in HIV-infected patients.

Other targets

A number of chemokine receptor antagonists have been identified including CXCR4 and CCR5 inhibitors. Concerns have been expressed that blockade of CCR5 will promote the outgrowth of CXCR4 tropic virus that has greater pathogenicity. Similarly, the value of CXCR4 inhibitors has been questioned because the majority of virus is CCR5 tropic. These issues demonstrate that the optimal strategy for development and use of chemokine receptor antagonists has yet to be defined.

Enhancement of current drug regimens

There is room to improve the potency, tolerability and simplicity of existing antiretroviral therapies. Multi-drug combination pills, containing two or three antiretrovirals, which can be administered once daily, should become available in the next few years. Such combination pills will be particularly useful in developing countries where resources are limited, but the morbidity and mortality from AIDS is high. In developed nations more intense monitoring of antiretroviral therapy is likely, including frequent assessment of HIV-1 RNA response to identify early signs of regimen failure [34], therapeutic drug monitoring to adjust dosing of protease inhibitors in treatment-experienced patients, and biomedical informatics to manage the increasing amount and complexity of clinical data and experience to optimize individual patient management.

Therapeutic drug monitoring

Therapeutic drug monitoring (TDM) is the laboratory measurement and interpretation of drug concentrations to individualize drug dosing. This can potentially boost the potency and durability of the antiretroviral regimen and diminish drug toxicity. The role of TDM in clinical care of HIV-1 continues to be evaluated and debated. TDM recommendations have already made their way into British and French antiretroviral treatment guidelines.

Although there is need for large, controlled trials of the clinical utility and cost effectiveness of TDM, there are solid arguments in its favor: i) a correlation between plasma PI and NNRTI concentrations and virologic response has been demonstrated in several studies [35–39]; ii) high plasma drug concentrations may be predictive of drug toxicity [40–42]; and iii) several-fold inter-individual variability may be observed in the trough concentrations of PIs [43] and NNRTIs. Indeed, the Athena trial, a large randomized trial of TDM *versus* standard of care in 600 previously naïve patients receiving indinavir or nelfinavir showed fewer treatment discontinuations and a greater proportion of patients with plasma HIV-1 RNA < 500 copies/ml at 12-month in the TDM group [44].

A more accurate expression of potency than drug concentration alone in treatment-experienced patients is the inhibitory quotient (IQ) which may be defined as the ratio of the actual trough (C_{min}) to the *in vitro* IC_{50}, EC_{50} or IC_{95} of the patient's pre-treatment HIV-1 isolate, determined by phenotypic assays or virtual phenotype testing. The value of different methods of determining IQ has not been defined. Nevertheless, in a study of 57 treatment-experienced patients treated with lopinavir/ritonavir and efavirenz [45], the IQ (lopinavir $C_{trough} \div$ baseline fold IC_{50} x serum-adjusted IC_{50} for wild-type HIV-1) showed a strong correlation with virologic response, whereas the lopinavir trough concentration alone was a poor predictor of response. A similar association has been reported between virologic response and IQ of indinavir [45, 46] and amprenavir [47].

Although efforts are underway to better define the potential role of TDM in routine antiretroviral therapy by ongoing trials, and some authorities recommend its use to monitor adherence, drug interactions and therapy in children, pregnancy, renal and hepatic dysfunction, TDM is likely to lose popularity as protease inhibitors and other drugs with better pharmacokinetic profiles and higher intrinsic IQs become available.

Biomedical informatics and decision support

Biomedical informatics is the scientific discipline that focuses on storage, retrieval, and optimal use of biomedical information for problem solving and medical decision-making [48]. This field is now being applied to management of HIV-1 infection. In a study by Tural et al. involving 326 patients, computerized expert advice, coupled with HIV-1 genotyping and clinical information, improved virologic response to antiretroviral therapy [49]. In the future, it is expected that the use of such decision-making tools for managing HIV-1 infection will increase in parallel with inevitable growth in the amount and complexity of information in the field and for individual patients. Application of such new technologies should facilitate individualized care of patients with HIV-1 infection. In addition, insight gained from models of care in industrialized nations should be applicable to patients in developing countries and help bridge the information gap between resource-poor and privileged parts of the world.

New antiretroviral treatment strategies

Intermittent therapy

Concerns about access, cost, toxicity, adherence, drug resistance and the inability of antiretroviral therapy to restore significant HIV-1-specific immunity or eradicate HIV-1 have led many to question the "gold standard" of continuous, lifelong therapy. Intermittent therapy is an approach in which antiretroviral therapy is switched off and on in a monitored setting, either triggered by changes in CD4+ T-cell counts or clinical symptoms. Intermittent therapy is not recommended for HIV-1 management at the time of this writing, but may well become more common in the future as limited resources force the parsimonious use of drugs to treat the greatest number of individuals possible.

Important risks of intermittent therapy include repopulation of latent reservoirs, acute antiretroviral syndrome and AIDS-related events, CD4 depletion and immune system deterioration, recurrent adverse drug effects, drug resistance, and increased risk of HIV-1 transmission. These risks will need to be compared carefully to the cost savings and toxicity reductions in long-term trials of intermittent therapy *versus* continuous therapy.

Immune-based therapies

The adaptive immune response to immunodeficiency virus infection comprises both cellular and humoral immunity [50]. Various epitopes expressed by the HIV-1 or on the surface of HIV-1-infected cells trigger neutralizing antibodies and cellular immune responses, respectively. The protective cytotoxic T lymphocyte (CTL) response [51] is stimulated by CD4+ cells by means of HIV-1-specific T-helper lymphocytes (THL) and cytokines. CD4+ cells respond to antigen-presenting cells (APCs) such as dendritic cells and macrophages. The immune response to HIV-1 is only partially effective, varies amongst individuals and ultimately fails in the majority of individuals [52].

Interest in immune-based therapies is largely driven by the desire to mimic the prolonged asymptomatic state in long-term nonprogressors [53–56].

However, several gaps in our knowledge must be filled before immune-based therapies can become the standard of care. Selection of escape variants from immune selective pressure is a major concern.

Therapeutic vaccines

Therapeutic vaccines can potentially slow the dynamics of the HIV-1 pandemic. By stimulating a potent HIV-1-specific immune response that reduces viremia, therapeutic vaccines could decrease morbidity for the individual and reduce transmissibility for the population at large. Different approaches are being used to develop HIV-1 immunogens that elicit broad immune responses, both humoral and cellular, that are cross reactive against a range of HIV-1 variants [57]. Many studies examining effects of therapeutic vaccines either alone or in combination with treatment interruptions or immune modulators are in progress.

Remune, a "whole" (clade A envelope, clade G gag, depleted of gp 120) killed HIV-1 vaccine, was the first therapeutic vaccine to reach large-scale clinical trials. Unfortunately, the outcome was disappointing. A 3-year multicenter, double-blind, placebo-controlled, randomized trial of 2527 patients conducted at 77 centers in the US failed to show a significant effect on plasma HIV-1 RNA, CD4 percentage and progression-free survival [58].

Several other therapeutic vaccines have shown tantalizing evidence of better HIV-1 specific immune function with control of viremia in Phase II trials, but larger trials are needed to assess the potential value of these vaccines at the population level.

Cytokines

The rationale for use of cytokines in HIV-1 infection include i) restoration of CD4 T-cells in those with incomplete recovery on antiretroviral therapy; ii) reduction of immune system activation; and iii) enhancement of HIV-1 specific immune function, alone or as a vaccine adjuvant.

Of all the cytokines, interleukin-2 (IL-2) has emerged as the most promising candidate. Results of two large Phase III clinical trials are expected to clarify the therapeutic role of IL-2 in HIV-1 infection in the next several years. The Evaluation of Subcutaneous Proleukin in a Randomized International Trial (ESPRIT) is a 5-year study that will enroll 4000 patients with CD4 counts >300 cells/uL, whereas the Study of IL-2 with low CD4 Counts on Active Anti-HIV-1 Therapy (SILCAAT) is a 4 to 6-year study of about 1400 patients with CD4 counts of 50–299 cells/uL. IL-2 is likely to be most useful in patients with inadequate restoration of CD4+ T-cell counts on antiretroviral therapy. By raising CD4+ T-cell counts, IL-2 could also prolong the interval between periods of antiretroviral therapy in patients receiving intermittent treatment strategy. The cost benefit and practicality of this approach needs to be better understood before it can be recommended.

Predictions for the future

The antiretroviral armament will continue to expand in coming years, allowing new combinations of drugs that are more effective and better tolerated. Old and new targets including HIV-1 entry molecules, integrase and accessory proteins (e.g., Vif) will be effectively inhibited. But more new antiretrovirals alone are likely to have only a modest effect on the HIV-1 pandemic. Resources for antiretroviral drugs will always be limited and life-long therapy is thus problematic. More cost-effective strategies such as intermittent therapy with simplified once-daily dosing regimens and a partially effective vaccine that reduces viremia and lowers infectiousness are needed to keep pace with the growing pandemic.

References

1 Levy JA (2001) What can be achieved with an HIV-1 vaccine? *Lancet* 357: 223–224

2 Furtado MR, Callaway DS, Phair JP, Kunstman KJ, Stanton JL, Macken CA, Perelson AS, Wolinsky SM (1999) Persistence of HIV-1-1 transcription in peripheral-blood mononuclear cells in patients receiving potent antiretroviral therapy. *N Engl J Med* 340(21): 1614–1622

3 Lewin SR, Vesanen M, Kostrikis L, Hurley A, Duran M, Zhang L, Ho DD, Markowitz M (1999) Use of real-time PCR and molecular beacons to detect virus replication in human immunodeficiency virus type 1-infected individuals on prolonged effective antiretroviral therapy. *J Virol* 73(7): 6099–6103

4 Zhang L, Ramratnam B, Tenner-Racz K, He Y, Vesanen M, Lewin S, Talal A, Racz P, Perelson AS, Korber BT et al. (1999) Quantifying residual HIV-1-1 replication in patients receiving combination antiretroviral therapy. *N Engl J Med* 340(21): 1605–1613

5 Gunthard HF, Frost SD, Leigh-Brown AJ, Ignacio CC, Kee K, Perelson AS, Spina CA, Havlir DV, Hezareh M, Looney DJ et al. (1999) Evolution of envelope sequences of human immunodeficiency virus type 1 in cellular reservoirs in the setting of potent antiviral therapy. *J Virol* 73(11): 9404–9412

6 Dornadula G, Zhang H, VanUitert B, Stern J, Livornese L Jr, Ingerman MJ, Witek J, Kedanis RJ, Natkin J, DeSimone J, Pomerantz RJ (1999) Residual HIV-1-1 RNA in blood plasma of patients taking suppressive highly active antiretroviral therapy. *JAMA* 282(17): 1627–1632

7 Sharkey ME, Teo I, Greenough T, Sharova N, Luzuriaga K, Sullivan JL, Bucy RP, Kostrikis LG, Haase A, Veryard C et al. (2000) Persistence of episomal HIV-1-1 infection intermediates in patients on highly active anti-retroviral therapy. *Nat Med* 6(1): 76–81

8 Finzi D, Hermankova M, Pierson T, Carruth LM, Buck C, Chaisson RE, Quinn TC, Chadwick K, Margolick J, Brookmeyer R et al. (1997) Identification of a reservoir for HIV-1-1 in patients on highly active antiretroviral therapy. *Science* 278(5341): 1295–1300

9 Chun TW, Stuyver L, Mizell SB, Ehler LA, Mican JA, Baseler M, Lloyd AL, Nowak MA, Fauci AS (1997) Presence of an inducible HIV-1-1 latent reservoir during highly active antiretroviral therapy. *Proc Natl Acad Sci U S A* 94(24): 13193–13197

10 Wong JK, Hezareh M, Gunthard HF, Havlir DV, Ignacio CC, Spina CA, Richman DD (1997) Recovery of replication-competent HIV-1 despite prolonged suppression of plasma viremia. *Science* 278(5341): 1291–1295

11 Richman D, Kessler H, Eron J, Thompson M, Raffi F, Jacobson J, Harris J, Mccreedy B, Bigley J and Rousseau F (2000) Anti-HIV-1 activity and tolerability of DAPD, a novel dioxolane guano-

sine RT inhibitor: initial results of a phase I/II 14-day monotherapy clinical trial. 7th Conference on Retroviruses and Opportunistic Infections. January –February 2000; San Francisco, CA [Abstract 668]

12 Eron JJ, Kessler H, Thompson M, Deeks S, Arduino R, Jacobson J, Sista N, Quinn J, Harris J, Bigley J, Rousseau F (2000) Clinical HIV-1 suppression after short-term monotherapy with DAPD. 40th Interscience Conference on Antimicrobial Agents and Chemotherapy September 17–20, 2000 Ontario, Canada [Abstract 34695]

13 Schinazi RF, McMillan A, Cannon D, Mathis R, Lloyd RM, Peck A, Sommadossi JP, St Clair M, Wilson J, Furman PA et al. (1992) Selective inhibition of human immunodeficiency viruses by racemates and enantiomers of cis-5-fluoro-1-[2-(hydroxymethyl)-1,3-oxathiolan-5-yl]cytosine. *Antimicrob Agents Chemother* 36(11): 2423–2431

14 van der Horst C, Sanne I, Wakeford C, Quinn J, Rousseau F (2001) Two Randomized, Controlled, Equivalence Trials of Emtricitibine (FTC) to Lamivudine (3TC). 8th Conference on Retroviruses and Opportunistic Infections. February 4–8, 2001; Chicago, Ill [Abstract 18].

15 Schinazi R, Ma L, Shi J, Liotta D, Faraj A, Sommadossi JP (1998) Anti-HIV-1 Activity, Biochemistry, And Pharmacokinetics Of β-D-2',3'-didehydro-2',3'-dideoxy-5-fluorocytidine (D-D4FC). 12th World AIDS Conf, June 28 – July 3, 1998, Geneva: [Abstract 557/41174]

16 Schinazi RF, Mellors J, Bazmi H, Diamond S, Garber S, Gallagher K, Geleziunas R, Klabe R, Pierce M, Rayner M, Wu JT et al. (2002) DPC 817: a cytidine nucleoside analog with activity against zidovudine- and lamivudine-resistant viral variants. *Antimicrob Agents Chemother* 46(5): 1394–1401

17 Calvez V, Tubiana R, Ghosn J, Wirden M, Marcelin AG, Westling C, Shoen H, Harmenberg J, Mardh G, Öberg B et al. (2002) MIV-30 reduces markedly viral load in patients with virological failure despite multiple-drug therapy: results from a 4-week Phase II study. *Antiviral Ther* 7:S4 [Abstract 3]

18 Vrang L, Zhang H, Palmer S, Kim E-Y, Merigan T, Öberg B (2002) *In vitro* effects of MIV-310 (alovudine, 3'-fluorodeoxythymidine, FLT) against HIV-1 mutants. *Antiviral Ther* 7:S18

19 Hernandez J, Amador L, Amantea M, Chao H, Hawley P, Paradiso L (2000) Short-course monotherapy with AG1549, a novel nonnucleoside reverse transcriptase inhibitor (NNRTI), in antiretroviral naïve patients. 7th Conference on Retroviruses and Opportunistic Infections. January 30 – February 4, 2000; San Francisco, CA.[Abstract 669]

20 Wolfe P, Hawley P, Boccia G, Clendeninn N, Paradiso L, Shaw T, Chi-Burris K (2001) Safety and efficacy of capravirine *versus* placebo in HIV-1-infected patients failing a nonnucleoside-reverse-transcriptase-inhibitor- containing regimen: results of a phase II, double-blind, placebo-controlled study. 8th Conference on Retroviruses and Opportunistic Infections. February 4–8, 2001; Chicago, Ill [Abstract 323]

21 Gruzdev B, Rakhmanova A, de Dier K, Comhaire S, Baede-van dijk P, van t Klooster G (2001) TMC125 is a highly potent non-nucleoside reverse transcriptase inhibitor (NNRTI) in antiretroviral therapy (ART)-naïve, HIV-1 infected subjects. 41st Interscience Conference on Antimicrobial Agents and Chemotherapy December 16–19, 2001 Chicago, IL [Abstract I668]

22 Gazzard B, Pozniak A, Arasteh K, Staszewski S, Rozenbaum W, Yeni P, van 't Klooster G, De Dier K, Peeters M, de Béthune MP, Graham N, Pauwels R (2002) TMC125, a next-generation NNRTI, demonstrates high potency after 7 days therapy in treatment-experienced HIV-1 infected individuals with phenotypic NNRTI resistance. 9th Conference on Retroviruses and Opportunistic Infections, February 24–28, 2002 Seattle, WA [Abstract 4].

23 Gong YF, Robinson BS, Rose RE, Deminie C, Spicer TP, Stock D, Colonno RJ, Lin PF (2000) *In vitro* resistance profile of the human immunodeficiency virus type 1 protease inhibitor BMS-232632. *Antimicrob Agents Chemother* 44(9): 2319–2326

24 Cahn P, Percival L, Phanuphak P, Sanne I, Kelleher T, Giordano M (2001) Phase II 24-week data from study AI424-008: comparative results of BMS-232632, stavudine, lamuvidine as HAART for treatment-naïve HIV-1-infected patients. 1st IAS Conference on HIV-1 Pathogenesis and Treatment – 2001 Buenos Aires, Argentina [Abstract 5]

25 Larder BA, Hertogs K, Bloor S, van den Eynde CH, DeCian W, Wang Y, Freimuth WW, Tarpley G (2000) Tipranavir inhibits broadly protease inhibitor-resistant HIV-1-1 clinical samples. *AIDS* 14(13): 1943–1948

26 Gathe J, Kohlbrenner VM, Pierone G, Arasteh K, Rubio R, LaLonde R, Piliero P, McCallister S, Garfinkel S, Chaves R et al. (2003) Tipranavir/ritonavir demonstrates potent efficacy in multiple protease inhibitor experienced patients: BI 1182.52. 10th Conference on Retroviruses and

Opportunistic Infections, February 10–14, 2003; Boston, MA [Abstract 179]
27 Arasteh K, Clumeck N, Pozniak A, Jaeger H, Pauw MD, Muller H, Peeters M, Hoetelmans R, Meyer SD, van der Sandt I et al. (2003) First clinical results on antiretroviral activity, pharmacokinetics, and safety of TMC114, an HIV-1 protease inhibitor, in multiple PI-experienced patients. 10th Conference on Retroviruses and Opportunistic Infections, February 10–14, 2003; Boston, MA [Abstract 8]
28 Lalezari JP, Henry K, O'Hearn M, Montaner JS, Piliero PJ, Trottier B, Walmsley S, Cohen C, Kuritzkes DR, Eron JJ Jr, et al. (2003) TORO 1 Study Group. Enfuvirtide, an HIV-1 fusion inhibitor, for drug-resistant HIV infection in North and South America. *N Engl J Med* 348(22): 2175–2185
29 Lazzarin A, Clotet B, Cooper D, Reynes J, Arasteh K, Nelson M, Katlama C, Stellbrink HJ, Delfraissy JF, Lange J et al. (2003) TORO 2 Study Group. Efficacy of enfuvirtide in patients infected with drug-resistant HIV-1 in Europe and Australia. *N Engl J Med* 348(22): 2186–2195
30 Lalezari JP, Beal JA, Ruane PJ, Cohen CJ, Jacobson EL, Sundin D, Leong WP, Lambert DM, Zhou J, Medinas R et al. (1999) HIV-1-1 isolates from patients treated with T-20 are sensitive to the second generation fusion inhibitor T-1249. *Antiviral Ther* 4(suppl 1): 8 [Abstract 10].
31 Eron J, Merigan T, Kilby M, Yangco B, Gleary J, Rusnak P, Dimassimo B, Smith R, Baker B, Duff F et al. (2001) A 14-day assessment of the safety, pharmacokinetics, and antiviral activity of T-129, a peptide inhibitor of membrane fusion. 8th Conference on Retroviruses and Opportunistic Infections. February 4–8, 2001; Chicago, Ill. [Abstract 14].
32 Miralles GD, Lalezari JP, Bellos N, Richmond G, Zhang Y, Murchison M, Spence R, Raskino C, DeMasi RA (2003) T-1249 demonstrates potent antiviral activity over 10 day dosing in most patients who have failed a regimen containing enfuvirtide (ENF): planned interim analysis of T1249-102, a phase I/II study. 10th Conference on Retroviruses and Opportunistic Infections, February 10–14, 2003; Boston, MA [Abstract 14lb]
33 Hazuda DJ, Felock P, Witmer M, Wolfe A, Stillmock K, Grobler JA, Espeseth A, Gabryelski L, Schleif W, Blau C, Miller MD (2000) Inhibitors of strand transfer that prevent integration and inhibit HIV-1-1 replication in cells. *Science* 287(5453): 646–50.
34 Polis MA, Sidorov IA, Yoder C, Jankelevich S, Metcalf J, Mueller BU, Dimitrov MA, Pizzo P, Yarchoan R, Dimitrov DS (2001) Correlation between reduction in plasma HIV-1-1 RNA concentration 1 week after start of antiretroviral treatment and longer-term efficacy. *Lancet* 358(9295): 1760–1765
35 Schapiro JM, Winters MA, Stewart F, Efron B, Norris J, Kozal MJ, Merigan TC (1996) The effect of high-dose saquinavir on viral load and CD4+ T-cell counts in HIV-1-infected patients. *Ann Intern Med* 124(12): 1039–1050
36 Mueller BU, Zeichner SL, Kuznetsov VA, Heath-Chiozzi M, Pizzo PA, Dimitrov DS (1998) Individual prognoses of long-term responses to antiretroviral treatment based on virological, immunological and pharmacological parameters measured during the first week under therapy. *AIDS* 12(15):F191–196
37 Gieschke R, Fotteler B, Buss N, Steimer JL (1999) Relationships between exposure to saquinavir monotherapy and antiviral response in HIV-1-positive patients. *Clin Pharmacokinet* 37(1): 75–86
38 Murphy RL, Sommadossi JP, Lamson M, Hall DB, Myers M, Dusek A (1999) Antiviral effect and pharmacokinetic interaction between nevirapine and indinavir in persons infected with human immunodeficiency virus type 1. *J Infect Dis* 179(5): 1116–1123
39 Marzolini C, Telenti A, Decosterd LA, Greub G, Biollaz J, Buclin T (2001) Efavirenz plasma levels can predict treatment failure and central nervous system side-effects in HIV-1-1-infected patients. *AIDS* 15(1): 71–75
40 Hsu A, Granneman GR, Witt G, Locke C, Denissen J, Molla A, Valdes J, Smith J, Erdman K, Lyons N et al. (1997) Multiple-dose pharmacokinetics of ritonavir in human immunodeficiency virus-infected subjects. *Antimicrob Agents Chemother* 41(5): 898–905
41 Gatti G, Di Biagio A, Casazza R, De Pascalis C, Bassetti M, Cruciani M, Vella S, Bassetti D (1999) The relationship between ritonavir plasma levels and side-effects: implications for therapeutic drug monitoring. *AIDS* 13(15): 2083–2089
42 Dieleman JP, Gyssens IC, van der Ende ME, de Marie S, Burger DM (1999) Urological complaints in relation to indinavir plasma concentrations in HIV-1-infected patients. *AIDS* 13(4): 473–478
43 Back DJ, Khoo SH, Maher B, Gibbons SE (2000) Current uses and future hopes for clinical pharmacology in the management of HIV-1 infection. *HIV-1 Med* 1 Suppl 2: 12–17
44 Burger DM, Hugen PWH, Droste J, Huitema ADR (2001) Therapeutic drug monitoring of nelfi-

navir 1250 mg bid in treatment-naive patients improves therapeutic outcome after 1 year: results from Athena. 2nd International Workshop on Clinical Pharmacology of HIV-1 Therapy; April 2–4, 2001; Noordwijk, The Netherlands. [Abstract 6.2b]
45 Kempf D, Hsu A, Isaacson J, Jiang P, Brun S, Renz C, Granneman GR, Sun E (2001) Evaluation of the inhibitory quotient as a pharmacodynamic predictor of the virologic response to protease inhibitor therapy. 2nd International Workshop on Clinical Pharmacology of HIV-1 Therapy; April 2–4, 2001 Noordwijk, The Netherlands [Abstract 7.1]
46 Kempf D, Hsu A, Jiang P, Rode R, Hertogs K, Larder B, Zolopa A, Shulman N, Havlir D, Gallant J et al. (2001) Response to ritonavir (RTV) intensification in indinavir (IDV) recipients is highly correlated with virtual inhibitory quotient. 8th Conference on Retroviruses and Opportunistic Infections. February 4–8, 2001; Chicago, Ill. [Abstract 523]
47 Duval X, Lamotte C, Race E, Descamps D, Damond F, Clavel F, Leport C, Dybul M, Chun TW, Yoder C et al. (2001) Short-cycle intermittent HAART. 8th Conference on Retroviruses and Opportunistic Infections; February 4–8, 2001; Chicago, Ill. [Abstract 354]
48 Shortliffe EH, Perreault LE (eds): (2001) *Medical informatics: computer applications in health care and biomedicine.* Springer, New York
49 Tural C, Ruiz L, Holtzer C, Schapiro J, Viciana P, Gonzalez J, Domingo P, Boucher C, Rey-Joly C, Clotet B (2002) Clinical utility of HIV-1-1 genotyping and expert advice: the Havana trial. *AIDS* 16(2): 209–218
50 Garber DA, Silvestri G, Fedanov A, Regos R, Barry A, Kozyr N, Larsen CP, McClure H, Montefiori D, Altmann JD et al. (2001) Use of T-cell costimulatory blockade during acute SIV infection of rhesus macaques to define the relative importance of cellular and humoral immune responses in control of viremia and disease progression. AIDS vaccine 2001; September 5–8, 2001, Philadelphia, PA [Abstract LB2]
51 Schmitz JE, Kuroda MJ, Santra S, Sasseville VG, Simon MA, Lifton MA, Racz P, Tenner-Racz K, Dalesandro M, Scallon BJ et al. (1999) Control of viremia in simian immunodeficiency virus infection by CD8+ lymphocytes. *Science* 283: 857–860
52 Klein MR, van Baalen CA, Holwerda AM, Kerhof Garde SR, Bende RJ, Keet IP, Eeftinck-Schattenkerk JK, Osterhaus AD, Schuitemaker H, Miedema F et al. (1995) Kinetics of Gag-specific cytotoxic T-lymphocyte responses during the clinical course of HIV-1-1 infection: a longitudinal analysis of rapid progressors and long-term aymptomatics. *J Exp Med* 181: 1365–1372
53 Pantaleo G, Menzo S, Vaccarezza M, Graziosi C, Cohen OJ, Demarest JF, Montefiori D, Orenstein JM, Fox C, Schrager LK et al. (1995) Studies in subjects with long-term nonprogressive human immunodeficiency virus infection. *N Engl J Med* 332(4): 209–216
54 Rinaldo C, Huang XL, Fan ZF, Ding M, Beltz L, Logar A (1995) High levels of anti-HIV-1-1 memory cytotoxic T-lymphocyte activity and low viral load are associated with lack of disease in HIV-1-1 infected long-term nonprogressors. *J Virol* 69: 5838–5842
55 Hogan CM, Hammer SM (2001) Host determinants in HIV-1 infection and disease. Part 1: cellular and humoral immune responses. [Review] *Annals of Internal Medicine* 134(9 Pt 1): 761–776
56 Hogan CM, Hammer SM (2001) Host determinants in HIV-1 infection and disease. Part 2: genetic factors and implications for antiretroviral therapeutics. [Review] *Annals of Internal Medicine* 34(10): 978–996
57 Peters BS (2002) The basis for HIV-1 immunotherapeutic vaccines. *Vaccine* 20: 688–705
58 Kahn JO, Cherng DW, Mayer K, Murray H, Lagakos S (2000) Evaluation of HIV-1 immunogen, an immunologic modifier, administered to patients infected with HIV-1 having 300 to 549 × 10(6)/L CD4 cell counts: A randomized controlled trial. *JAMA* 284(17): 2193–2202

Index

MIX
Papier aus verantwortungsvollen Quellen
Paper from responsible sources
FSC® C105338

If you have any concerns about our products,
you can contact us on
ProductSafety@springernature.com

In case Publisher is established outside the EU,
the EU authorized representative is:
Springer Nature Customer Service Center GmbH
Europaplatz 3, 69115 Heidelberg, Germany

Printed by Libri Plureos GmbH
in Hamburg, Germany